Springer Series on

ATOMIC, OPTICAL, AND PLASMA PHYSICS 39

Springer Series on
ATOMIC, OPTICAL, AND PLASMA PHYSICS

The Springer Series on Atomic, Optical, and Plasma Physics covers in a comprehensive manner theory and experiment in the entire field of atoms and molecules and their interaction with electromagnetic radiation. Books in the series provide a rich source of new ideas and techniques with wide applications in fields such as chemistry, materials science, astrophysics, surface science, plasma technology, advanced optics, aeronomy, and engineering. Laser physics is a particular connecting theme that has provided much of the continuing impetus for new developments in the field. The purpose of the series is to cover the gap between standard undergraduate textbooks and the research literature with emphasis on the fundamental ideas, methods, techniques, and results in the field.

27 **Quantum Squeezing**
By P.D. Drumond and Z. Ficek

28 **Atom, Molecule, and Cluster Beams I**
Basic Theory, Production and Detection of Thermal Energy Beams
By H. Pauly

29 **Polarization, Alignment and Orientation in Atomic Collisions**
By N. Andersen and K. Bartschat

30 **Physics of Solid-State Laser Physics**
By R.C. Powell
(Published in the former Series on Atomic, Molecular, and Optical Physics)

31 **Plasma Kinetics in Atmospheric Gases**
By M. Capitelli, C.M. Ferreira, B.F. Gordiets, A.I. Osipov

32 **Atom, Molecule, and Cluster Beams II**
Cluster Beams, Fast and Slow Beams, Accessory Equipment and Applications
By H. Pauly

33 **Atom Optics**
By P. Meystre

34 **Laser Physics at Relativistic Intensities**
By A.V. Borovsky, A.L. Galkin, O.B. Shiryaev, T. Auguste

35 **Many-Particle Quantum Dynamics in Atomic and Molecular Fragmentation**
Editors: J. Ullrich and V.P. Shevelko

36 **Atom Tunneling Phenomena in Physics, Chemistry and Biology**
Editor: T. Miyazaki

37 **Charged Particle Traps**
Physics and Techniques of Charged Particle Field Confinement
By V.N. Gheorghe, F.G. Major, G. Werth

38 **Plasma Physics and Controlled Nuclear Fusion**
By K. Miyamoto

39 **Plasma-Material Interaction in Controlled Fusion**
By D. Naujoks

Vols. 1–26 of the former Springer Series on Atoms and Plasmas are listed at the end of the book

D. Naujoks

Plasma-Material Interaction in Controlled Fusion

With 54 Figures and 11 Tables

Springer

Dr. Dirk Naujoks
Max-Planck-Institut für Plasmaphysik
Teilinstitut Greifswald
17491 Greifswald, Germany
E-mail: naujoks@ipp.mpg.de

ISSN 1615-5653

ISBN-10 3-540-32148-9 Springer Berlin Heidelberg New York

ISBN-13 978-3-540-32148-4 Springer Berlin Heidelberg New York

Library of Congress Control Number: 2006927044

Springer is a part of Springer Science+Business Media.

springer.com

Printed in The Netherlands

Typesetting by the authors and SPi using a Springer LaTeX macro package
Cover concept by eStudio Calmar Steinen
Cover design: *design & production* GmbH, Heidelberg

Printed on acid-free paper SPIN: 11531203 57/3141/SPI - 5 4 3 2 1 0 to Natalie

Preface

Nuclear fusion has the potential to provide a major part of mankind's energy needs for many millennia. On the way to controlled thermonuclear fusion on our planet, the principal goals—from the physical point of view—are firstly to obtain a sufficiently stable plasma, secondly, to heat this plasma to ignition temperature, and finally, to avoid excessive interaction of the hot plasma with the solid wall of the containing vessel. With respect to the foreseen use of a fusion reactor as an energy-producing device, an ideal plasma confinement is unattainable nor is it desired. The generated energy together with the helium particles (fusion "ash") must be removed from the central region and conducted to the energy exchanging facilities (blanket), as well as to the gas exhausting and purifying systems—at the rate they are produced. The so-called first wall, the border between the hot plasma with sun-like parameters (and beyond) and the "cool earth", should be able to withstand the high energy and particle fluxes with little or no maintenance.

During the last few decades a large number of dedicated experimental results as well as theoretical and simulation studies have been performed—thanks to the effort of scientists from many countries participating in this truly international project of controlled fusion. Several aspects of plasma–surface interaction have been reviewed in various publications [1–5] as well as in proceedings of conferences such as the series of *Plasma Surface Interaction* and *Fusion Reactor Materials* conferences. A comprehensive "Data Compendium" related to atomic processes taking place in plasma–surface interactions and material questions is given in special supplements issued by the journal *Nuclear Fusion* [6–9].

I tried to provide an in-depth look at the multi-faceted aspects of plasma–surface interaction in controlled fusion, to give a comprehensive analysis of the main processes and the main parameters ruling them, together with an assessment of the most critical questions and open points that demand further investigation. I hope this can assist the reader by performing their own estimations and assessments of relevant physical processes and problems. For further studies, references to selected papers are given. A comprehensive review of the enormous experimental work done in this field is out of the scope of this book, but can be found in the references given above. A more detailed quantitative analysis can be acquired by applying simulation techniques, which are presented shortly, used together with special data compilations.

Since the involved processes are identical, the book might also be of interest in the fast-paced field of surface modification by means of plasma technology. Whether thin layers are deposited on materials in order to improve the surface characteristics or whether plasma ions are implanted into the depth using biased targets, the underlying physics is the same as in fusion experiments.

I would like to thank Prof. V. N. Afanas'ev for conducting my initial steps into scientific work in the field of particle interaction with solids, who has encouraged me to combine experimental, theoretical, and simulation studies wherever it seems possible. He suffered, but dealt gracefully with my poor Russian during my stay in Moscow.

I am much obliged to Dr. R. Behrisch, who has introduced me to the world of plasma–surface interaction. He allowed me to benefit from his wide experience and pushed me to make things clear and simple without unessential elaborating—not always with success I am afraid. I have also learned from him to *fight* for each discharge against the persistent fear of the operators of how a rather small surface probe would harm the device by exposing it into the plasma. I am thankful to Prof. G. Fussmann, who surprised me with analytical descriptions of effects, which I thought were studied only by computer simulation, and who showed me, in detail, that the complexity of particle–surface interaction is quite negligible compared to the situation in another topic—in plasma physics. The suggestions and corrections made by both of them are very appreciated and helped to improve the manuscript significantly.

Thanks to all colleagues I worked with in the inspiring and challenging atmosphere of the international fusion community.

I am grateful to Dr. Ascheron for his encouragement and support to publish this book with Springer, and in particular, to Ms. Blanck for her excellent technical assistance. Many thanks to Ms. Dewitz for most of the drawings and illustrations.

Greifswald, May 2006 *Dirk Naujoks*

Contents

Part III Operation Limits and Criteria

1

Introduction

In plasma–material interactions the fourth state of matter, the plasma, meets the first, the solid state, generally leaving out the states in between, i.e., the gaseous and the liquid state. Serious problems arise. The solid is subjected to power load and to bombardment of energetic particles from the plasma. Due to the contact, the plasma loses energy and particles. At the same time, it is cooled by low-energy plasma particles in the boundary layer and by wall atoms entering the plasma. Such situations are found in the cosmos at the surface of objects with no atmosphere, such as on the moon, which is loaded by the solar wind, or at space vehicles and dust particles in the sun system (for example in the rings of Saturn) and interstellar space. The ignition of electric arcs and discharges on the surfaces of satellites can cause partial or full failure of their sensitive electronics. Before the first astronauts set their feet on the moon surface there had been some fears that the moon could be covered by thick layers of dust produced via erosion processes due to particle impact. A special investigation was launched to estimate the particle fluxes and to get an idea of the corresponding sputtering yields. As we all know, Neil Armstrong had "both feet on the ground". The dust layer on the moon is rather thin—as was estimated.

On our planet, plasma is produced in electric gas discharges in a large number of applications and in experiments dedicated to controlled fusion. The vessel walls in such experimental devices are bombarded with energetic ions, electrons, and neutrals causing the release of wall material atoms that penetrate as impurities into the plasma. Due to this erosion, the lifetime of wall components is limited. Radiation damage is inflicted on the wall material by high-energy neutrons created in the fusion reactions in an ignited fusion plasma.

Fusion represents a very difficult challenge from the viewpoint of developing the required materials in order to meet the demands for economic, safe and environmentally acceptable fusion power reactors. Besides long pulse operation, high power and particle fluxes, short transient events such as disruptions, ELMs, hot spots, and arcing could damage the wall considerably.

The description of the strong, entirely mutual, interaction of a plasma with a solid will require significant improvements in the detailed understanding of the different processes taking place in the plasma–solid transition zone:

- The power may be reflected or absorbed. It diffuses into the bulk and will be removed at the rear side by cooling. The surface temperatures may rise to high values leading to melting and sublimation.
- The bombarding particles may be backscattered into the plasma. They may also be implanted into the wall material and diffuse further into the bulk or towards the surface, where they can be thermally desorbed back into the plasma. The particles may be trapped in the solid permanently, and, therefore, would be lost in the recycling process.
- Particle bombardment causes a release of surface atoms from the solid, i.e., sputtering. This results in an introduction of wall atoms into the plasma, where they are converted to impurity ions, resulting in fuel dilution and strong energy losses due to radiation.
- Electrons hitting the surface of the solid are mostly implanted, but a fraction may also be backscattered, or secondary electrons may be released.
- The loss of electrons from the plasma causes the build-up of an electric potential between the plasma and the solid, the Langmuir sheath potential. In this electrical field the impinging ions increase their velocity, and subsequently, their ability to sputter target atoms. Electric arcs may be ignited causing an additional flux of atoms from the solid into the plasma.

This book is divided into three parts. In Part I, fusion as a possible future energy source is shortly discussed, on the basis of the burn and exhaust criteria (Sect. 3.2), and the main fusion concepts, i.e., the magnetic and inertial plasma confinement, are outlined. The general design of the first wall in fusion experiments is considered.

In Part II, the interaction of a plasma attached to a solid is discussed in terms of particle (Chap. 6), electric (Chap. 7), and power coupling (Chap. 8), after a chapter introducing the main plasma characteristics (Chap. 5). The main processes of plasma–surface interaction, as partly listed above, are described using rather simple analytical relations to show the parameters of importance and to make the underlying physics clear.[1] In addition, numerical methods are presented for simulation of plasma–material interaction processes for the complex conditions usually met in experiments (Sect. 6.10, Sect. 7.6, Sect. 9.3.9). The serious problems caused by impurities in fusion plasma are described in Chap. 9.

In Part III, a set of criteria, e.g., the impurity criteria (Chap. 10), as well as plasma operation limits (Chap. 11) and material operation limits (Chap. 12) are presented. The choice of materials for a future fusion reactor is critically discussed, and a summary of issues of concern and open questions is given (Chap. 13).

---

[1]The SI unit system has been used throughout this compilation (energy and temperature are given also in eV)

Part I

Fusion as Energy Source

2

Energy Problem and Related Safety Aspects

The lifestyle of modern human civilization and the demographic development require an ever-increasing expenditure of energy. Today, the standard of living is directly linked to energy consumption, and there is little hope that this connection could be broken in the future. The global energy usage is increasing at about 2%[1] p.a. resulting in an exponential growth over the years [10]. In dynamical systems that evolve under such conditions, sooner or later, saturation or collapse occurs due to exhaustion of resources.

Currently, most electricity is provided by burning hydrocarbons or carbon-based fossil fuels. Smaller contributions are delivered by hydroelectricity, and by nuclear fission. Despite considerable financial encouragement, the renewable sources—wind, waves, direct solar heating, geothermal energy, photovoltaics—and a more efficient use of energy contribute only little on a global scale [11]. They all suffer from high costs, low energy densities, and low availability. On top of that, efficient use is possible only at remote locations far away from the centers of population and industry.

For example, bioethanol is a possible substitute for petrol for road transport. The main energy crops used to produce bioethanol are sugar-beet and wheat crops. Lignocellulostic biomass (poplar and willow trees), waste straw and sawdust materials can also be used to produce bioethanol with a production rate of approximately 30 to 40 barrels (1 barrel = 159 l) per year and hectare. For comparison, the Organization of Petroleum Exporting Countries (OPEC) alone, which pumps about a third of the world's crude supply, increased their daily oil output in 2005 to about 30 million barrels. In order to substitute the global daily amount of about 90 million barrels, one would need an area of about 25 000 km^2. To substitute the yearly production of oil, a farmland of 9 million km^2 is required. The area available for agriculture is worldwide only about 10 million km^2 and is almost completely used for food production.

[1]In 2004 the increase of energy consumption exceeded the 4% level. The energy demand in China grew by 15% last year

Today, a one-family house can be supplied with energy—with some financial effort—using photovoltaic sources, and it is easy to ignite the Olympic fire with the help of the sun and a concave mirror. However, providing industry, transport, and large cities with sufficient energy is up to now well beyond the means of renewable energy sources. It is feared that this situation will not improve in the near future.

Efforts in energy saving and increased efficiency of energy systems in the technologically developed world are counteracted by the annual rise of energy consumption. It is a frightening scenario indeed, if the large population in developing countries such as China and India reach the world average of energy consumption. A massive increase in use of fossil fuels for power generation is going to take place in the developing world. China has already today the highest concentration of atmospheric sulfur dioxide. The consequences are not only the fact that fossil fuels, accumulated over millions and millions of years, are burned up in a few hundred years, but also acid rain, accumulation of CO_2 in the atmosphere, and the associated greenhouse effect. Invaluable raw materials, which are subjected to an ever-increasing demand by the industry, among them the chemical and pharmaceutical industry, are reduced to simple molecules such as carbon oxides and water vapor. The ongoing depletion of world fuel resources has led and will increasingly lead to political and military confrontations.

The use of fission power as a conceivable alternative is limited due to critical questions of safety, public acceptance, disposal of the radioactive waste, and proliferation of nuclear technology. Fusion power is a promising candidate, besides solar and fission power, for long-term energy supply. Each of these options should be developed to its full potential allowing coming generations to select the solution most suited for their needs.

Since the fuel is fed on continuously in fusion devices (like fossil fuel plants but in contrast to fission reactors), the energy density in the reaction zone of a fusion reactor is comparably small and so is the hazard potential. Already small disturbances, e.g., a slightly higher impurity concentration, will lead to a quench of the burning fusion plasma. Chain reactions are not possible, since the fusion process is not based on neutron multiplication reactions.

The deuterium fuel and the direct end product of fusion—the inert gas helium – are neither toxic nor radioactive. They do not produce atmospheric pollution, nor do they contribute to the greenhouse effect. In contrast, fission produces a multitude of highly radioactive elements. Nevertheless, a few kilograms of radioactive tritium will be on site of a fusion reactor. Its handling, however, will be facilitated by the fact that it is bred in the outer region of the reactor (in the blanket) and no transport over large distances is required. Studies have shown that routine losses of tritium to the biosphere can be made less than prescribed by regulatory guidelines, and that at maximum only a few hundred grams of the tritium inventory could be released under severe accident conditions [12].

The neutrons from the fusion reactions are the main source of both penetrating radiation (requires shielding of primary and secondary neutrons and gamma rays) and radioactive contamination due to activation (containing and waste storage problem). The inventory of radioactive materials produced by neutron activation can be compared with the overall inventory in an equivalent pressurized water reactor. The toxicity, in terms of the gross potential to cause biological damage to man, is found to be significantly less than that in a fission reactor and many orders of magnitude less after a few years. The only significant source of afterheat, i.e., energy released from the decay of radioactive material, is associated with the structural material of the blanket, which should be cooled during both planned and emergency shutdowns to avoid damage. The use of low activation materials and carefully purified materials can reduce this inventory further, thus fully realizing the potential superiority of fusion over fission in this respect [12].

3

Fusion Fuel

In comparison to chemical reactions, where the released energies are in the range of a few eVs, fusion operates in the MeV range. That is why a rather small amount of fusion fuel produces a huge amount of energy, a million times more than in simple fuel combustion. The most competitive advantage of fusion is the almost inexhaustible fuel resource (Sect. 3.1). The problem is to initiate fusion, i.e., to overcome the repulsive force of the Coulomb interaction, and to sustain the fusion reactions on a longer time scale (Sect. 3.2).

3.1 Fusion Reactions

Altogether more than 80 different fusion reactions are known. The fusion reaction with the largest cross-section at the lowest energy is the deuterium–tritium (DT) reaction

$$\mathrm{D} + \mathrm{T} \quad \rightarrow \quad {}^{4}\mathrm{He}\,(3.517\,\mathrm{MeV}) + \mathrm{n}\,(14.069\,\mathrm{MeV})\,. \tag{3.1}$$

The released energy $\Delta E = \Delta m c^2$ is due to the mass defect

$$\Delta m = m_{\mathrm{D}^+} + m_{\mathrm{T}^+} - (m_{^4\mathrm{He}^{2+}} + m_{\mathrm{n}}) \doteq 3.1 \times 10^{-29}\,\mathrm{kg} \tag{3.2}$$

according to the equivalence of mass and energy as result of Einstein's special theory of relativity. Only this fusion reaction is being regarded for a first fusion reactor. Its effective cross-section ($T < 100\,\mathrm{keV}$ [13])

$$\begin{aligned} \ln(\langle \sigma v \rangle_{\mathrm{DT}})\,[\mathrm{cm}^3/\mathrm{s}] = {} & \frac{-21.38}{T_{\mathrm{e}}^{0.29}} - 25.2 - 7.1 \cdot 10^{-2} T_{\mathrm{e}} + 1.94 \cdot 10^{-4} T_{\mathrm{e}}^2 \\ & + 4.92 \cdot 10^{-6} T_{\mathrm{e}}^3 - 3.98 \cdot 10^{-8} T_{\mathrm{e}}^4 \end{aligned} \tag{3.3}$$

(for $T = 23\,\mathrm{keV}$ $\langle \sigma v \rangle_{\mathrm{DT}} = 5.14 \cdot 10^{-22}\,\mathrm{m}^3/\mathrm{s}$) is about 100 times larger, between 1 and 100 keV, than that of the deuterium–deuterium (DD) reactions

$$\mathrm{D + D} \rightarrow {}^3\mathrm{He}\,(0.817\,\mathrm{MeV}) + \mathrm{n}\,(2.450\,\mathrm{MeV}) \tag{3.4}$$

$$\mathrm{D + D} \rightarrow \mathrm{T}\,(1.008\,\mathrm{MeV}) + \mathrm{p}\,(3.024\,\mathrm{MeV})\,. \tag{3.5}$$

The temperature in (3.3) is given in keV. The simple relation [14]

$$\langle \sigma v \rangle_{\mathrm{DT}} = 1.1 \times 10^{-24}\,(T[\mathrm{keV}])^2\ \mathrm{m}^3/\mathrm{s} \tag{3.6}$$

is valid near $T = 10\,\mathrm{keV}$. More exact formulae, valid in a wider range of temperatures, can be found in [15].

On the other hand, the neutrons produced in the DD reactions are less energetic and therefore less harmful to the reactor structure materials. However, in initial pure D plasmas tritium will also be produced, and therefore, so will highly energetic neutrons according to (3.1).

The energy of the α-particles (helium nucleus) is transferred by collisions to the fuel, i.e., to D and T, thus helping to establish a self-sustaining reaction. The fast neutrons easily penetrate the fuel zone and have to be slowed down in a specially designed wall element (blanket), where their energy is converted into heat and then into electricity applying conventional water–steam technology.

High energies of the reacting particles are required to overcome the Coulomb barrier ($\simeq 0.4\,\mathrm{MeV}$), but, thanks to the tunnel effect [16], not as high as estimated from classical calculations. The probability for a particle of mass m and velocity v to cross a potential barrier can be roughly estimated by

$$f_{\mathrm{G}} = \exp\left(-\frac{\Delta}{\Delta_\lambda}\right) \tag{3.7}$$

where for the characteristic decay length Δ_λ, the de Broglie wavelength $\Delta_\lambda = h/(mv)$ of the incoming particle can be taken. The width of the barrier Δ, i.e., the distance to cross the barrier, is obtained by calculating the distance between two particles with their nuclear charges Z_1 and Z_2 at which the potential energy is equal to the kinetic energy of the incoming particle leading to

$$\frac{Z_1 Z_2 e^2}{4\pi\epsilon_{\mathrm{o}}\Delta} = \frac{mv^2}{2} \quad \rightarrow \quad \Delta = \frac{Z_1 Z_2 e^2}{2\pi\epsilon_{\mathrm{o}} m v^2}\,. \tag{3.8}$$

Using this expression, the tunnel probability is proportional to

$$f_{\mathrm{G}} \propto \exp\left(-\frac{Z_1 Z_2 e^2}{\epsilon_{\mathrm{o}} h v}\right) \tag{3.9}$$

an expression close to the famous Gamov factor $\exp(-\pi Z_1 Z_2 e^2/(\epsilon_{\mathrm{o}} h v))$ [16]. Without the tunnel effect, plasma temperatures of several billion K would be necessary in comparison to the significantly lower temperatures of several 100 million K envisaged in fusion reactors.

Deuterium, a non-radioactive isotope of hydrogen, is very abundant and may be extracted from seawater. Deuterium is found in natural hydrogen compounds to the extent of 0.014 to 0.015 percent, i.e., roughly 1 D in every 6500 H atoms. The estimated resource in the oceans is 4.6×10^{13} tons, sufficient for ten billion years based on today's average annual world consumption [17]. It is possible to use the different boiling points of heavy water (101.4°C) and normal water (100°C) for separation, but this would require a lot of energy due to the high heat of vaporization of water. More efficient methods include distillation of liquid hydrogen and various chemical exchange processes which exploit different affinities of deuterium and hydrogen for various compounds. These include the ammonia/hydrogen system, which uses potassium amide as the catalyst, and the hydrogen sulfide/water system.

Tritium is an isotope with a relatively short period of radioactivity (half-life of 12.3 years), so it does not occur in nature and should be bred on site from lithium via either the reaction

$$^{6}\mathrm{Li} + \mathrm{n} \quad \rightarrow \quad ^{4}\mathrm{He}\,(2.05\,\mathrm{MeV}) + \mathrm{T}\,(2.73\,\mathrm{MeV}) \,, \tag{3.10}$$

which can be achieved by slow neutrons, or via the reaction

$$^{7}\mathrm{Li} + \mathrm{n} \quad \rightarrow \quad ^{4}\mathrm{He} + \mathrm{T} + \mathrm{n} - 2.47\,\mathrm{MeV} \tag{3.11}$$

using fast neutrons. The latter reaction does not consume a neutron. Each fusion neutron could produce one new tritium atom. Lithium-6 makes up 7.4% of natural lithium. The average quantity of lithium in the Earth's crust is around 50 ppm. It is more abundant than tin or lead and even ten times more abundant than uranium (3 to 4 ppm) [17]. Reserves of Li are estimated at 12 million tons, if taken from seawater, an additional potential of 230 000 million tons would be available. A self-cooled lithium blanket appears to be the simplest and most reliable approach to a breeding blanket in a reactor. In order to balance losses in the blanket, a small additional neutron source, e.g., via the reaction

$$^{9}\mathrm{Be} + \mathrm{n} \quad \rightarrow \quad 2\,^{4}\mathrm{He} + 2\mathrm{n} - 1.57\,\mathrm{MeV} \tag{3.12}$$

is necessary since only one neutron is produced in each fusion reaction. In the new ITER design such a breeding blanket is not foreseen [5, 18]. The basic function of the planned blanket system is to provide thermal and nuclear shielding to the vessel and external components. A partial conversion of this shielding blanket to a breeding one is envisaged for a later state of operation.[1]

The fuel resources, limited by lithium, are estimated at several thousand years if the lithium is of telluric origin and at several million years if the lithium is taken from seawater. The realization of fusion power on earth would mean a virtually limitless source of energy at comparatively low fuel cost. The enormous impact this would have on our civilization makes controlled fusion the most important scientific challenge man has ever faced [19].

---

[1] see also http://www.iter.org

Apart from fusion fuel other materials needed for construction of thermonuclear reactors might run out. A fusion reactor consists of a large number of high-technology components containing tons of rare, non-ferrous materials. For example, the world production capacity for superconducting wires such as NbTi (Nb_3Sn and MgB_2 wires are the newest superconducting material under development) will be almost fully occupied by providing just the next fusion device ITER. The extensive use of high-temperature superconductors (HTS) based on rather rare materials such as yttrium, bismuth, thallium or mercury is therefore also questionable, not only because of their structural problems. Carbon fiber composites (CFC), ceramic fiber matrices and other advanced materials requiring an extraordinary technological effort would be needed in bulk.

Helium is a unique industrial gas. However, only a handful of sources in the world produces it. The high cost of extraction restricts helium use to relatively few high-technology applications. By far the largest use of helium is as liquid coolant for various superconductors. This use accounts for about 30% of global demand. There is growing concern over the ability to meet future supply needs.

Hydrogen has one major advantage over helium in that it is freely available at relatively low cost. Hydrogen could serve equally well for many helium applications, but it is flammable, therefore explosive. And though hydrogen's boiling point is only a few K higher than helium, it cannot be used as a cheaper substitute for cooling low-temperature superconductors.

3.2 Ignition and Burn Criteria

For a continuously burning DT fusion plasma, appropriate parameters (particle and energy confinement times, plasma density, and temperature) have to be achieved in order to be able to exhaust the produced ^{4}He and the energy [20,21]. Fulfilling the particle and energy balances, only a limited operational window for a fusion reactor is available [22–24], which is even more narrow due to plasma impurities and operation limits such as the β and density limits (see Chap. 11).

In the following, the concentrations of helium and impurity ions are denoted by $f_{\mathrm{He}} = n_{\mathrm{He}}/n_{\mathrm{e}}$ and $f_i = n_i/n_{\mathrm{e}}$, respectively, relating the corresponding densities to the electron density. Taking charge neutrality

$$\sum_i q_i n_i = n_{\mathrm{D}} + n_{\mathrm{T}} + 2n_{\mathrm{He}} + \sum q_i n_i = n_{\mathrm{e}} \tag{3.13}$$

and equal densities of deuterium and tritium, $n_{\mathrm{D}} = n_{\mathrm{T}} = n_{\mathrm{DT}}$, we obtain

$$n_{\mathrm{DT}} = \frac{n_{\mathrm{e}}}{2}\left(1 - 2f_{\mathrm{He}} - \sum q_i f_i\right) \tag{3.14}$$

where q_i is the charge state of ions of kind i and f_i is their concentration ($f_i = n_i/n_{\mathrm{e}}$). In particular, q_i is the charge state of impurity species i and

$q_{He} = 2$. The dilution of the DT fusion fuel by helium and impurity ions is characterized by the additional factor $(1 - 2f_{He} - \sum q_i f_i)$, which reduces the fusion rate. Under steady state conditions, the fusion rate (equalizing the production rate of α-particles) must balance the exhaust rate of helium

$$n_D\, n_T\, \langle \sigma v \rangle_{DT} = \frac{n_{He}}{\tau^*_{He}} \tag{3.15}$$

where $\langle \sigma v \rangle_{DT}$ is the fusion rate coefficient. The particle confinement time, τ_{He}, is defined by

$$\tau_{He} = \frac{\int_{volume} n_{He}\, dV}{\int_{surface} \Gamma_{He}\, dS} \tag{3.16}$$

where Γ is the helium flux density out of the volume V enfolded by the surface S. The He exhaust itself can be parameterized by introducing the effective confinement time [25]

$$\tau^*_{He} = \frac{\tau_{He}}{1 - R_{cyc}} = \frac{\tau_{He}}{\epsilon} \tag{3.17}$$

where R_{cyc} is the recycling coefficient and $\epsilon = 1 - R_{cyc}$ is the exhaust efficiency determined by the scrape-off layer (SOL) physics and pumping capabilities. During a typical discharge, a helium particle leaves and re-enters the plasma many times before it can be pumped out. The effective confinement time is usually considerably longer than the particle confinement time, $\tau^*_{He} \gg \tau_{He}$, since R_{cyc} is close to unity (Fig. 3.1).

When external fuel sources such as a gas inlet are turned off, the density is observed to decay with a characteristic time τ^*_p, i.e., as $\exp(-t/\tau^*_p)$ [2], since edge fueling by recycling continues and the confinement/replacement time is given by

$$\tau_p = (1 - R_{cyc})\tau^*_p\ . \tag{3.18}$$

The difference between τ_p and τ^*_p becomes critical by analyzing the burn criterion (see below). In fusion experiments τ_p is usually on the order of tens of milliseconds, while τ^*_p is about one second.

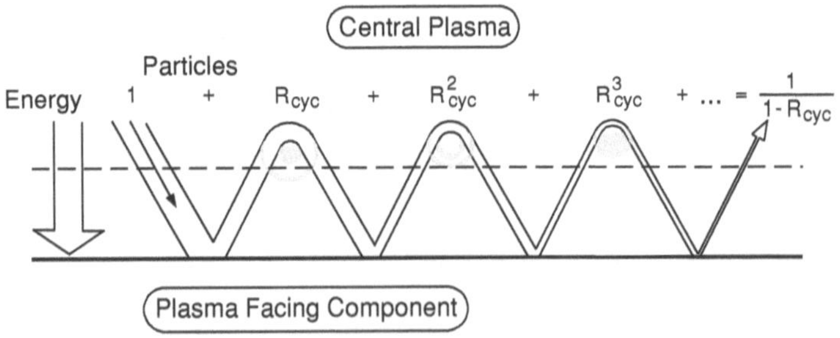

Fig. 3.1. The recycling process. The energy is almost fully absorbed by the material, whereas particles such as He can leave the central plasma region and return many times

For a fusion reactor, the produced helium ash corresponding to the produced fusion power has to be removed as well as the additional fueling in the case of neutral beam injection (NBI) of highly energetic deuterium, if it is used as a plasma heating method.

Unfortunately, the pumping system will probably exhaust the hydrogen isotopes with an efficiency comparable to that of helium. Thus, the required throughput of helium will be accompanied by a large throughput of DT particles due to the lower concentration. The pumping requirements are therefore quite substantial.

Combining (3.14) and (3.15) gives the so-called exhaust criterion

$$n_e \tau_{\mathrm{He}}^* = \frac{4 f_{\mathrm{He}}}{\left(1 - 2 f_{\mathrm{He}} - \sum q_i f_i\right)^2 \langle \sigma v \rangle_{\mathrm{DT}}} \,. \tag{3.19}$$

The plasma is transparent for the highly energetic neutrons produced in the fusion reactions. They penetrate more or less freely through the first wall with its thickness of a few centimeters. Their kinetic energy will be converted into thermal energy in thick blanket structures located behind the wall. Contrarily, the part of helium fusion power P_α ($E_\alpha = 3.52\,\mathrm{MeV}$, see (3.1)) which is not lost through radiation must be confined in the central plasma long enough to replace convective and diffusive heat losses in order to maintain the plasma temperature

$$\begin{aligned} P_\alpha - P_{\mathrm{brems}} - P_{\mathrm{rad}} = \frac{W_{\mathrm{E}}}{\tau_{\mathrm{E}}} = \frac{3}{2} \Big(& n_e k_{\mathrm{B}} T_e \\ & + 2 n_{\mathrm{DT}} k_{\mathrm{B}} T_{\mathrm{DT}} + n_{\mathrm{He}} k_{\mathrm{B}} T_{\mathrm{He}} \\ & + \sum n_i k_{\mathrm{B}} T_i \Big) \Big/ \tau_{\mathrm{E}} \end{aligned} \tag{3.20}$$

where W_{E} is the energy content of the plasma, τ_{E} is the energy confinement time, P_{brems} are the power losses due to bremsstrahlung and P_{rad} are the additional power losses due to impurity radiation. In striking contrast to particle confinement, τ_{E} is predominantly determined by transport processes, since most of the energy leaving the plasma is not recycled back from the wall.

The helium fusion power is given by

$$P_\alpha = n_{\mathrm{DT}}^2 \langle \sigma v \rangle_{\mathrm{DT}} E_\alpha = \frac{n_e^2}{4} \left(1 - 2 f_{\mathrm{He}} - \sum f_i q_i\right)^2 \langle \sigma v \rangle_{\mathrm{DT}} E_\alpha \,. \tag{3.21}$$

Since fully ionized ions do not radiate (besides bremsstrahlung), neither DT nor helium contribute to P_{rad}, but only partially stripped impurity ions. Recombination can be neglected in the hot central plasma. Introducing radiation functions L_i, which are functions of electron temperature T_e and charge state of the impurity ion species i, P_{rad} becomes

$$P_{\mathrm{rad}} = n_e \sum n_i L_i(T_e) \,. \tag{3.22}$$

P_{brems} can be written (see (9.9)) as (with $q_{\text{DT}} = 1$ and $q_{\text{He}} = 2$)

$$P_{\text{brems}} = c_{\text{br}} n_{\text{e}} \left(2n_{\text{DT}} + 4n_{\text{He}} + \sum_i q_i^2 n_i \right) \sqrt{k_{\text{B}} T_{\text{e}}} = c_{\text{br}} n_{\text{e}}^2 Z_{\text{eff}} \sqrt{k_{\text{B}} T_{\text{e}}}$$
$$= c_{\text{br}} n_{\text{e}}^2 \left(1 + 2f_{\text{He}} + \sum \left[q_i^2 - q_i\right] f_i\right) \sqrt{k_{\text{B}} T_{\text{e}}} \qquad (3.23)$$

since $2n_{\text{DT}}/n_{\text{e}} = 1 - 2f_{\text{He}} - \sum q_i f_i$ (see (3.14)), and

$$c_{\text{br}} = \frac{16\sqrt{2\pi}\gamma_{\text{G}}}{3\sqrt{3}(4\pi\epsilon_{\text{o}})^3 m_{\text{e}}^{3/2} c^3 \hbar} = 3.84 \times 10^{-29}\, \text{Wm}^2\text{s}/\sqrt{\text{kg}} \qquad (3.24)$$

where $\gamma_{\text{G}} \simeq 1$ is the Gaunt factor, which was introduced by Kramers to take account for discrepancies between the classical and quantum mechanical treatments. Z_{eff} is the effective charge state of the plasma ions

$$Z_{\text{eff}} = \frac{\sum\limits_{\text{all ions}} q_i^2 n_i}{\sum\limits_{\text{all ions}} q_i n_i} = \sum_{\text{all ions}} q_i^2 n_i \Big/ n_{\text{e}} \,. \qquad (3.25)$$

As far as bremsstrahlung is concerned, in an ignited DT fusion reactor, medium and light impurities can be tolerated up to an effective space charge of about 2, whereas for DD and DHe^3 plasma ignition is possible only for very clean plasmas [23].

Substituting (3.21), (3.22), and (3.23) into (3.20), we obtain with the assumption of $T = T_{\text{e}} = T_{\text{DT}} = T_{\text{He}} = T_i$

$$n_{\text{e}}\tau_{\text{E}} = (3/2)\, k_{\text{B}} T \left(2 - f_{\text{He}} + \sum f_i \left[1 - q_i\right]\right) \Big/$$
$$\left[\left(1 - 2f_{\text{He}} - \sum f_i q_i\right)^2 \langle\sigma v\rangle_{\text{DT}} E_\alpha/4 \right.$$
$$\left. - c_{\text{br}} \left(1 + 2f_{\text{He}} + \sum \left[q_i^2 - q_i\right] f_i\right) \sqrt{k_{\text{B}} T} - \sum f_i L_i \right] . \qquad (3.26)$$

The exhaust criterion (3.19) and the burn criterion (3.26) can be combined in order to eliminate the electron density by defining the ratio [22]

$$\gamma = \frac{\tau_{\text{He}}^*}{\tau_{\text{E}}} \,, \qquad (3.27)$$

which results in a cubic equation for a certain impurity concentration f_i for given γ, T, and f_{He}

$$\gamma = \frac{8f_{\text{He}}}{3} \left[\left(1 - 2f_{\text{He}} - \sum f_i q_i\right)^2 \langle\sigma v\rangle_{\text{DT}} E_\alpha/4 - c_{\text{br}} \left(1 + 2f_{\text{He}}\right.\right.$$
$$\left.\left. + \sum \left[q_i^2 - q_i\right] f_i\right) \sqrt{k_{\text{B}} T} - \sum f_i L_i \right] \Big/ \left[\left(1 - 2f_{\text{He}} - \sum q_i f_i\right)^2\right.$$
$$\left. \times \left(2 - f_{\text{He}} + \sum f_i [1 - q_i]\right) \langle\sigma v\rangle_{\text{DT}}\, k_{\text{B}} T \right] . \qquad (3.28)$$

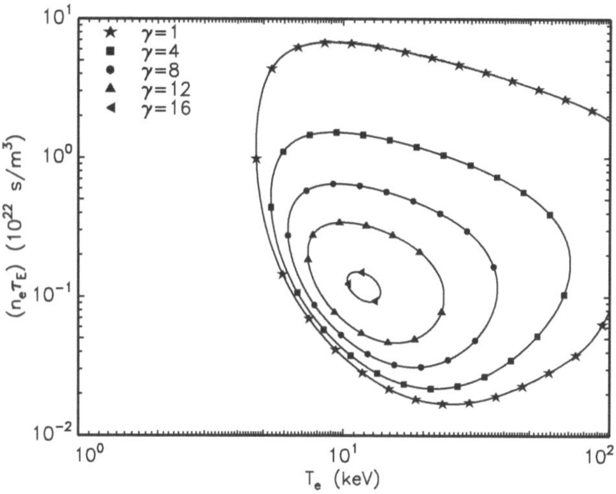

Fig. 3.2. Operation space of a fusion reactor considering only helium as an impurity, according to (3.29). The chosen values of γ are indicated. Here, the fusion rate coefficient is taken according to (3.3)

In the case of $f_i = 0$ (only helium, no other impurities), this reduces to

$$\gamma = \frac{2f_{\mathrm{He}}}{3} \frac{\langle\sigma v\rangle_{\mathrm{DT}} E_\alpha - 4c_{\mathrm{br}}\sqrt{k_{\mathrm{B}}T}(1+2f_{\mathrm{He}})/(1-2f_{\mathrm{He}})^2}{(2-f_{\mathrm{He}})\,\langle\sigma v\rangle_{\mathrm{DT}}\,k_{\mathrm{B}}T}\,, \tag{3.29}$$

which is also a cubic equation in regard to f_{He}. Due to charge neutrality

$$2n_{\mathrm{DT}} + 2n_{\mathrm{He}} + q_{\mathrm{imp}}n_{\mathrm{imp}} = n_{\mathrm{e}} \tag{3.30}$$

(here only one impurity species with charge state q_{imp} is considered) an additional condition arises

$$f_{\mathrm{imp}} = \frac{n_{\mathrm{imp}}}{n_{\mathrm{e}}} \le \frac{1-2f_{\mathrm{He}}}{q_{\mathrm{imp}}} \tag{3.31}$$

reflecting the extreme case of $n_{\mathrm{DT}} = 0$ in the case of the equal sign. In a plasma without any impurities, other than helium, the limiting condition is $f_{\mathrm{He}} \le 1/2$.

Usually, (3.29) has two real solutions, $f_{\mathrm{He},1}$ and $f_{\mathrm{He},2}$, for a given γ value and a certain temperature in the available range of $0 \le f_{\mathrm{He}} \le 1/2$. Varying the temperature T, the obtained solutions of (3.29) can be substituted into the burn criterion (3.26). The resulting values of $n_{\mathrm{e}}\tau_{\mathrm{E}}$ build the workspace of a fusion reactor as a function of T for a certain value of γ (Fig. 3.2).[2]

As seen, there are no solutions at very low temperatures. A real point of concern is the fact that the operation space is tied up with increasing

[2]Often, a similar presentation by means of the so-called triple product $n_{\mathrm{e}}\,\tau_{\mathrm{E}}\,T_i$ as a function of temperature is used in the literature [22, 24]

values of γ. For $\gamma \geq 16$, there is no solution at all (Fig. 3.2). Assuming linked transport mechanisms for particle and energy transport, i.e., $\tau_{\mathrm{He}} \simeq \tau_{\mathrm{E}}$, the exhaust efficiency ϵ (see (3.17) and (3.27) should be larger than $\epsilon > 1/16 \simeq 0.06$.

Confinement times are strongly dependent on the source location (the sink is assumed to be fixed at the edge). The values for τ_{He} and τ_{E} can be quite different, even if the transport coefficients are numerically the same, since energy sources are usually in the plasma center while particle sources (neutral ionization from recycled particles) are located near the edge.

The ratio $\gamma = \tau^*_{\mathrm{He}}/\tau_{\mathrm{E}}$ was measured to be approximately 8 in the fusion experiment DIII-D, using helium gas puffing as well as 75 keV neutral helium beam injection simulating a central source of helium [26]. In ASDEX-Upgrade, a ratio of about 7 has been demonstrated in helium exhaust studies [27, 28].

From the statistical analysis of confinement results obtained in different fusion devices, expressions of the energy confinement time τ_{E} are found as a function of geometrical and physical parameters. The expression [18]

$$\tau_{\mathrm{E}}^{\mathrm{tokamak}} = 0.0562\, I_{\mathrm{p}}^{0.93} B_{\mathrm{T}}^{0.15} P^{-0.69} n_{\mathrm{e}}^{0.41} M^{0.19} R^{1.97} \varepsilon^{0.58} \kappa^{0.78}\ [\mathrm{s}] \qquad (3.32)$$

is valid in tokamaks with H-Mode and ELMs. The parameters and units are the plasma current I_{p} in MA, the toroidal magnetic field B_{T} in T, the total power crossing the separatrix P in MW, the electron density in $10^{19}\,1/\mathrm{m}^3$, the average ion mass M in amu, the major radius R in m, the inverse aspect ratio $\varepsilon = a/R$, and the elongation κ. For stellarators, the following relation is proposed [29]

$$\tau_{\mathrm{E}}^{\mathrm{stellarator}} = 0.048\, a^{2.39} R^{1.22} B_{\mathrm{T}}^{0.88} P^{-0.77} n_{\mathrm{e}}^{0.69} \iota^{0.04}\ [\mathrm{s}] \qquad (3.33)$$

with the rotational transform $\iota = 2\pi/q_{\mathrm{s}}$ and the minor radius a. The parameters are given here with the same units as in (3.32). The energy confinement time is usually calculated from the measured plasma energy W_{E} and heating power P according to

$$\tau_{\mathrm{E}} = \frac{W}{P - \mathrm{d}W/\mathrm{d}t}\,. \qquad (3.34)$$

4

Fusion Concepts

Although the sun successfully demonstrates that fusion works, its confinement principle cannot be applied, since gravitational forces are far too weak for terrestrial plasmas. There are two basically different approaches to controlled thermonuclear fusion on our planet: magnetic and inertial confinement fusion.

4.1 Inertial Plasma Confinement

In the latter method, the surface of a small pellet containing the fusion fuel is rapidly heated by high-energy lasers or particle beams. By a rocket-like inward reaction the pellet implodes. The fusion fuel is compressed to super high densities and is adiabatically heated until the pellet core is brought to ignition. The confinement time needed to comply with the $(n_e \, \tau_E)$-criterion (see Sect. 3.2) and should be smaller than the time during which the inertia of the pellet material is sufficient to hold the pellet together (inertial confinement). In the direct drive approach, several beams are producing a highly symmetric illumination with an optimized pulse shape. Using an indirect drive, the beam radiation is first converted to soft x-rays that isotropically fill a cavity containing in its center the fuel pellet.

The key issues of the inertial confinement fusion are the pellet design, the efficiency and repetition rate of the laser or particle beams, and the cyclic energy deposition. Additional problems arise during the repetitive ignition of pellets one after the other falling down into the reaction vessel. Already after the first two or three successful ignitions, the resulting "dirty" conditions in the reaction vessel would not allow a further localized energy input with the required high accuracy. The long cleaning phases between the ignitions and the required reloading time of the drivers are the problems to encounter on the way to, at least, quasistationary conditions [30–32]. Plasma and particle interaction processes with the wall are quite similar to those in the magnetic confinement concept (see below), while being here of less concern [33].

Laser physicists in Europe have proposed a new "fast ignition" facility HiPER. In the conventional approach to inertial confinement the lasers that compress the fuel pellet also heat it. Fast ignition relies on the usage of different lasers for these two phases. The HiPER facility would consist of a long-pulse laser with an energy of 200 kJ for compression and a short-pulse laser with an energy of 70 kJ for heating [34].

4.2 Magnetic Plasma Confinement

In magnetic confinement, the currents of the charged particles, ions and electrons, are impeded from flowing across the magnetic surfaces, i.e., in the radial direction. Soon after starting with linear devices, which had a mirror configuration of the magnetic field coils, it became clear that the magnetic field lines have to be bent into a torus in order to avoid the significant parallel losses. To provide toroidal plasma equilibrium and stability, the magnetic field lines have to pass around both the poloidal way and the toroidal way to form magnetic surfaces [14, 19, 35, 36]. Without the poloidal field, the magnetic field gradient would cause charge separation between the top and bottom of the plasma. The resulting electric field would drive the plasma outwards to the wall. The helical structure of the magnetic field configuration gives stability by shorting out charge imbalances. The amount of the helical twist is given by the so-called safety factor q_s, which becomes approximately

$$q_s \simeq \frac{r}{R} \frac{B_{\text{toroidal}}}{B_{\text{poloidal}}} \tag{4.1}$$

in the case of circular cross-section plasmas, where r and R are the minor and major radius, respectively.

Since the temperature for ignition is defined and the reacting plasma must have a sufficiently low pressure (see below, β-criterion), the plasma density is rather low (seven to eight orders of magnitude lower than the particle density in solids). In order to fulfill the $(n\,\tau)$-criterion (see Sect. 3.2) the time τ, which characterizes the energy confinement, should typically be a few seconds.

The most promising concepts of magnetic confinement fusion are the tokamak and stellarator systems.

4.3 Stellarator Concept

In the stellarator, initiated in 1950 by Spitzer, the vacuum magnetic field for plasma confinement is produced entirely by currents flowing in external coils (Fig. 4.1). In the absence of a net toroidal plasma current, the magnetic configuration can be maintained in steady state provided only that currents in the coils are maintained and that particle and energy exhaust can be ensured

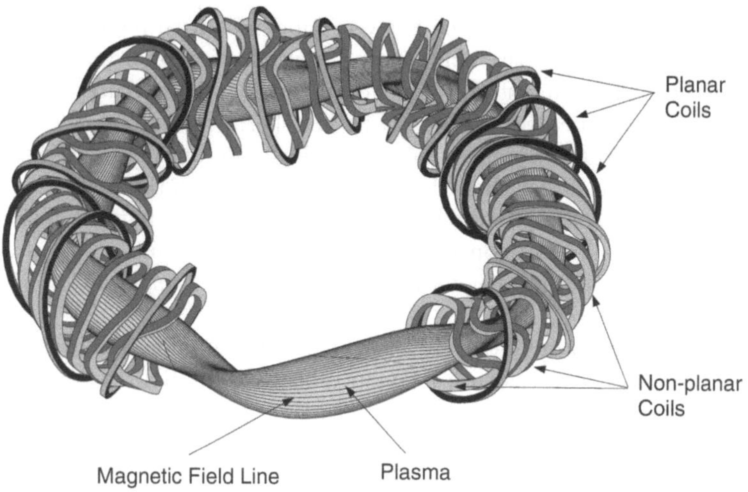

Fig. 4.1. Schematic of a five-period optimized stellarator with modular coils

together with refueling. No external current drive system is needed. The system is necessarily asymmetric in the toroidal direction. Despite the increased engineering complexity, stellarators do not show disruptive instabilities due to a quench of the plasma current as can happen in tokamaks, and this could prove decisive in the long term. The absence of a net plasma current as in tokamaks minimizes the free energy available for driving instabilities.

With the Wendelstein line, the Max Planck Institute for Plasma Physics (IPP) is exploring the reactor potential of advanced stellarators [37–40], a concept for confining toroidal plasmas with magnetic fields generated exclusively by a set of external, planar, and non-planar coils. Progress in experiments and theory has eliminated major concerns about low stability limits and enhanced transport losses due to the large magnetic ripple in stellarators. The engineering concept of modular coils [37] allows the realization of an optimized magnetic configuration with considerable reactor potential. As a consequence of optimization, the configuration possesses an inherent magnetic separatrix with three-dimensional particle and energy fluxes at the plasma boundary. The three-dimensional nature of the boundary topology gives rise to a rather complex plasma–wall interaction pattern. Transport simulations [41] have shown that the plasma outflow is rather concentrated in regions where the flux surfaces show the strongest curvature in the poloidal cross-section. This region corresponds to a helical ridge, which starts at the lower apex of the elliptical cross-section and stretches along the outside to the upper elliptical apex through each field period [42, 43]. Correspondingly, visible ribbons of deposition have been observed along the vessel wall [44].

The development of a proper exhaust technology is another important task. One of the interesting properties of the optimized design is the adaptation to divertor operation (see below), important with respect to power and particle exhaust, impurity control, and minimization of erosion and impurity production. The exhaust will depend on the island structures at the edge of the plasma and the degree of ergodization there [45].

The aspect ratio, which is the ratio of the major plasma radius to the minor plasma radius, is larger than that of tokamaks, hence, the power loading onto the first wall is correspondingly reduced. On the other hand, a fusion reactor based on the stellarator concept will be large with a radius of about 25 m. The use of superconducting magnetic coils, for example using NbTi conductors, is inevitable.

Without net toroidal current, heating relies upon non-ohmic methods such as electron and ion cyclotron resonance heating (ECRH, ICRH) and neutral beam injection (NBI).

4.4 Tokamak Concept

In the tokamak, proposed by Sakharov in 1952, the poloidal magnetic field component is produced by a toroidal current, flowing in the plasma itself. This current is induced by transformer action. The primary winding changes the magnetic flux and the resulting toroidal electric field with a loop voltage of a few volts drives the plasma current. Furthermore, the plasma is ohmically heated by the induced current. The toroidal field is produced by currents flowing in uniformly distributed poloidal coils encircling the plasma torus. As a result, the plasma is free to flow on certain closed and nested surfaces—the axis-symmetric magnetic surfaces (Fig. 4.2).

If a tokamak plasma is heated only by ohmic heating, the ohmic heating power density ηj^2 decreases and the bremsstrahlung loss increases with increasing temperature. Therefore, the upper limit of the electron temperature can be estimated from $\eta j^2 = P_{\mathrm{brems}}$. This limit is less than a few keV. Additional heating, e.g., neutral beam heating, is necessary in a tokamak system to approach the condition for ignition. The tokamak confinement system has been well developed and experiments demonstrated high fusion triple products: $n_{\mathrm{e}}\tau_{\mathrm{E}}T_{\mathrm{DT}}$ (Sect. 3.2).

With the ITER tokamak, the upcoming project of international cooperation, the scientific and technological feasibility of fusion energy should be demonstrated using the magnetic confinement concept [18]. The device is conceived for controlled ignition and extended burn of DT plasmas. During the design engineering activities (EDA phase, starting in 1992), its technical concept has been developed and progressively evolved fulfilling the revised programmatic objectives and the cost reduction target adopted by the ITER parties in 1998. While having reduced technical objectives, the new ITER design (ITER–FEAT for Fusion Energy Advanced Tokamak) will provide an integrated demonstration of a burning fusion plasma addressing confinement,

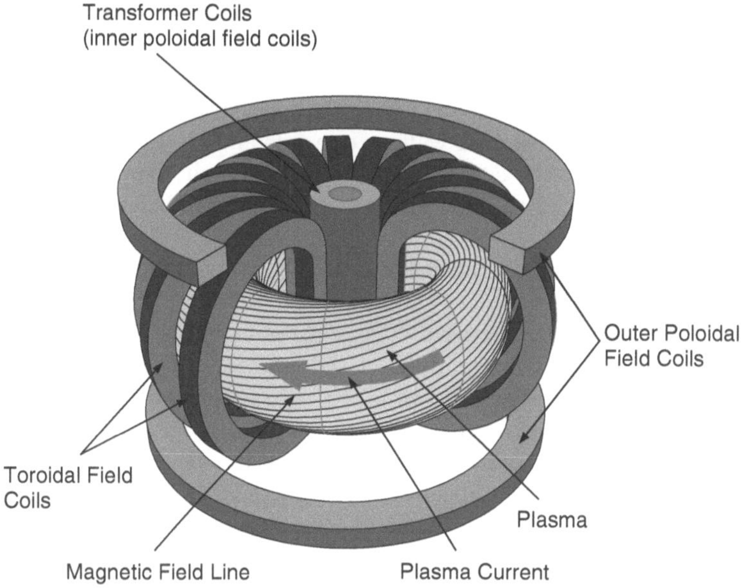

Fig. 4.2. Schematic of a tokamak. The transformer induces the toroidal current. The outer poloidal field coils are used for plasma positioning and shaping

stability, exhaust of helium ash, and impurity control as well as the availability of essential fusion technologies. In 2005, the site was selected. ITER is to be constructed in France, at Cadarache. After the construction time of about ten years, the device is anticipated to operate over an approximately 20-year period in order to execute its challenging program [18]. ITER is also an experiment for investigating plasma–wall interaction issues to prepare a viable solution for future steady state reactors such as DEMO.

4.5 Design of the First Wall

Due to the limited confinement of energy and particles and the demands connected with the aim to convert the thermal energy into electricity in a future fusion reactor, the design of the first wall elements requires special care and attention. It has to perform three main functions: (1) to sustain the impact of the energetic particles and of radiation without releasing many impurities and without a large degradation of its thermophysical properties, (2) to transfer the heat load of about 0.5 to 1 MW/m^2 coming from the plasma to a cooling medium and to withstand the resulting thermal stress, while allowing the neutron flux to reach the blanket modules where the kinetic energy of the neutrons is used to heat up a coolant as in a primary loop of a conventional heat-power plant, (3) be able to withstand high heat loads during events such as disruptions, ELMs or generation of runaway electrons and the related electromagnetic loads, (4) minimizing the tritium retention in accordance with the general objective of a low tritium inventory.

The crucial problem is to save power exhaust from the core plasma without intolerable effects to the first wall on one side and to the plasma itself on the other. Radiation at the edge can help to transfer the power onto sufficiently large areas, as well as the power distribution over sufficiently many particles by recycling.

Impurity sources from the wall can largely contribute to the contamination of the core plasma because of the large wall area and the ineffective redeposition of once emitted particles. The sources of impurities can also be very localized as observed, for example, at the limiters of ICRH antennae.

One can try to restrict the area of plasma–wall interaction (Chap. 6) to dedicated regions by exposing so-called limiters into the plasma or to divert the edge plasma by means of special magnetic coils into a so-called divertor. In both schemes, the plasma is streaming along the magnetic field lines towards the exposed surfaces, where it is neutralized. Most of the plasma electrons are expelled by the electric field being automatically established in the thin sheath above surface—the electric sheath (Chap. 7). The plasma ions are accelerated up the ion sound speed by reaching the sheath entrance and recombine at the surface. To ensure the equality of ion and electron fluxes, the surface is negatively charged with respect to the plasma potential. The loss of plasma is partly compensated by dissociation and subsequent ionization of emitted molecules but mainly balanced by the diffusive transport out of the undisturbed core region into the edge plasma.

The thickness of the so-called scrape-off layer (SOL) formed by the sink action of limiters or divertor plates (Fig. 4.3) can easily be estimated by equating the cross-field particle flux into the SOL of length L_c and poloidal extent l [46, 47]

$$D_\perp \left.\frac{\mathrm{d}n}{\mathrm{d}r}\right|_{\mathrm{LCFS}} L_\mathrm{c}\, l \approx D_\perp \frac{n_\mathrm{LCFS}}{\lambda_\mathrm{SOL}} L_\mathrm{c}\, l \tag{4.2}$$

to the particle flux reaching a wall element

$$l \int\limits_{\mathrm{LCFS}}^{wall} n_\mathrm{e}(r) c_\mathrm{s}(r)\, \mathrm{d}r \approx 0.5\, n_\mathrm{LCFS}\, c_\mathrm{s}\, \lambda_\mathrm{SOL}\, l \tag{4.3}$$

where L_c is the shortest connection length to an absorbing wall element (limiter, divertor, or wall) and $c_\mathrm{s} = \sqrt{k_\mathrm{B}(T_\mathrm{e}+T_\mathrm{i})/m_\mathrm{i}}$ is the isothermal ion acoustic velocity. The plasma density n_LCFS far from the wall element is reduced roughly to half of the value near above the surface.

The cross-field diffusion in the boundary layer, characterized by the diffusion coefficient $D_\perp \simeq 1\,\mathrm{m}^2/\mathrm{s}$, is slow enough in comparison with the parallel motion to give rise to a sharp plasma boundary with a short plasma decay length

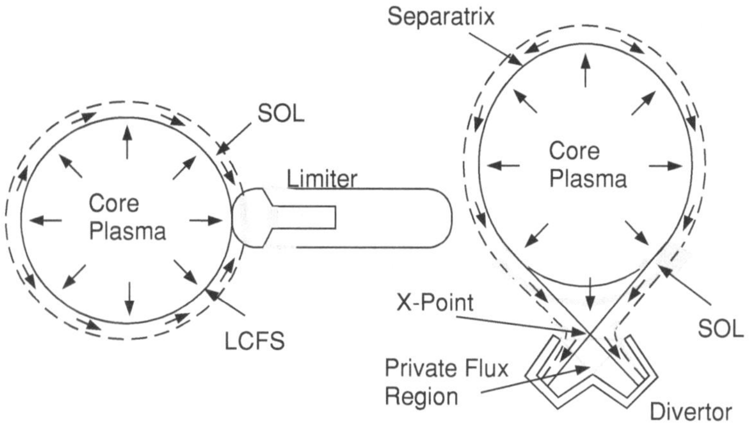

Fig. 4.3. Limiter and divertor configuration shown in the poloidal cross-section of a tokamak with indicated flow pattern

$$\lambda_{\mathrm{SOL}} = \sqrt{\frac{2D_{\perp}L_{\mathrm{c}}}{c_{\mathrm{s}}}} = \sqrt{2D_{\perp}\,\tau_{\mathrm{SOL}}} \tag{4.4}$$

on the order of several centimeters. It is worth noting that an exponential decay of the plasma parameters in the scrape-off layer can also be explained, in another way, by an increasing cross-field transport across the SOL, for example, by increasing $D_{\perp}$ with increasing radius.

The extreme anisotropy of heat and particle transport due to the dominant parallel flows in the SOL lead to very localized regions of high load onto the limiter or divertor targets. This constraint can only be faced by establishing an extremely glancing incidence of the magnetic field with an angle of about one degree. However in practice, wall tolerances and magnetic error fields cause a variation of the local angle typically of the same order. For such conditions, sweeping of the magnetic field configuration in the divertor, i.e., changing the strike point position of the separatrix at the divertor plates in time, could spread the power deposition over larger areas.

4.5.1 Limiter

Limiters are very useful for concentrating the plasma surface interaction on specially designed and geometrically optimized structures (Fig. 4.3). The main disadvantage of limiters is the high possibility that eroded particles once ionized will reach the central plasma by cross-field diffusion, since ionization will probably occur already inside the confined region.

Even a proper alignment, i.e., establishing an almost parallel incidence of the magnetic field lines at the top of the limiter, would not reduce the particle flux density to these areas (and subsequently the erosion from there) down to zero as one would hope applying the simply geometrical $\sin\alpha$-law, where α is the angle between the magnetic field line and the surface. The substantial heat and particle fluxes are dominated in this case by direct, cross-field transport [48].

However, using strong gas injection, for example neon, a cold radiative plasma mantle could be established in TEXTOR [49,50]. Such a concept allows for dissipation of the energy on the whole plasma vessel, thereby reducing the load onto the limiter. Nevertheless, the crucial limitation for radiative edge cooling is also given by the impurity concentration in the center.

4.5.2 Divertor

With additional magnetic coils, usually placed at the top and/or bottom of tokamaks, the magnetic topology in the plasma boundary is changed in a way that previously closed magnetic flux surfaces are open and particles, which pass beyond the separatrix, are directed to target plates far away from the hot central region (Fig. 4.3). Although the radial distance of the divertor region to the separatrix is usually less than a meter, the particles have to move a long distance up to hundreds of meters along the magnetic field lines before reaching the targets. This opens the possibility to actively control the plasma parameters in the divertor region without affecting the plasma core parameters and, therefore, the confinement. The divertor should decouple the plasma parameter at the edge from those at the target plates [51–53].

It has already been pointed out earlier by Düchs et al. [54] that steady state operation is not feasible without a divertor when considering that the following goals have to be achieved: (1) to exhaust the power at an acceptable erosion rate, (2) to control the impurity content of the main plasma by retention of impurities produced at the divertor targets, (3) to pump the He ash out of the confined region.

Given that particle and energy fluxes can be transferred to the wall components either at a high temperature and low plasma density (called low recycling regime) or at low temperature and high density (high recycling). The most promising divertor concept seems to be the losses-in-series using the combination of all available energy loss mechanisms from the core region down to the divertor to disperse the power by atomic processes as the only way to reduce the heat load onto the plates [55,56]. The power loss channels are: (1) impurity radiation in the SOL and in the divertor, (2) momentum and energy transfer by charge-exchange (and elastic ion-neutral collisions) neutrals to the divertor chamber wall, a concept called "neutral cushion", characterized by a flame-like envelope of the ionization front ($T_e \simeq 5\,\mathrm{eV}$), (3) volume recombination by three-body recombination, which becomes effective at low plasma temperatures ($< 2\,\mathrm{eV}$), and (4) ionization and hydrogen radiation near the targets, i.e., high recycling. Impurity radiation mainly occurs in regions where the electron temperature is about $10\,\mathrm{eV}$ and higher, hydrogen ionization occurs near $5\,\mathrm{eV}$, and plasma recombination at $1\,\mathrm{eV}$ and lower. Only by employing all of the listed effects, e.g., using gas puffs and feedback-controlled injection of recycling impurities such as Ar or Ne [50], a reduction of the heat load down to an acceptable value of 5 to 10 $\mathrm{MW/m^2}$ can be realized.

Such additional measures as the increase of "wetted area", by tilting the target poloidally (which is limited due to alignment and plasma position control) and the spatial sweeping of strike zones (which needs space and makes tight baffling of neutrals difficult) could only help to some degree but not solve the problem. There is no material and/or target design in sight to stand a higher load under steady state conditions. Thus, a successful divertor design is based on the optimization of both sides of the problem: an adequate choice of the material and optimized plasma regimes.

Reducing the plasma temperature at the target plates offers the favorable possibility of decreasing the erosion yield. For some materials, e.g., high-Z elements, the energy of impinging ions could then be even smaller than the threshold energy for sputtering. For the same energy flux, it is much better—with respect to erosion—to have high densities and low temperature near the target plates. The primary indication of such a detachment is a decrease up to an order of magnitude in the ion saturation current, recorded, for example, by Langmuir probes embedded in the target plate.

Released impurities will be ionized near the plates. Their parallel streaming is effected by (1) friction forces, which tend to bring their flow in-line with the streaming velocity of the background plasma ions, by (2) thermal forces $\propto \partial T_\mathrm{i}/\partial s$ which tend to drive the impurities back from the plates, and by (3) electric forces which occur because of the electric potentials needed to establish ambipolarity of ion and electron fluxes.

In particular the thermal forces imply a sufficient plasma flow towards the divertor plates that drives the impurity ions back to their place of birth. The ratio of the probabilities for impurities to enter the main plasma, either when emitted from the divertor plates, or when they are sputtered at the main chamber wall allows for the assessment of the retention capability of the divertors [57].

Especially under high-density conditions, where the re-ionization of the fuel neutrals is close to the target, the unfortunate case can occur that certain flux tubes will experience an excess of ionization, resulting in flow reversal in that flux tube. In such a case, the number of fuel neutrals ionized in the tube exceeds the outflow of plasma (fuel) to the plate [58–60].

Not only is the retention of impurities important, but also the fuel neutrals should be kept inside the divertor regions by means of a closed divertor design, which is essential in order to allow efficient pumping of H and He and to prevent charge-exchange sputtering at the main plasma wall. The neutrals must recirculate in the whole divertor chamber. Efficient momentum removal and thermalization of hot neutrals require the wall to be as close as possible to the plasma. Possible solutions are a wide gas box with transparent walls close to the plasma or, as a possible alternative, the application of vertical targets in the "slot divertor" [61]. In particular, charge-exchange neutrals are only helpful if they can escape, travel to a region of high ion temperature, and return as cold ions to the target. In a slot divertor, the neutrals transfer their energy to the side walls. If they are ionized in front of the target, they will give their energy to the same local region and only marginally contribute to a reduction of power flux densities.

The He pump capability of the divertor depends, on one hand, whether it will be possible to achieve a sufficient accumulation of He in the divertor plasma, and on the other hand, requires a pump which recycles the hydrogen isotopes such as D, T, and retains He.

In addition to the normal load, the divertor components have to withstand electromagnetic loads due to disruptions leading to eddy currents up to several MA conducted through the divertor structure and resulting in loads of several hundred tons.

A special divertor concept has been developed for stellarators and will be tested in future experiments. The complex plasma geometry characterized by island structures at the edge offers relaxed possibilities to enable the required particle and power exhaust. It has been shown that in the case of optimized stellarators the critical "leading-edge" problem, i.e., excessive heat and particle deposition due to vertical impact of field lines on the targets, can be solved [45] in spite of the complex three-dimensional geometry.

Part II

The Plasma-Material Interface

5

The Plasma State

The equilibrium between the thermal (or random kinetic) energy of the particles and the interparticle binding potential determines the state of matter. A plasma is a quasineutral gas of charged and neutral particles, while the transition from a gas to a plasma occurs gradually with increasing temperature. In a neutral gas, all macroscopic forces are transmitted to each particle by collisions. A plasma exhibits in addition to binary collisions collective behavior, i.e., the motion of charged particles generates electric fields and currents, which, in turn, give rise to magnetic fields. Owing to the particle motion in these self-generated fields (and external fields) the plasma is able to sustain a great variety of wave phenomena and show its unique property: long-range interaction.

The term plasma is used to describe media containing free charged particles, which remain macroscopically neutral. Three conditions must be satisfied to call an ionized gas a plasma:

1. The Debye length λ_D (Sect. 5.2) must be smaller than the plasma dimension.
2. The number of particles N_D in a sphere with radius of λ_D should be large in ideal plasmas.
3. The frequency of typical plasma oscillations ω_pl (Sect. 5.3) should be larger than the reciprocal value of the mean time between collisions with neutral atoms.

The three parameters, λ_D, N_D, and ω_pl, which characterize a plasma and its behavior, are in close relationship to each other [62, 63]. In particular, a shift of $N_\mathrm{D} = n_\mathrm{e}(4/3)\pi\lambda_\mathrm{D}^3$ electrons, usually several tens of thousands, over a distance of about $\lambda_\mathrm{D} = \sqrt{\epsilon_\mathrm{o} k_\mathrm{B} T_\mathrm{e}/(n_\mathrm{e} e^2)}$ (keeping the positively-charged particles at their initial positions) leads to an increase of potential energy in that region, which is comparable to the kinetic energy of one electron $\propto k_\mathrm{B} T_\mathrm{e}$. The plasma frequency ω_pl is roughly the reciprocal value of the time

t_{pl} required for an electron to fly over a distance of λ_{D}

$$\omega_{\mathrm{pl}} \simeq \frac{1}{t_{\mathrm{pl}}} = \frac{v_{\mathrm{e}}}{\lambda_{\mathrm{D}}} \approx \frac{\sqrt{k_{\mathrm{B}}T_{\mathrm{e}}/m_{\mathrm{e}}}}{\sqrt{\epsilon_{\mathrm{o}}k_{\mathrm{B}}T_{\mathrm{e}}/(n_{\mathrm{e}}e^2)}} = \sqrt{\frac{n_{\mathrm{e}}e^2}{\epsilon_{\mathrm{o}}m_{\mathrm{e}}}} \tag{5.1}$$

and defines how fast a plasma (and especially the electrons) can react on disturbances of charge neutrality. The extent of disturbed regions, i.e., local regions where charge imbalance is allowed in a plasma, is given by the Debye length.

Every forced disturbance is met by the plasma with an adequate rearrangement of the charged particles. This collective adjustment is much faster and more effective than the momentum and energy change due to binary collisions. In fact, the so-called self-collision time τ_{c} (see (5.50)), a measure of the collisions between charged particles, is much larger than $\tau_{\mathrm{pl}} = 1/\omega_{\mathrm{pl}}$

$$\frac{t_{\mathrm{c}}}{t_{\mathrm{pl}}} = \frac{2\pi\epsilon_{\mathrm{o}}^2 m_{\mathrm{e}}^2 v_{\mathrm{e}}^3}{e^4 n_{\mathrm{e}} \ln \Lambda}\frac{v_{\mathrm{e}}}{\lambda_{\mathrm{D}}} = \left(\frac{\epsilon_{\mathrm{o}}(m_{\mathrm{e}}\, v_{\mathrm{e}}^2)/3}{e^2 n_{\mathrm{e}}}\right)^2 \frac{9 \cdot 2\pi\, n}{\lambda_{\mathrm{D}}\, \ln \Lambda} = \frac{9 \cdot 2}{\ln \Lambda}\pi\lambda_{\mathrm{D}}^3 n_{\mathrm{e}} \approx N_{\mathrm{D}} \tag{5.2}$$

since N_{D} is a large number in a low-density, ideal plasma. The collective behavior of plasmas is mainly governed by the electrons having the same charge as the ions, and therefore feeling the same electric force, but are less inertia-controlled due to their smaller mass.

5.1 Ionization Degree and Coupling Constant

At non-zero temperatures any gas contains, along with the neutrals, a certain number of ionized atoms and electrons. Their concentration is small if the temperature is considerably below the ionization energy. For a plasma with no refueling or recycling and no transport losses, the neutral density would rapidly drop to a low value, which is the equilibrium value obtained by balancing the recombination rate with the ionization rate at a certain electron temperature.

The degree of ionization is defined as the ratio of the electron density to the total density of the electrons and neutral atoms

$$\frac{n_{\mathrm{e}}}{n_{\mathrm{n}} + n_{\mathrm{e}}} \tag{5.3}$$

not to be confused with the ionization ratio $n_{\mathrm{e}}/n_{\mathrm{n}}$. When the average energy of the particles approaches the ionization energy, the gas almost completely turns into an ionized plasma. In thermal equilibrium, the particle velocity distribution is a Maxwellian distribution (A.1). Often, a partial equilibrium is observed, where the temperatures determining the distributions for the electrons and the ions are different.

Even at very low concentrations of charged particles, their interactions dominates the behavior of an ionized gas. The cross-section for binary collisions of charged particles (see Sect. 5.4)

$$\sigma = \frac{e^4 \ln \Lambda}{4\pi\epsilon_o^2 m^2 v^4} \simeq \frac{e^4 \ln \Lambda}{36\pi\epsilon_o^2 (k_\mathrm{B} T_\mathrm{e})^2} = \frac{4 \times 10^{-17}}{T_\mathrm{e}[\mathrm{eV}]^2}\,\mathrm{m}^2 \tag{5.4}$$

is for $T_\mathrm{e} \leq 1\,\mathrm{eV}$ several orders of magnitude larger than the cross-section for ion-neutral collisions $\sigma_n \simeq 10^{-20}\,\mathrm{m}^2$. The cross-section according to (5.4) becomes, however, very small at high temperatures and therefore close to σ_n. A plasma (with negligible losses) is almost fully ionized at electron temperatures of a few eVs, and the concentration of neutral atoms is therefore low.

The main difference between the behavior of a plasma and a gas is due to the effects exerted by the electric and magnetic fields on the motion of the charged particles. The coupling constant Γ_coupl, i.e., the ratio of the average Coulomb energy of particles separated by the mean distance to the thermal energy

$$\Gamma_\mathrm{coupl} = \frac{Q^2}{4\pi\epsilon_o d}\frac{1}{k_\mathrm{B} T} \tag{5.5}$$

characterizes a system of interacting charged particles, where Q is the charge of the particles. The average distance between the particles is defined by the density n

$$d = \left(\frac{3}{4\pi n}\right)^{1/3} . \tag{5.6}$$

The boundary between ideal and non-ideal plasmas is given by $\Gamma_\mathrm{coupl} = 1$. When the ratio (5.5) exceeds a critical value of about 170 solidification occurs. In gaseous plasmas, this condition would require, for example, a density of

$$n > \left(170\,\frac{4\pi\epsilon_o 3^{1/3} k_\mathrm{B} T}{e^2 (4\pi)^{1/3}}\right)^3 = 4 \times 10^{32} (T[\mathrm{eV}])^3 \frac{1}{\mathrm{m}^3} = 4 \times 10^{29}\,\frac{1}{\mathrm{m}^3} \tag{5.7}$$

for a plasma temperature of 0.1 eV. In dusty plasmas, this condition can be satisfied. Small particles, grains embedded in a plasma are negatively charged by the incoming electrons, as every solid object does during plasma contact (see Chap. 7). The collected charge on a grain of only 0.1 μm radius can be very large (several thousand elementary charges) and the distance between the grains is small in a dense plasma. In this case, the electrostatic interaction energy between grains can become comparable to the thermal energy of their random motion, and the criterion $\Gamma_\mathrm{coupl} > 170$ can be met. Thus, the grains can form regular structures or lattices—so-called Coulomb or plasma crystals, as theoretically predicted in 1986 by Ikezi [64]. The critical value of the coupling parameter depends on plasmas of the shielding effect acting on charged particles embedded in a uniform background of neutralizing charge. It increases nearly exponentially with d/λ_D [65].

Dusty plasmas as an ensemble of charged dust particles immersed in a plasma occur in planetary ring systems, interstellar clouds, and tails of comets [66, 67] but also on earth in plasma technology applications such as etching, plasma assisted deposition, or sputtering. The resulting contamination problems can only be reduced by a better understanding of the complex behavior of dusty plasmas with their tendency to self-organize into ordered structures. Analyzing (the rather large) plasma crystals by both simulations and experiments offers the chance for detailed studies of solidification in three dimension, melting, and phase transitions [68, 69].

5.2 Debye Length

In the case of no magnetic field, the Debye length turns out to be the characteristic length describing the screening of a single test charge Q in a plasma by the surrounding electrons and ions. The density distributions of the electrons $n_\mathrm{e}(r)$ and ions $n_\mathrm{i}(r)$ as a function of distance r from this test charge are described by

$$n_\mathrm{e}(\boldsymbol{r}) = n_\mathrm{pl} \exp[e\phi(\boldsymbol{r})/(k_\mathrm{B}T_\mathrm{e})] \tag{5.8}$$
$$n_\mathrm{i}(\boldsymbol{r}) = n_\mathrm{pl} \exp[-e\phi(\boldsymbol{r})/(k_\mathrm{B}T_\mathrm{i})] \tag{5.9}$$

where n_pl is the undisturbed plasma density far away ($n_\mathrm{pl} = n_\mathrm{e}(\boldsymbol{r} \to \infty) = n_\mathrm{i}(\boldsymbol{r} \to \infty)$), T_i and T_e the temperature of ions and electrons, respectively, and $\phi(r)$ is the electric potential distribution. Using the Poission equation

$$\nabla \cdot \boldsymbol{E}(\boldsymbol{r}) = \rho(\boldsymbol{r})/\epsilon_\mathrm{o}\,, \quad \boldsymbol{E}(\boldsymbol{r}) = -\nabla\phi(\boldsymbol{r}) \tag{5.10}$$

where the charge density is given by

$$\rho(\boldsymbol{r}) = e[n_\mathrm{i}(\boldsymbol{r}) - n_\mathrm{e}(\boldsymbol{r})] + Q\delta(\boldsymbol{r}) \tag{5.11}$$

we obtain

$$\nabla^2\phi(\boldsymbol{r}) + \frac{en_\mathrm{pl}}{\epsilon_\mathrm{o}} \{\exp[-e\phi(\boldsymbol{r})/(k_\mathrm{B}T_\mathrm{i})] - \exp[e\phi(\boldsymbol{r})/(k_\mathrm{B}T_\mathrm{e})]\} = -\frac{Q}{\epsilon_\mathrm{o}}\,\delta(\boldsymbol{r}) \tag{5.12}$$

with the Dirac function $\delta(\boldsymbol{r})$. With the condition that the kinetic energy is much higher than the potential disturbance caused by the test charge

$$e\phi(\boldsymbol{r}) \ll k_\mathrm{B}T_\mathrm{e,i} \tag{5.13}$$

(5.12) simplifies to

$$\nabla^2\phi(\boldsymbol{r}) - (1/\lambda_\mathrm{D}^2)\phi(\boldsymbol{r}) = -\frac{Q}{\epsilon_\mathrm{o}}\,\delta(\boldsymbol{r}) \tag{5.14}$$

with

$$\lambda_{\mathrm{D}} = \sqrt{\frac{\epsilon_{\mathrm{o}} k_{\mathrm{B}} T_{\mathrm{e}} k_{\mathrm{B}} T_{\mathrm{i}}}{n_{\mathrm{pl}} e^2 (k_{\mathrm{B}} T_{\mathrm{e}} + k_{\mathrm{B}} T_{\mathrm{i}})}} \,. \tag{5.15}$$

The Debye length does not depend on the mass of the involved particles, i.e., inertia does not play a role here, but is affected by the smaller temperature of both plasma constituents—the smaller the kinetic energy of one species, the smaller the region of potential disturbance as well. In spherical coordinates, the solution of (5.14) is

$$\phi(r) = \frac{1}{4\pi\epsilon_{\mathrm{o}}} \frac{Q}{r} \exp(-r/\lambda_{\mathrm{D}}) \tag{5.16}$$

which is valid for $r \neq 0$ and satisfies the condition $\phi(r) \to 0$ for $r \to \infty$, and $\phi(r) = Q/(4\pi\epsilon_{\mathrm{o}} r)$ for $r \to 0$. Accordingly, we obtain for the density distribution $(\exp(\pm e\phi(r)/k_{\mathrm{B}}T) \approx 1 \pm e\phi(r)/k_{\mathrm{B}}T)$

$$\begin{aligned} \rho(\boldsymbol{r}) &= -n_{\mathrm{pl}} e^2 \phi(\boldsymbol{r}) \left(\frac{1}{k_{\mathrm{B}} T_{\mathrm{i}}} + \frac{1}{k_{\mathrm{B}} T_{\mathrm{e}}} \right) + Q\delta(\boldsymbol{r}) \\ &= -\frac{Q}{4\pi r \lambda_{\mathrm{D}}^2} \exp(-r/\lambda_{\mathrm{D}}) + Q\delta(\boldsymbol{r}) \,. \end{aligned} \tag{5.17}$$

The test charge Q causes a spatial rearrangement of the charged particles in a way that the disturbance is as small as possible. Of course, one species alone is able to realize the screening, the other might be fixed. Out of the sphere with radius λ_{D}, the effect of the test charge almost vanishes.

The process of charge screening, however, should not be considered as a static one as might be suggested by the above analysis. The charged particles oscillate around a test charge driven by continuously rising and decaying electric fields and inertia. The distribution (5.17) is obtained only be averaging over several oscillation periods.

5.3 Plasma Frequency

Having small density fluctuations in a fully ionized plasma

$$n_{\mathrm{e}}(\boldsymbol{r}, t) = n_{\mathrm{pl}} + n'_{\mathrm{e}}(\boldsymbol{r}, t) \quad \text{and} \quad n_{\mathrm{i}}(\boldsymbol{r}, t) = n_{\mathrm{pl}} + n'_{\mathrm{i}}(\boldsymbol{r}, t) \tag{5.18}$$

the continuity equations of both species can be linearized

$$\frac{\partial n'_{\mathrm{e,i}}}{\partial t} + n_{\mathrm{pl}} (\nabla \cdot \boldsymbol{v}_{\mathrm{e,i}}) = 0 \tag{5.19}$$

as well as the momentum equations assuming constant temperatures

$$\frac{\partial \boldsymbol{v}_{\mathrm{e,i}}}{\partial t} = \frac{Q_{\mathrm{e,i}}}{m_{\mathrm{e,i}}} \boldsymbol{E} - \frac{k_{\mathrm{B}} T_{\mathrm{e,i}}}{m_{\mathrm{e,i}} n_{\mathrm{pl}}} \nabla n'_{\mathrm{e,i}} - \nu_{\mathrm{e,i}} \boldsymbol{v}_{\mathrm{e,i}} \tag{5.20}$$

with the charges $Q_\mathrm{i} = q_\mathrm{i}e$ and $Q_\mathrm{e} = -e$. Neglecting collisions and the pressure term, since a small density disturbance is already sufficient to generate large electric fields, allows for the removal of the temperature dependence. For singly charged ions ($q_\mathrm{i} = 1$), the electric field is given by

$$\nabla \cdot \boldsymbol{E} = \frac{\rho}{\epsilon_\mathrm{o}} = \frac{e}{\epsilon_\mathrm{o}}(n_\mathrm{i} - n_\mathrm{e}) = \frac{e}{\epsilon_\mathrm{o}}(n'_\mathrm{i} - n'_\mathrm{e}) \,. \tag{5.21}$$

Taking the divergence of (5.20) and combining (5.19) with (5.21), we obtain for each species

$$\frac{\partial^2 n'_\mathrm{e,i}}{\partial t^2} = -\frac{Q_\mathrm{e,i} n_\mathrm{pl}}{m_\mathrm{e,i}} \nabla \cdot \boldsymbol{E} = -\frac{Q_\mathrm{e,i} n_\mathrm{pl}}{m_\mathrm{e,i}} \frac{e}{\epsilon_\mathrm{o}}(n'_\mathrm{i} - n'_\mathrm{e}) \,. \tag{5.22}$$

Substituting $n'_\mathrm{pl} = n'_\mathrm{i} - n'_\mathrm{e}$ yields

$$\frac{\partial^2}{\partial t^2}(n'_\mathrm{i} - n'_\mathrm{e}) = -\frac{e^2 n_\mathrm{pl}}{m_\mathrm{i}\epsilon_\mathrm{o}}(n'_\mathrm{i} - n'_\mathrm{e}) - \frac{e^2 n_\mathrm{pl}}{m_\mathrm{e}\epsilon_\mathrm{o}}(n'_\mathrm{i} - n'_\mathrm{e}) \tag{5.23}$$

and thus

$$\frac{\partial^2 n'_\mathrm{pl}}{\partial t^2} + \omega_\mathrm{pl}^2 n'_\mathrm{pl} = 0 \,. \tag{5.24}$$

The plasma density disturbance n'_pl varies harmonically according to the plasma frequency

$$\omega_\mathrm{pl} = \sqrt{\frac{n_\mathrm{pl} e^2}{\epsilon_\mathrm{o}} \left(\frac{1}{m_\mathrm{e}} + \frac{1}{m_\mathrm{i}} \right)} = \sqrt{\frac{n_\mathrm{pl} e^2 (m_\mathrm{e} + m_\mathrm{i})}{\epsilon_\mathrm{o} m_\mathrm{e} m_\mathrm{i}}} \,. \tag{5.25}$$

In the case of $m_\mathrm{e} \ll m_\mathrm{i}$, (5.25) reduces to the well-known electron plasma frequency

$$\omega_\mathrm{pe} = \sqrt{n_\mathrm{pl} e^2 / (m_\mathrm{e} \epsilon_\mathrm{o})} \tag{5.26}$$

often simply called the plasma frequency.

5.4 Collisions in Plasmas

A particle does not perform a collision during a time Δt with the probability

$$f_\mathrm{P}(\Delta t) = \exp(-\Delta t / \tau) \tag{5.27}$$

where $\tau = 1/\nu$ is the average time between two collisions and ν the collision frequency. This relation is readily obtained from the balance equation for particles that have not participated in collisions after a time t

$$\frac{\mathrm{d}N}{\mathrm{d}t} = -\frac{N(t)}{\tau} \quad \rightarrow \quad N(t) = N_\mathrm{o} \exp(-t/\tau) \tag{5.28}$$

where N is the number of particles and $N_\mathrm{o} = N(t = 0)$. The collision time τ, the mean free path λ, and the velocity of the particle v are simply related to each other, via $\lambda = \tau v$. In other words, by crossing a plasma slab of thickness Δx the incoming flux density $\Gamma_{\rm in}$ is reduced by collisions

$$\Gamma_{\rm out} = \Gamma_{\rm in} - \Gamma_{\rm in}\, n\, S\, \Delta x \,\frac{\sigma}{S} \tag{5.29}$$

where n is the particle density in the plasma, S is the surface area of the plasma slab, and σ is the cross-section. The ratio σ/S is a measure of the collision probability, and the term $n\,S\,\Delta x$ gives simply the number of particles in the plasma slab. For the change of the flux density, we obtain

$$\frac{\Gamma_{\rm out} - \Gamma_{\rm in}}{\Delta x} = \frac{\Delta \Gamma}{\Delta x} = \frac{{\rm d}\Gamma}{{\rm d}x} = -n\sigma\Gamma \tag{5.30}$$

with the solution $\Gamma = \Gamma_{\rm o} \exp(-x/\lambda)$ where $\lambda = 1/(n\sigma)$ is the mean free path, $\tau = \lambda/v$ is the mean time between two collisions, and $\nu = n\sigma v$ is the collision frequency.

In partially-ionized plasmas, collisions with neutral atoms may be important. The corresponding cross-section is simply given by

$$\sigma = \pi r_{\rm o}^2 \simeq 10^{-20}\,{\rm m}^2 \tag{5.31}$$

where $r_{\rm o}$ is the gas-kinetic radius. In a fully ionized plasma, the motion of each charged particle is mainly governed by the long-range Coulomb interaction with the force $F = e^2/(4\pi\epsilon_{\rm o} r^2)$, where r is the distance between both (singly charged) particles. A strong deflection ($\geq 90^{\rm o}$) is always accompanied by a change of momentum, i.e., the forward directed momentum is lost,

$$\begin{aligned} m\frac{\Delta v}{\Delta t} &= F \\ \frac{\Delta(mv)}{\Delta t} \approx \frac{mv}{\Delta t} &= \frac{e^2}{4\pi\epsilon_{\rm o} r^2}\,. \end{aligned} \tag{5.32}$$

The time of interaction Δt can be estimated with $\Delta t = r/v$ yielding

$$r = \frac{e^2}{4\pi\epsilon_{\rm o} m v^2} \tag{5.33}$$

and the cross-section for large deflections is

$$\sigma = \pi r^2 = \frac{e^4}{16\pi\epsilon_{\rm o}^2 m^2 v^4}\,. \tag{5.34}$$

In fact, r is equal to the impact parameter required for a $90^{\rm o}$ deflection and twice the value of distance Δ at which the potential energy is equal to the kinetic energy of the incoming particle

$$\frac{e^2}{4\pi\epsilon_{\rm o}\Delta} = \frac{mv^2}{2} \quad \rightarrow \quad \Delta = \frac{e^2}{2\pi\epsilon_{\rm o} m v^2} = 2r = 2\rho_{\rm p}^{90^{\rm o}}\,. \tag{5.35}$$

However, small-angle scattering contributes mainly to the cross section of binary collisions. Therefore, the effective cross-section is by a factor of $4\ln\Lambda$ larger than given by (5.34). The Coulomb logarithm $\ln\Lambda$ varies typically from 5 for cold and dense plasmas to about 20 in the case of hot fusion plasmas. The corresponding collision frequency is then (see (5.34))

$$\nu = n\sigma v = \frac{ne^4 \ln\Lambda}{4\pi\epsilon_o^2 m^2 v^3} \quad \text{or} \quad \nu = \frac{ne^4 \ln\Lambda}{4\pi\epsilon_o^2 \sqrt{m}\,(3k_B T)^{3/2}} \tag{5.36}$$

(with $mv^2/2 = 3k_B T/2$). The symbol ν is used for electrons with mass m_e higher than ions with mass M by the factor

$$\frac{\nu_{ei}}{\nu_{ii}} = \sqrt{\frac{M}{m_e}} \simeq 40\ . \tag{5.37}$$

Usually, the collision frequencies are obtained by taking into account the velocity distributions of both the "test" and "field" particles (Sect. A.1) [14].

The exchange of momentum in collisions results in friction, when different particles are involved, and in viscosity in the case of collisions between particles of the same species. The friction force

$$\begin{aligned} F_R &= (\text{flux density}) \times (\text{cross} - \text{section}) \times (\text{momentum transfer}) \\ &= (\text{collision frequency}) \times (\text{momentum transfer}) \\ &= (nv) \cdot \sigma \cdot (mv) \end{aligned} \tag{5.38}$$

equals the change of momentum in time. Therefore we obtain

$$F_R = 4\ln\Lambda\, nm\pi \left(\rho_p^{90^\circ}\right)^2 v^2 \tag{5.39}$$

where $\rho_p^{90^\circ}$ is the impact parameter for a 90° deflection. In the equations of plasma physics, this force appears, for example, as (in N/m^3)

$$\boldsymbol{F}_R = n_e m_e \nu_{ei} (\boldsymbol{v}_i - \boldsymbol{v}_e) = m_e \nu_{ei} \frac{\boldsymbol{j}}{e} \tag{5.40}$$

expressing the force acting on electrons due to their collisions with the plasma ions. Bearing in mind that each plasma species has its own velocity distribution, the term $(\boldsymbol{v}_i - \boldsymbol{v}_e)$ is the velocity difference of both species. Note that $\nu_{ei} \neq \nu_{ie}$ but the friction forces—according to Newton's law "actio" equals "reactio"—are the same, i.e., $F_{ei} = -F_{ie}$.

Several physical effects in plasmas besides friction such as mobility, diffusion, and conductivity are closely connected to collision processes. From the force balance

$$en\boldsymbol{E} = \nabla p + m\,n\,\nu\,\boldsymbol{v} \tag{5.41}$$

we obtain for an isothermal plasma the solution

$$\boldsymbol{v} = \frac{e\,n}{m\,n\,\nu}\boldsymbol{E} - \frac{k_B T}{m\,\nu}\frac{\nabla n}{n} = \mu_b\,\boldsymbol{E} - D\frac{\nabla n}{n} \tag{5.42}$$

with the mobility coefficient $\mu_\text{b} = e/(m\nu)$ and the diffusion coefficient $D = k_\text{B}T/(m\nu)$. This description simplifies the process of acceleration in the electric field between two single collisions and the slowing down process during collisions.

In the case of no pressure gradient, the current density according to (5.42)

$$j = env = \frac{e^2\,n}{m\,\nu}\boldsymbol{E} = \sigma_\text{c}\boldsymbol{E} \tag{5.43}$$

is characterized by the coefficient σ_c

$$\sigma_\text{c} = \frac{e^2\,n}{m\,\nu} = \frac{4\pi\epsilon_0^2\,(3k_\text{B}T)^{3/2}}{e^2\ln\Lambda\sqrt{m}} \quad \text{or in general} \quad \sigma_\text{c} = \frac{e^2 n}{m\sum_k \nu_{eV}} \tag{5.44}$$

called plasma conductivity, where the collision frequency is taken from (5.36). It depends strongly on plasma temperature T but not on plasma density, since both the collision frequency and the current density are proportional to the particle density. The exact value of the plasma conductivity including also the effect of the effective charge is given by [70]

$$\sigma_\text{c} = 1.92\times 10^4\left(2 - Z_\text{eff}^{-1/3}\right)\frac{T_\text{e}^{3/2}[eV]}{Z_\text{eff}\ln\Lambda}\frac{1}{\Omega\text{m}}\,. \tag{5.45}$$

A clean H-plasma with $T_\text{e} = 100\ eV$ has about the same resistivity as stainless steel ($7\times10^{-7}\,\Omega\,\text{m}$). At a temperature of about 1 keV, its conductivity comes close to the conductivity of copper ($2\times10^{-8}\,\Omega\,\text{m}$). With higher plasma temperatures, the conductivity increases up to even higher values, when a plasma current of several MA is driven by a loop voltage of only a few volts in tokamak experiments.

Several relaxation times can be introduced. The "slowing down time" is defined by

$$t_\text{s} = -\frac{v}{\langle\Delta v/\Delta t\rangle} \tag{5.46}$$

with the average velocity change per unit time $\langle\Delta v/\Delta t\rangle$. For small velocities, t_s is near constant, and for higher velocities $t_\text{s}\propto v^3$. The "deflection time" is given by

$$t_\text{d} = \frac{v^2}{\langle(\Delta v_\perp)^2/\Delta t\rangle} \tag{5.47}$$

with the average change of the square of the velocity component perpendicular to the initial direction of the test particle $\langle(\Delta v_\perp)^2/\Delta t\rangle$. The change of energy in a collision can be expressed as

$$\begin{aligned}\Delta E &= \frac{M}{2}\left[(v+\Delta v_\parallel)^2 + \Delta v_\perp^2\right] - \frac{1}{2}Mv^2\\ &= \frac{M}{2}\left[2v\Delta v_\parallel + \Delta v_\parallel^2 + \Delta v_\perp^2\right] \approx Mv\Delta v_\parallel\,.\end{aligned} \tag{5.48}$$

Hence, we have for the "energy exchange time"

$$t_E = \frac{E^2}{\langle(\Delta E)^2\rangle} = \frac{M^2v^4}{4M^2v^2\langle(\Delta v_\parallel)^2/\Delta t\rangle} = \frac{v^2}{4\langle(\Delta v_\parallel)^2/\Delta t\rangle} \tag{5.49}$$

with the average change of the square of the velocity component parallel to the initial direction $\langle(\Delta v_{a\parallel})^2/\Delta t\rangle$. One important parameter is the "self-collision time" t_c. It determines the time required to establish thermal equilibrium by collisions after a disturbance. For test and field particles of the same kind moving with the average thermal velocity, we have approximately [71]

$$t_c \approx t_E \approx t_D \approx \frac{2\pi\epsilon_o^2M^2v^3}{Z^4e^4n\ln\Lambda} \approx \frac{\epsilon_o^2\sqrt{M}(k_BT)^{3/2}}{Z^4e^4n} \; . \tag{5.50}$$

For a hydrogen plasma with a temperature of $k_BT = 1\,\mathrm{keV}$ and a density of $n = 10^{19}\,\mathrm{m}^{-3}$, the self-collision time is about $t_c = 0.001\,\mathrm{s}$. In the case of $k_BT = 100\,\mathrm{eV}$, we obtain $t_c = 3 \times 10^{-5}\,\mathrm{s}$. This characteristic time increases strongly with temperature, and it is much larger for ions than for the light electrons.

With an average distance between the particles of about

$$\Delta_l = \frac{1}{n_{\mathrm{pl}}^{1/3}} = 1.15 \times 10^{-7}\,\mathrm{m} = 1150\,\text{Å} \tag{5.51}$$

for a plasma density of $n_{\mathrm{pl}} = 10^{20}\,1/\mathrm{m}^3$, it is useful to distinguish binary collisions, i.e., to neglect here the interaction of one particle with all particles in the near and far neighborhood, since the impact parameter $\rho_p^{90°}$ for a 90° deflection of an electron with an energy of $E = 10\,\mathrm{eV}$ at a proton

$$\rho_p^{90°} = \frac{e^2(m_e + m_p)}{8\pi\epsilon_o m_p E} = 7.2 \times 10^{-11}\,\mathrm{m} = 0.72\,\text{Å} \tag{5.52}$$

is small in comparison to Δ_l. The exact derivation of (5.52) is given in Sect. 6.1 (see (6.34)). Even for a 1 degree deflection, the corresponding impact parameter is rather small, i.e., $\rho_p^{1°} = \rho_p^{90°}\cot(\alpha/2) = 8.2 \times 10^{-9}\,\mathrm{m} = 82\,\text{Å} < \Delta_l$.

For larger impact parameters, $\rho_p > \Delta_l$, the model of binary collisions becomes questionable. Nevertheless, up to a value of $\rho_p = \lambda_D$ ($\lambda_D = \sqrt{\epsilon_o k_B T_e/(n_{\mathrm{pl}}e^2)} = 2 \times 10^{-6}\,\mathrm{m}$ for $k_BT_e = 10\,\mathrm{eV}$, i.e., $\lambda_D \gg \Delta_l \gg \rho_p^{90°}$) multiple collision events can be thought as a sequence of several binary collisions. With this assumption, the collision cross-section is usually derived by integrating in the interval $\rho_p = [0, \lambda_D]$. This integration leads to the additional factor $4\ln\Lambda$ mentioned above.

In contrast to solids, the charged particles in plasmas are not fixed at rigid lattice positions. They react sensitively in electric and magnetic fields by rearrangement, charge separation, and induced currents. The influence of the self-generated fields back on the particle motion itself leads to the flexible and chaotic appearance of plasmas, which try to evade any manipulation.

Each single particle *feels* the surrounding plasma particles and moves in their field. Such terms as mean free path or collision time are not fully adequate to deal with the true many-particle interactions in plasmas.

Collective processes dominate on a spatial scale equal and larger than the Debye length. Introducing a density description (Sect. 5.7) allows the determination, via the Poisson equation, of the electric field based on a certain spatial distribution of charged particles. Inside the Debye sphere, the local electric fields have to be considered as well. However, these fields are small in the case of a symmetrical arrangement of the particles. The more particles are included in the Debye sphere, the less local fluctuations play a role.

According to the described methodology collective effects, i.e., many-particle collisions, are more or less artificially separated from binary collisions even though the physics behind them, i.e., the Coulomb interaction, is the same. A plasma is often called collisionless when the absence of only the binary collisions with impact parameters smaller than the Debye length is meant.

5.5 Transport Processes in Plasmas

Transport effects arise whenever inhomogeneities, for example in density or energy, occur in plasmas. Several parameters help to classify transport phenomena. The degree of ionization decides which kind of collisions are important. The direction of the magnetic field is also the direction of maximum transport, since every process of momentum and energy transfer is much more efficient along the magnetic field lines than across them. The ratio of the thermal energy density of electrons and ions to the magnetic pressure determines the degree of confinement, which is limited due to arising instabilities at larger ratios. The ratio of the ion gyro-radius to the characteristic length scale of the gradients of the plasma parameter gives the regions of validity of the different simplifying theoretical models of plasma transport [14, 35, 36].

Any real plasma will have a density gradient. The plasma particles diffuse toward regions of low density, thereby tending to flatten out the density gradient. If the plasma is only weakly ionized, then the charged particles collide primarily with neutral atoms rather than with one another. In ionized plasmas collisions of charged particles dominate. Like-particle collisions, e.g., ion–ion collisions, do not lead to cross-field diffusion, since the gyration centers of two particles being involved in a collision move exactly equal but opposite distances across the magnetic field. For each particle that moves outward, there is another that moves inward. The transport due to unlike-particle collisions, e.g., electron–ion collisions, is ambipolar, since the displacements of the gyrations centers of both particles are equal for each collision [36].

The diffusion process can be described introducing the concept of a "random walk". In the direction of the magnetic field, the more or less free transport of the charged particles is reduced by collisions. The transport perpendicular to the magnetic field lines is, on the other hand, enhanced by collisions. During collisions the particles change their positions in a random walk across the field. The step length in this random-walk process is about the gyro-radius, which is usually much smaller than the mean free path between two collisions that characterized the random motion along the magnetic field lines. Global drifts of charged particles (Sect. A.1) usually do not lead to radial transport, rather to rotation of a plasma column.

The net fluxes of electrons and ions are adjusted to retain the high degree of charge neutrality as required in plasmas. This process of adjustment involves the generation of electric fields as soon as a slight charge imbalance occurs. Diffusion involves the transport of momentum in the plasma. The flux of particles to the density gradient is characterized by the diffusion coefficient. Thermal conduction is connected to the transport of kinetic energy. The thermal conductivity relates the heat flux to the temperature gradient.

Besides binary collisions, collective effects dominate the transport in magnetized plasmas. Local drifts caused by microinstabilities enhance the transport across the magnetic field by several orders of magnitude in comparison with the transport effects due to binary collisions. Analyzing experimental data of different fusion experiments, Fussmann concludes that the anomalous large transport of impurities does not depend on the impurity mass, charge, or velocity [72]. This serves as a clear indication that electric drift effects are responsible for the enhanced transport. The origins of anomalous diffusion are assumed to lie in fluctuating electric fields, which exceed the thermal level.

5.5.1 Transport by Binary Collisions

In the classical picture, the gyrating charged particles are fixed to the magnetic field lines. Radial displacements are only possible by collisions. The corresponding diffusion coefficient for cross-field transport is, in the frame of the random-walk model, given by the collision frequency ν and the square of the mean step length, which is about the gyro-radius. For electrons, we have then $(m_\mathrm{e} v_\mathrm{e}^2/2 = 3k_\mathrm{B}T_\mathrm{e}/2)$

$$D_\perp^{\mathrm{classical}} \simeq \nu\, \rho_{\mathrm{ce}}^2 = \nu \left(\frac{v_\mathrm{e}}{\omega_{ce}}\right)^2 \simeq \nu\, \frac{3\, m_\mathrm{e} k_\mathrm{B} T_\mathrm{e}}{e^2\, B^2}\,. \tag{5.53}$$

The prefactor $1/(2d)$ in (6.109) has been omitted for simplification. Only unlike-particle collisions lead to transport, and we have with (5.36)

$$\begin{aligned} D_\perp^{\mathrm{classical}} &= \nu_{\mathrm{ei}}\, \rho_\mathrm{e}^2 = \frac{1}{4\pi\epsilon_\mathrm{o}^2} \frac{e^4\, n \ln\Lambda}{\sqrt{m_\mathrm{e}}\, (3k_\mathrm{B}T_\mathrm{e})^{3/2}} \frac{3\, m_\mathrm{e}\, k_\mathrm{B} T_\mathrm{e}}{e^2\, B^2} \\ &= \frac{e^2\, n \ln\Lambda\, \sqrt{m_\mathrm{e}}}{4\pi\epsilon_\mathrm{o}^2\, B^2 \sqrt{3k_\mathrm{B}T_\mathrm{e}}} \\ &\simeq 3\times 10^{-22}\, \frac{n\,[1/\mathrm{m}^3]}{(B\,[\mathrm{T}])^2 \sqrt{T_\mathrm{e}\,[\mathrm{eV}]}} \left[\frac{\mathrm{m}^2}{\mathrm{s}}\right]. \end{aligned} \tag{5.54}$$

For example, taking $B = 2\,\mathrm{T}$, $T_\mathrm{e} = 10\,\mathrm{eV}$ and $n = 10^{19}\,1/\mathrm{m}^3$, (5.54) yields a very small diffusion coefficient of $D_\perp^{\mathrm{classical}} = 2.4 \times 10^{-4}\,\mathrm{m}^2/\mathrm{s}$. Important to note is that the same result is obtained when considering the plasma ions (if $T_\mathrm{i} = T_\mathrm{e}$), since

$$\nu_{\mathrm{ei}}\rho_\mathrm{e}^2 = \nu_{\mathrm{ie}}\rho_\mathrm{i}^2 \tag{5.55}$$

due to $\nu_{\mathrm{ie}} = (m_\mathrm{e}/M_\mathrm{i})\,\nu_{\mathrm{ie}}$.

In the fluid picture, the simplified Ohm's law [35]

$$\boldsymbol{E} + \boldsymbol{v} \times \boldsymbol{B} = \eta_\parallel \boldsymbol{j}_\parallel + \eta_\perp \boldsymbol{j}_\perp \tag{5.56}$$

is used together with the pressure balance

$$\boldsymbol{j} \times \boldsymbol{B} = \nabla p \tag{5.57}$$

where $\boldsymbol{j}$ is the current density, η the resistivity, and p the plasma pressure $p = (k_\mathrm{B}T_\mathrm{e} + k_\mathrm{B}T_\mathrm{i})n$. Taking the cross product of (5.56) with $\boldsymbol{B}$ and replacing $\boldsymbol{j} \times \boldsymbol{B}$ from (5.57) yields the fluid flow velocity across the magnetic field

$$\boldsymbol{v}_\perp = -\eta_\perp \frac{\nabla p}{B^2} + \frac{\boldsymbol{E} \times \boldsymbol{B}}{B^2} \tag{5.58}$$

where the first term describes the transport caused by collisions, and the second term is the electric drift term. For the collision driven flux density, we obtain

$$n v_\perp = -n\,\frac{\eta_\perp (k_\mathrm{B}T_\mathrm{e} + k_\mathrm{B}T_\mathrm{i})\nabla n}{B^2} = -D_\perp^{\mathrm{classical}}\,\nabla n \tag{5.59}$$

the familiar result (5.54), since the resistivity can be represented as $\eta = m_\mathrm{e}\nu_{\mathrm{ei}}/(e^2 n) \propto 1/(k_\mathrm{B}T_\mathrm{e})^{3/2}$.

5.5.2 Neoclassical Diffusion

In toroidal devices such as tokamaks and stellarators, the magnetic field has to be twisted in order to eliminate gradient-B and curvature drifts. While a charged particle moves along a magnetic field line, it comes into regions with higher magnetic field strength near the inner wall of the torus and regions with weaker magnetic field strength near the outer wall at larger radii. A magnetic mirror configuration is therefore established and some particles are trapped. They do not circulate around the torus, but follow trajectories whose projections on the torus cross-section form banana-shaped orbits. By performing collisions, the particles are able to *hop* from one banana orbit to another. The resulting diffusion across the magnetic field is then characterized by the width of the banana orbit, which is usually much larger than the gyro-radius. This effect, called neoclassical diffusion, enhances the transport by a factor of 10 to 100 in comparison to classical diffusion. These values are, nevertheless, still too small to explain the large transport observed in fusion experiments. In addition, the neoclassical theory leads to some predictions and dependencies which could not be confirmed in experiments [73–75].

5.5.3 Anomalous Transport

The expected strong dependence of transport on the magnetic field strength B according to the classical theory, i.e., $D_\perp \propto 1/B^2$, in the beginning of fusion research gave rise to hope of quick realization of controlled fusion. Soon, an anomalously large diffusion, which exceeded the calculated one by four orders of magnitude, was observed in the first experiments, i.e., $D_\perp^{\text{experiment}} \simeq D_\text{B} = (10^3\text{–}10^4)\, D_\perp^{\text{classical}}$. Furthermore, the experimental diffusion coefficient scales were more like $1/B$. Developing an arc discharge with magnetic field for uranium isotope separation, Bohm gave (without derivation) the following semi-empirical formula to describe this anomalous diffusion [76]

$$D_\perp = \frac{1}{16}\frac{k_\text{B}T_\text{e}}{e\,B} \equiv D_\text{B}\ . \tag{5.60}$$

This relation may be expressed using the random-walk approach as

$$D_\text{B} \simeq v_\text{e}\rho_\text{ce} = \omega_\text{ce}\,\lambda_\text{D}\,\rho_\text{ce} = \sqrt{\frac{k_\text{B}T_\text{e}}{m_\text{e}}\frac{\sqrt{k_\text{B}T_\text{e}\,m_\text{e}}}{e\,B}} = \frac{k_\text{B}T_\text{e}}{e\,B} \tag{5.61}$$

where ω_ce is the electron gyro-frequency and λ_D is the Debye length. Equation (5.61) does not provide, however, a reasonable, physical explanation of (5.60), since a step length of $\sqrt{\lambda_D\rho_\text{ce}}$ is of no physical relevance.

The scaling of D_B with $k_\text{B}T_\text{e}$ and B can be shown to be the natural one, whenever the losses are caused by $\boldsymbol{E} \times \boldsymbol{B}$ drifts due to fluctuating electric fields E [77]. The local particle flux density is then proportional to the drift velocity $v_\perp$

$$\Gamma_\perp = n v_\perp \propto n\frac{E}{B}\ . \tag{5.62}$$

Because of Debye shielding, the maximum potential in the plasma is given by $e\phi_\text{max} \approx k_\text{B}T_\text{e}$. If Δ is a characteristic scale length of the plasma, the maximum electric field is then

$$E_\text{max} \approx \frac{k_\text{B}T_\text{e}}{e\,\Delta} \tag{5.63}$$

leading to

$$\Gamma_\perp \approx \frac{n}{\Delta}\frac{k_\text{B}T_\text{e}}{e\,B} \approx -\frac{k_\text{B}T_\text{e}}{e\,B}\nabla n = -D_\text{B}\nabla n\,. \tag{5.64}$$

With the prefactor $1/16$, the relation (5.60) agrees surprisingly well with various experiments. Well applicable to the diffusion in the rather cold edge plasma of fusion experiments with $D_\text{B} \simeq 1\,\text{m}^2/\text{s}$, it is clearly not valid for the hot core region, since there it would suggest much too high values of the diffusion coefficient.

Although the anomalous transport is often treated in the frame of a diffusion description, it is actually governed by convective terms ("zonal flows") [78, 79] and the behavior of vortex structures ("blobs") in magnetized plasmas [80]. These turbulent eddies have a radial extent of 1–2 cm and a typical lifetime of about 0.5–1 ms. By poloidal shear flows, i.e., $v_{\text{poloidal}} = \boldsymbol{E}_{\text{radial}} \times \boldsymbol{B}_{\text{toroidal}}$, the eddies can be radially decorrelated resulting in suppression of cross-field transport. The Kelvin-Helmholtz instability as well as the Rayleigh convection instability occur not only in fluids but also in magnetized plasmas due to non-uniformity of plasma flow (shear flow) and non-uniformity of temperature, respectively.

Despite the effort and progress made in the field of plasma transport theories [81–87], reliable predictions can hardly be given up to now. In estimations and computer simulations the transport coefficients are usually given as constant input parameters, for example $D_\perp = 0.1$–$1\,\text{m}^2/\text{s}$.

The exact solution of the Vlasov equation (see below) by numerical methods or by using a PIC simulation opens, in principle, the opportunity to analyze the "anomalous" transport, since all the necessary physics is included. Such a task, however, would clearly overload existing computer systems.

5.6 The Vlasov Equation

The use of the Boltzmann kinetic equation represents one of the most comprehensive approaches to describe the collective dynamics in many-particle systems such as plasmas. In a very useful approximation, the Vlasov equation

$$\frac{df}{dt} = \frac{\partial f}{\partial t} + \boldsymbol{v} \cdot \frac{\partial f}{\partial \boldsymbol{r}} + \boldsymbol{a} \cdot \frac{\partial f}{\partial \boldsymbol{v}} = 0 \tag{5.65}$$

the collision term on the right-hand side of the Boltzmann equation is omitted, but the internal electric and magnetic fields are included in the force term

$$\boldsymbol{a} = \frac{\boldsymbol{F}_{\text{ext}} + \boldsymbol{F}_{\text{int}}}{m} = \frac{\boldsymbol{F}_{\text{ext}}}{m} + \frac{Q}{m} \left(\boldsymbol{E}_{\text{int}} + [\boldsymbol{v} \times \boldsymbol{B}_{\text{int}}] \right) . \tag{5.66}$$

The Vlasov equation describes the evolution of the particle distribution function $f = f(\boldsymbol{r}, \boldsymbol{v}, t)$ in time, where the force F is given by externally applied fields and by internal electromagnetic fields, i.e., $\boldsymbol{E}_{\text{int}}$ and $\boldsymbol{B}_{\text{int}}$. In many applications, the magnetic field is almost constant and given by external sources. In these cases, the plasma currents are too small to generate considerable internal magnetic fields, and the plasma pressure is much smaller than the magnetic pressure $B^2/(2\mu_\text{o})$. However, the internal electric field $\boldsymbol{E} = -\nabla\phi$ cannot be neglected. It should be obtained using the Poisson equation

$$\nabla^2 \phi = -\frac{\rho}{\epsilon_\text{o}} \tag{5.67}$$

where the electric potential ϕ is determined by the charge density ρ

$$\rho(\boldsymbol{r}) = Q \int f(\boldsymbol{r}, \boldsymbol{v}, t)\, \text{d}\boldsymbol{v} . \tag{5.68}$$

Analytical solutions of the self-consistent set of equations (5.65–5.68) are available for only very simplified situations. In most of the cases of interest, extensive numerical calculations are required.

In principle, all collision processes—including even binary collisions—can be described within that approach. This is at first glance somewhat surprising, since the collision term of the Boltzmann equation has been disregarded. However, the internal fields are obtained by a macroscopic smoothing of the local fields between the colliding particles. It is rather a question of spatial resolution, which is applied in the process of density calculation knowing the position of particles (Sect. 5.7), than a principal one. The characteristic length usually applied in the smoothing procedure is the Debye length, so one can argue that in this case collisions with impact parameters smaller than λ_D are neglected. More importantly, the collective processes, i.e., the many-particle interactions, which dominate the plasma behavior are still well-treated.

The evolution of the particle distribution function can be described by following the motion of the particles through phase space. The set of equations (5.65–5.68) can be solved, therefore, by evaluating the trajectories

$$m\frac{\mathrm{d}\boldsymbol{v}}{\mathrm{d}t} = Q\left(\boldsymbol{E} + [\boldsymbol{v}\times\boldsymbol{B}]\right) \tag{5.69}$$

of all particles, since the points on the enclosing surface of any volume in phase space move according to $\mathrm{d}\boldsymbol{r}/\mathrm{d}t = \boldsymbol{v}$. For the next time step, we have

$$\boldsymbol{v}' = \boldsymbol{v} + \int_t^{t'} \frac{Q}{m}\left(\boldsymbol{E} + [\boldsymbol{v}\times\boldsymbol{B}]\right)\,\mathrm{d}t \quad \text{and} \quad \boldsymbol{r}' = \boldsymbol{r} + \int_t^{t'} \boldsymbol{v}\,\mathrm{d}t\,. \tag{5.70}$$

The particle density is obtained directly from the positions of the particles and not via the distribution function according to (5.68). Knowing the positions and velocities of all particles for all moments of time, the distributions function is well-defined. Such a numerical simulation (see Sect. 7.6) is fully equivalent to the self-consistent solution of the Vlasov equation together with the Poisson equation.

5.7 The Poisson Equation

By defining a certain interaction law, for example the Coulomb interaction law, the force acting on one particle is given by the spatial positions of all other particles with respect to the position of the particle under consideration.

A point-like charge $q_i e$ generates the electric field

$$\boldsymbol{E}_i(\boldsymbol{r}) = \frac{q_i e}{4\pi\epsilon_\mathrm{o}}\frac{\boldsymbol{R}_i}{R_i^3} \tag{5.71}$$

at a position given by the radius vector $\boldsymbol{r}$, where $\boldsymbol{R}_i = \boldsymbol{r} - \boldsymbol{r}_i$ and $\boldsymbol{r}_i$ is the radius vector of the test charge $q_i e$. The flux of the electric field $\boldsymbol{E}_i$ through a closed surface is

$$\Gamma_{\mathrm{E},i} = \oint_S (\boldsymbol{n} \cdot \boldsymbol{E}_i) \mathrm{d}S \tag{5.72}$$

where $\boldsymbol{n}$ is the normal vector of the surface element $\mathrm{d}S$. From position $\boldsymbol{r}_i$ the surface element is seen under the solid angle element (Fig. 5.1)

$$\mathrm{d}\Omega = \mathrm{d}S(\boldsymbol{n} \cdot \boldsymbol{R}_i)/(R_i)^3 \,. \tag{5.73}$$

Considering (5.71), we obtain from (5.72)

$$\Gamma_{\mathrm{E},i} = \frac{q_i e}{4\pi\epsilon_o} \oint_S \mathrm{d}\Omega = \begin{cases} 4\pi \frac{q_i e}{4\pi\epsilon_o} & \text{for} \quad \boldsymbol{r}_i \in V \\ 0 & \text{for} \quad \boldsymbol{r}_i \notin V \end{cases} \tag{5.74}$$

where the volume V is enclosed by the surface S. The field generated by several charges

$$\boldsymbol{E}(\boldsymbol{r}) = \sum_i \boldsymbol{E}_i = \sum_i \frac{q_i e}{4\pi\epsilon_o} \frac{\boldsymbol{R}_i}{R_i^3} \tag{5.75}$$

is the superposition of the various individual fields. The general flux of the electric field yields

$$\Gamma_{\mathrm{E}} = \oint_S (\boldsymbol{n} \cdot \boldsymbol{E}) \mathrm{d}S = \sum_i \Gamma_{\mathrm{E},i} = \frac{1}{\epsilon_o} \sum_{i \in V} q_i e = \frac{Q_{\mathrm{t}}}{\epsilon_o} \tag{5.76}$$

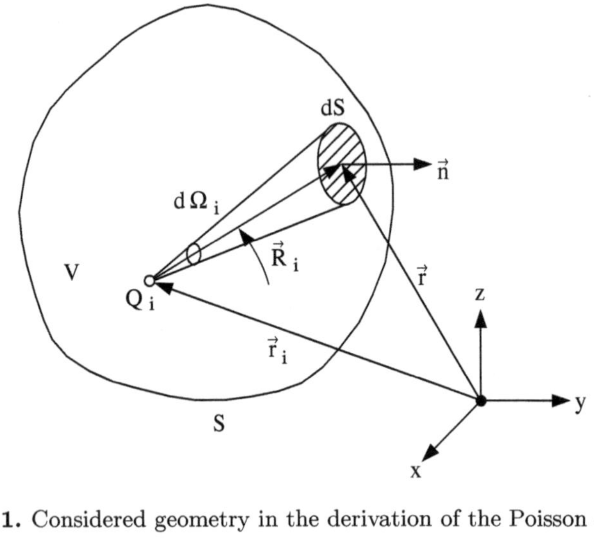

Fig. 5.1. Considered geometry in the derivation of the Poisson equation

where Q_t is the total charge enclosed in the volume V. Dividing the space now into small volume elements ΔV_k with their centers $\boldsymbol{r}_k$, a charge density can be defined

$$\rho(\boldsymbol{r}) = \frac{1}{\Delta V_k} \sum_{i \in \Delta V_k} q_i e \tag{5.77}$$

leading in the limit $\Delta V_k \to 0$

$$Q_\mathrm{t} = \int_V \rho \,\mathrm{d}V \,. \tag{5.78}$$

Under the assumption of $\boldsymbol{R}_i = \boldsymbol{r} - \boldsymbol{r}_i \approx \boldsymbol{r} - \boldsymbol{r}_k = \boldsymbol{R}_k$, (5.75) is replaced by

$$\begin{aligned} \boldsymbol{E}(\boldsymbol{r}) &= \sum_k \sum_{i \in \Delta V_k} \frac{q_i e}{4\pi\epsilon_\mathrm{o}} \frac{\boldsymbol{R}_i}{R_i^3} \approx \sum_k \frac{\boldsymbol{R}_k}{R_k^3} \sum_{i \in \Delta V_k} \frac{q_i e}{4\pi\epsilon_\mathrm{o}} \\ &= \sum_k \frac{\boldsymbol{R}_k}{R_k^3} \frac{\rho(\boldsymbol{r}_k)}{4\pi\epsilon_\mathrm{o}} \Delta V_k \end{aligned} \tag{5.79}$$

where the summation can be transformed into an integration

$$\boldsymbol{E}(\boldsymbol{r}) = \int \frac{\rho(\boldsymbol{r}')}{4\pi\epsilon_\mathrm{o}} \frac{\boldsymbol{R}}{R^3} \mathrm{d}V' \tag{5.80}$$

with $\boldsymbol{R} = \boldsymbol{r} - \boldsymbol{r}'$. A similar presentation can be given for the electric potential

$$\phi(\boldsymbol{r}) = \frac{1}{4\pi\epsilon_\mathrm{o}} \int_V \frac{\rho(\boldsymbol{r}')\,\mathrm{d}V'}{|\boldsymbol{r} - \boldsymbol{r}'|} \,. \tag{5.81}$$

Taking (5.76) and (5.78) together with the Gauss law

$$\oint_S (\boldsymbol{n} \cdot \boldsymbol{E}) \mathrm{d}S = \int_V \mathrm{div} \boldsymbol{E} \,\mathrm{d}V \tag{5.82}$$

yields the Poisson equation

$$\mathrm{div} \boldsymbol{E} = \frac{\rho}{\epsilon_\mathrm{o}} \quad \text{and} \quad \nabla^2 \phi = -\frac{\rho}{\epsilon_\mathrm{o}} \tag{5.83}$$

since $\boldsymbol{E} = -\nabla\phi$. Of course, (5.81) is a solution of (5.83). To prove that, it is inserted into the Poisson equation

$$\frac{1}{4\pi\epsilon_\mathrm{o}} \nabla^2 \int_V \frac{\rho(\boldsymbol{r}')\mathrm{d}V'}{|\boldsymbol{r} - \boldsymbol{r}'|} = \frac{1}{4\pi\epsilon_\mathrm{o}} \int_V \rho(\boldsymbol{r}') \nabla^2 \left(\frac{1}{|\boldsymbol{r} - \boldsymbol{r}'|} \right) \mathrm{d}V' \,. \tag{5.84}$$

Using the relation

$$\nabla^2 \left(\frac{1}{|\boldsymbol{r} - \boldsymbol{r}'|} \right) = -4\pi\delta(\boldsymbol{r} - \boldsymbol{r}') \tag{5.85}$$

this leads again to (5.83)

$$\nabla^2\phi = \frac{1}{4\pi\epsilon_o}\int\limits_V \rho(\boldsymbol{r}')[-4\pi\delta(\boldsymbol{r}-\boldsymbol{r}')]\mathrm{d}V' = -\rho(\boldsymbol{r})/\epsilon_o\,. \tag{5.86}$$

Without losing universality, the interaction between particles can be described in term of fields instead in terms of forces. The transition from point-like particles to a density description marks the transition from the Coulomb force to the Poisson equation (5.67) that relates the distribution of charged particles to the electric field. One particle can be thought of as being represented by a localized "charge cloud". The calculations via fields is as accurate as the spatial resolution. Just inside the charge cloud and in its direct neighborhood inaccuracies do occur. The interaction between the charge clouds is taken exactly into account at larger distances.

The numerical solution of the Poisson equation based on efficient algorithms is much faster than the calculation of the interaction between all particles (Sect. 7.6.1). In addition, the number of representative particles, the charge clouds, used in the simulation is usually several orders of magnitude smaller than the number of particles in real plasmas.

6

Particle Coupling

The interaction of plasma particles with solid surfaces can be classified according to the particles under consideration: (1) bombardment by atomic particles, in particular neutral atoms formed by charge exchange, molecules, and ions, (2) electron bombardment, (3) interaction with electromagnetic radiation, and (4) neutron irradiation.

While neutrons cause a lot of damage in the bulk material (Sect. 12.5), the sputtering coefficient is smaller than 10^{-4}, and this effect can be neglected. The effect of electrons [88] and photons on irradiated surfaces reduces primarily to heating. The resulting temperature gradient creates thermal stresses, which impose a limit on the maximum thickness of the material through which the heat flux is removed (see Sect. 8.3).

Due to the impact of the particles, a number of interaction and transformation processes are induced in the material leading to a change of its structure and composition, and to emission of particles.

A number of processes have to be considered:

1. Erosion processes (sputtering, evaporation, etc.) (Sect. 6.5)
2. Backscattering of ions and electrons from the surface (Sect. 6.8)
3. Emission of electrons (Sect. 6.9)
4. Ion implantation and release of gaseous deuterium, tritium, and helium (Sect. 6.4)
5. Diffusion of gas atoms and defects, the formation and decay of mobile and immobile complexes of implanted atoms and defects, and the solubility of the gas in the material (Sect. 6.3)
6. Segregation (Sect. 6.3)
7. Desorption of impurities during ion and electron impact
8. Blistering (Sect. 6.5.5)
9. Changes in a surface layer subjected to ion irradiation, changes in the chemical and phase composition (Sect. 6.3)
10. Chemical processes at the surface

11. Inelastic processes during the interaction of ions with the surface (changes in the charge state [89], ion, electron and photon emission) (Sect. 6.6)

Each particle that strikes a material surface is subjected to collisions (predominantly elastic) with lattice atoms and predominantly inelastic collisions with the electrons of the bulk material [90–93] (Sect. 6.2). During these interactions, the particles change their initial direction of motion and their charge state, lose their kinetic energy (Sects. 6.2.2 and 6.2.3) and momentum. They come to rest by reaching energies in the eV range, i.e., less than the binding energies. At these low energies, the particle is in energetic equilibrium with the target. This process is called implantation (Sect. 6.3).

The momentum transfer from the impinging particles to the lattice atoms can lead to developing cascades of knock-on atoms, which may also leave the surface by surmounting the surface potential determined by the sublimation energy. These sputtered particles contribute to erosion and plasma contamination. Surface erosion (see Sect. 6.5) primarily determines the lifetime of components and the source of impurities, which increase the plasma radiation. Almost all particles emitted by sputtering leave the surface as neutrals.

The basic model considering transport of particles in matter is the assumption of binary collisions (Sect. 6.1). As discussed in the case of collisions in plasmas (Sect. 5.4), this assumption holds roughly—in the case of solids—as long as the impact parameter for large deflections is smaller than the interatomic distance of about 2–3 Å. Putting in an impact parameter equal to that distance, i.e., $\rho_{\mathrm{p}}^{90^\circ}$ =2–3 Å, the energy of particles should be larger than

$$E_{\mathrm{o}} > \frac{Z_1 Z_2 e^2 (M_1 + M_2)}{8\pi\epsilon_{\mathrm{o}} M_2 \rho_{\mathrm{p}}^{90^\circ}} \simeq Z_1 Z_2 \cdot 3\,\mathrm{eV} \tag{6.1}$$

when additionally $M_1/M_2 \ll 1$ is assumed. Another limit is the binding energy in the crystal lattice (several eVs), since the assumption of two-particle collisions is valid for free particles. Resonance scattering as well as nuclear reactions have to be considered separately and, of course, do not belong to the class of elastic collisions.

6.1 Binary Collisions

For a particle of mass M_1 with velocity v_{o} that collides with a particle of mass M_2 at rest (Fig. 6.1), the laws of energy and momentum conservation state

$$\frac{M_1 v_{\mathrm{o}}^2}{2} = \frac{M_1 v_1^2}{2} + \frac{M_2 v_2^2}{2} \tag{6.2}$$

$$M_1 v_{\mathrm{o}} = M_1 v_1 \cos\theta_1 + M_2 v_2 \cos\theta_2 \tag{6.3}$$

$$0 = M_1 v_1 \sin\theta_1 - M_2 v_2 \sin\theta_2 \tag{6.4}$$

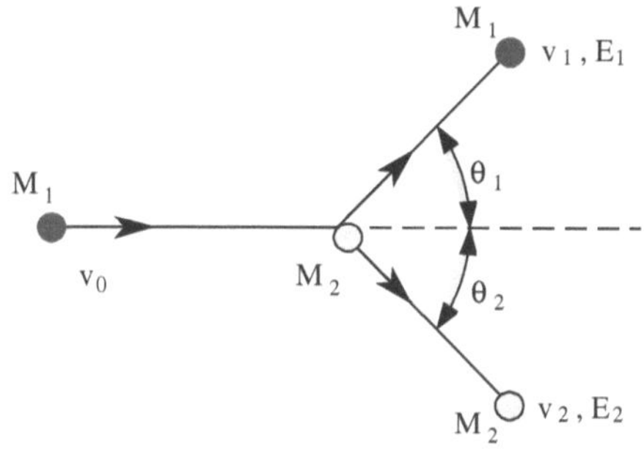

Fig. 6.1. Binary collision in the laboratory system (L-system)

taking into account that the trajectories of both particles during the collision lie in one plane (Fig. 6.1). Relations for $\sin\theta_2$ and v_2 can be deduced from (6.3) and (6.4), respectively, and inserted into (6.2) yielding

$$v_o^2 = v_1^2 + \left(\frac{M_1}{M_2}\right)^2 \left[v_1^2 + v_o^2 - 2v_1 v_o \cos\theta_1\right] \tag{6.5}$$

and

$$v_1 = v_o \frac{M_1}{M_1 + M_2} \left[\cos\theta_1 \pm \sqrt{\left(\frac{M_2}{M_1}\right)^2 - \sin^2\theta_1}\,\right] . \tag{6.6}$$

We have further

$$\begin{aligned} v_1 &= v_o \frac{\sin\theta_2}{\sin(\theta_1 + \theta_2)} \\ v_2 &= v_o \frac{2M_1}{M_1 + M_2} \cos\theta_2 \end{aligned} \tag{6.7}$$

and

$$\begin{aligned} E_1 &= E_o \frac{\sin^2\theta_2}{\sin^2(\theta_1 + \theta_2)} \\ E_2 &= E_o \frac{4M_1 M_2}{(M_1 + M_2)^2} \cos^2\theta_2 \end{aligned} \tag{6.8}$$

where $E_1 = M_1 v_1^2/2$ and $E_2 = M_2 v_2^2/2$. The obtained energy of the particle that was initially at rest is equal to the energy loss suffered by the incoming particle, i.e., $E_2 = \Delta E$.

The introduction of the center-of-mass system (CM-system) facilitates the analysis of collision processes. In this coordinate system, the total momentum is zero, by definition. We have

$$M_1 \boldsymbol{u}_{o1} + M_2 \boldsymbol{u}_{o2} = 0 \quad \text{and} \quad M_1 \boldsymbol{u}_1 + M_2 \boldsymbol{u}_2 = 0 \tag{6.9}$$

where $\boldsymbol{u}_{o1}$, $\boldsymbol{u}_{o2}$ are the velocities in the CM-system before collision; $\boldsymbol{u}_1$, $\boldsymbol{u}_2$ are the velocities after collision. The center of gravity is at rest in the CM- system.

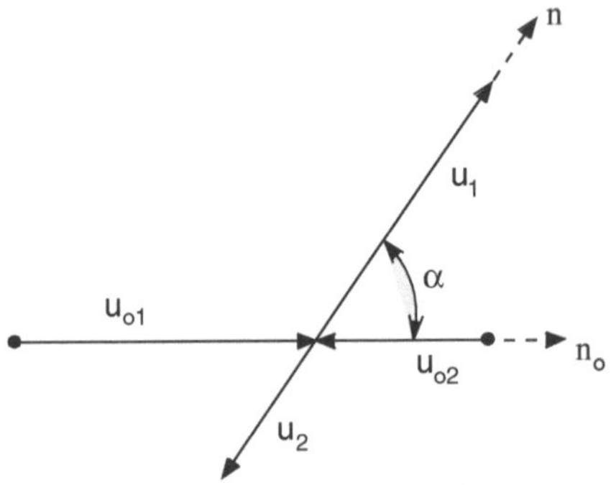

Fig. 6.2. Binary collision in the center-of-mass system ($\boldsymbol{n}_\mathrm{o}$ is the direction vector before and $\boldsymbol{n}$ the direction vector after collision)

Introducing the velocity of the CM-system $\boldsymbol{v}_\mathrm{c}$ with respect to the laboratory system (L-system), the velocities in both coordinate systems are simply related to each other

$$\boldsymbol{u}_{\mathrm{o}1} = \boldsymbol{v}_{\mathrm{o}1} - \boldsymbol{v}_\mathrm{c} = \boldsymbol{v}_\mathrm{o} - \boldsymbol{v}_\mathrm{c} \,, \quad \boldsymbol{u}_{\mathrm{o}2} = \boldsymbol{v}_{\mathrm{o}2} - \boldsymbol{v}_\mathrm{c} = -\boldsymbol{v}_\mathrm{c} \tag{6.10}$$

where $\boldsymbol{v}_{\mathrm{o}2} = 0$ and $\boldsymbol{v}_{\mathrm{o}1} = \boldsymbol{v}_\mathrm{o}$. After the collision, we have $\boldsymbol{u}_1 = \boldsymbol{v}_1 - \boldsymbol{v}_\mathrm{c}$ and $\boldsymbol{u}_2 = \boldsymbol{v}_2 - \boldsymbol{v}_\mathrm{c}$ (see Fig. 6.2). Using (6.10) in (6.9) gives the velocity of the center of mass

$$\boldsymbol{v}_\mathrm{c} = (M_1\boldsymbol{v}_{\mathrm{o}1} + M_2\boldsymbol{v}_{\mathrm{o}2})/(M_1 + M_2) = M_1\boldsymbol{v}_\mathrm{o}/(M_1 + M_2) \,. \tag{6.11}$$

With this relation together with (6.10), the velocities of the particle before the collision can be expressed in the CM-system by

$$\boldsymbol{u}_{\mathrm{o}1} = \frac{M_2\boldsymbol{v}_\mathrm{o}}{M_1 + M_2} \,, \quad \boldsymbol{u}_{\mathrm{o}2} = -\frac{M_1\boldsymbol{v}_\mathrm{o}}{M_1 + M_2} \,. \tag{6.12}$$

The velocity vectors are turned only by an angle α during the collision (Fig. 6.2)

$$\boldsymbol{u}_1 = u_{\mathrm{o}1}\boldsymbol{n} = \frac{M_2 v_\mathrm{o}\boldsymbol{n}}{M_1 + M_2} \,, \quad \boldsymbol{u}_2 = u_{\mathrm{o}2}\boldsymbol{n} = -\frac{M_1 v_\mathrm{o}\boldsymbol{n}}{M_1 + M_2} \,. \tag{6.13}$$

Their absolute values remain the same. Note that $\boldsymbol{u}_{\mathrm{o}1}$ and $\boldsymbol{u}_{\mathrm{o}2}$ as well as $\boldsymbol{u}_1$ and $\boldsymbol{u}_2$ are in opposition to each other. The transition back to the L-system is simply done by adding the center-of-mass velocity (Fig. 6.3)

$$\boldsymbol{v}_1 = \boldsymbol{u}_1 + \boldsymbol{v}_\mathrm{c} \,, \quad \boldsymbol{v}_2 = \boldsymbol{u}_2 + \boldsymbol{v}_\mathrm{c} \,. \tag{6.14}$$

As illustrated in the figure, the following relation holds that

$$u_1 \sin\alpha = v_1 \sin\theta_1 \,, \quad u_1 \cos\alpha + v_\mathrm{c} = v_1 \cos\theta_1 \tag{6.15}$$

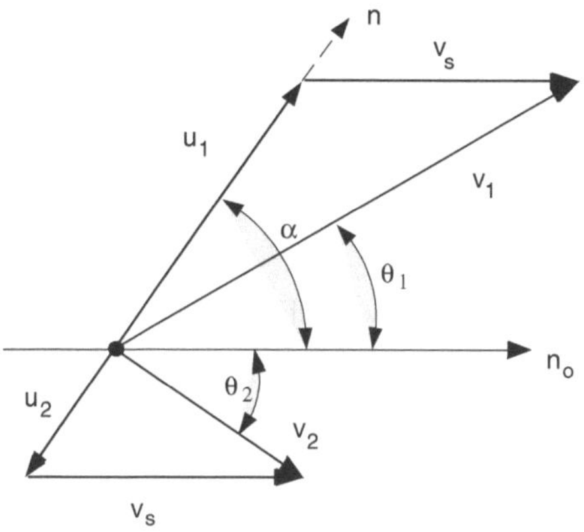

Fig. 6.3. Velocities in the L-system and CM-system after collision

leading to $\tan\theta_1 = \sin\alpha/(\cos\alpha + v_c/u_1)$. Since $v_s/u_1 = M_1/M_2$ as shown above, we obtain

$$\tan\theta_1 = \frac{\sin\alpha}{\cos\alpha + M_1/M_2} . \tag{6.16}$$

In a similar way, the scattering angle of the second particle can be deduced

$$\theta_2 = \frac{\pi - \alpha}{2} . \tag{6.17}$$

6.1.1 Scattering Angle α

The conservation laws are not sufficient to determine the scattering angle. Its value must be calculated by assuming a certain interaction potential $U(r)$ (not to be confused with the electric potential, since the interaction potential is given in energy units). The Lagrange function of the considered two-body system is

$$L = \frac{M_1\dot{\boldsymbol{r}}_1^2}{2} + \frac{M_2\dot{\boldsymbol{r}}_2^2}{2} - U(|\boldsymbol{r}_1 - \boldsymbol{r}_2|) \tag{6.18}$$

with the relative radius vector $\boldsymbol{r} = \boldsymbol{r}_1 - \boldsymbol{r}_2$. If the center of the force field is in the position of the center of gravity, i.e., $M_1\boldsymbol{r}_1 + M_2\boldsymbol{r}_2 = 0$, then

$$\boldsymbol{r}_1 = \frac{M_2}{M_1 + M_2}\boldsymbol{r}, \qquad \boldsymbol{r}_2 = -\frac{M_1}{M_1 + M_2}\boldsymbol{r} \tag{6.19}$$

and the Lagrange function

$$L = \frac{M_1 M_2}{M_1 + M_2}\frac{\dot{\boldsymbol{r}}^2}{2} - U(r) = \mu\frac{\dot{\boldsymbol{r}}^2}{2} - U(r) \tag{6.20}$$

describes the dynamics of one particle, which moves in a field with a fixed center. Thus, the two-body problem is reduced to the motion of one particle with the reduced mass $\mu = M_1 M_2/(M_1 + M_2)$ in a central field. The corresponding conservation laws of energy and angular momentum M_φ are given in the polar coordinate system by

$$E_\mathrm{r} = \frac{\mu}{2}\left(\dot{r}^2 + r^2\dot{\varphi}^2\right) + U(r) = \text{const.} \tag{6.21}$$

$$M_\varphi = \mu r^2 \dot{\varphi} = \text{const.} \tag{6.22}$$

where E_r is the relative energy of the system. Replacing $\dot{\varphi}$ in (6.21) as given by (6.22) yields

$$\dot{r} \equiv \frac{\mathrm{d}r}{\mathrm{d}t} = \sqrt{\frac{2}{\mu}[E_\mathrm{r} - U(r)] - \frac{M_\varphi^2}{\mu^2 r^2}} \tag{6.23}$$

or, after separation of r and t,

$$t = \int \frac{\mathrm{d}r}{\sqrt{\frac{2}{\mu}[E_\mathrm{r} - U(r)] - \frac{M_\varphi^2}{\mu^2 r^2}}} + \text{const.} \tag{6.24}$$

If (6.22) is written in the form

$$\mathrm{d}\varphi = \frac{M_\varphi}{\mu r^2}\,\mathrm{d}t \tag{6.25}$$

then $\mathrm{d}t$ can be substituted by (6.23), and after performing an integration, we obtain the trajectory equation

$$\varphi = \int \frac{(M_\varphi/r^2)\mathrm{d}r}{\sqrt{2\mu[E_\mathrm{r} - U(r)] - (M_\varphi^2/r^2)}} + \text{const.} \tag{6.26}$$

The scattering angle φ varies monotonically in time and does not change its sign. In the case of

$$U(r_\mathrm{min}) + \frac{M_\varphi^2}{2\mu r_\mathrm{min}^2} = E_\mathrm{r}\,, \tag{6.27}$$

the radial velocity component is zero and the minimum distance r_min to the center of the force field is reached. The particle does not stop but turns back (Fig. 6.4). Due to the symmetric trajectory, the scattering angle α can be expressed as

$$\alpha = |\pi - 2\varphi_\mathrm{o}|\,. \tag{6.28}$$

Instead of the constants E_r and M_φ, it is more convenient to apply the initial velocity v_o of the incoming particle and the impact parameter ρ_p. We have

$$E_\mathrm{r} = \mu v_\mathrm{o}^2/2 = E_\mathrm{o} M_2/(M_1 + M_2)\,, \quad M_\varphi = \mu \rho_\mathrm{p} v_\mathrm{o}\,. \tag{6.29}$$

According to (6.26), the angle φ_o is given by the definite integral

$$\varphi_\mathrm{o} = \int\limits_{r_\mathrm{min}}^{\infty} \frac{\frac{\rho_\mathrm{p}}{r^2}\mathrm{d}r}{\sqrt{1 - \frac{\rho_\mathrm{p}^2}{r^2} - \frac{2U(r)}{\mu v_\mathrm{o}^2}}} \tag{6.30}$$

with the limits of r_min and infinity.

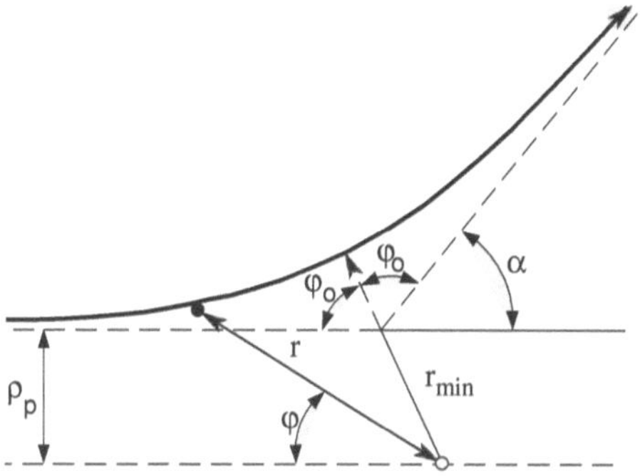

Fig. 6.4. Trajectory in the CM-system

6.1.2 Scattering in the Coulomb Field, $U(r) = C/r$

For an interaction potential of the kind $U(r) \propto 1/r$, the integration in (6.30) can be executed analytically, and we obtain for the scattering angle in the CM-system ($\int dx/(x\sqrt{x^2-a^2}) = (1/a)\arccos(a/x)$, $\cos x = 1/(\sqrt{1+\tan^2 x})$)

$$\alpha = \pi - 2\int_{r_{\min}}^{\infty} \frac{\rho_{\mathrm{p}}\,\mathrm{d}r}{r\sqrt{r^2-\rho_{\mathrm{p}}^2 - C\,r/E_{\mathrm{r}}}} = 2\arctan\left(\frac{C}{2\rho_{\mathrm{p}}E_{\mathrm{r}}}\right) \tag{6.31}$$

with $E_{\mathrm{r}} = \mu v_{\mathrm{o}}^2/2 = M_2E_{\mathrm{o}}/(M_1+M_2)$ and $C = Z_1Z_2e^2/(4\pi\epsilon_{\mathrm{o}})$, where Z_1 and Z_2 are the atomic numbers of the particles. Another expression often used is

$$\sin^2\left(\frac{\alpha}{2}\right) = \frac{1}{1+(2\rho_{\mathrm{p}}E_{\mathrm{r}}/C)^2} \tag{6.32}$$

using $\tan^2 x = \sin^2 x/(1-\sin^2 x)$. The impact parameter $\rho_{\mathrm{p}}^{90^\circ}$ corresponding to a scattering angle of 90 degrees is readily obtained, since for $\alpha = \pi/2 = 2\arctan(1)$ the condition

$$\frac{C}{2\rho_{\mathrm{p}}^{90^\circ}E_{\mathrm{r}}} = 1 \tag{6.33}$$

must be fulfilled; thus,

$$\rho_{\mathrm{p}}^{90^\circ} = \frac{Z_1Z_2e^2(M_1+M_2)}{8\pi\epsilon_{\mathrm{o}}M_2E_{\mathrm{o}}}\,. \tag{6.34}$$

6.1.3 Cross-Section

The scattering of many particles moving with the same velocity is described by introducing the ratio

$$\mathrm{d}\sigma = \frac{\mathrm{d}N}{\Gamma}\,[\mathrm{m}^2] \tag{6.35}$$

called cross-section, where Γ is the particle flux density. It relates the number of particles $\mathrm{d}N$ scattered by an angle in the interval $[\alpha, \alpha+\mathrm{d}\alpha]$ per unit time to the number of particles crossing a unit area of the (homogeneous) beam during the same time interval. In the simple geometrical interpretation, the cross-section, more exactly, the total cross-section, defines the extension of an "active" area. Only by hitting this area, can a certain reaction occur. The number of particles with impact parameter ρ_p out of the interval $[\rho_\mathrm{p}, \rho_\mathrm{p}-\mathrm{d}\rho_\mathrm{p}]$ is equal to the number of particles scattered by an angle α lying in the interval $[\alpha, \alpha+\mathrm{d}\alpha]$

$$\mathrm{d}N = 2\pi\rho_\mathrm{p}\,\Gamma\mathrm{d}\rho_\mathrm{p}\ . \tag{6.36}$$

Using this in (6.35) leads to

$$d\sigma = 2\pi\rho_\mathrm{p}\,\mathrm{d}\rho_\mathrm{p} \quad \text{or} \quad d\sigma = 2\pi\rho_\mathrm{p}(\alpha)\left|\frac{\mathrm{d}\rho_\mathrm{p}(\alpha)}{\mathrm{d}\alpha}\right| d\alpha\ . \tag{6.37}$$

In order to avoid negative cross-sections, the modulus is applied, since the derivative $\mathrm{d}\rho_\mathrm{p}/\mathrm{d}\alpha$ has, in most cases, gives a negative value. More frequently, the cross-section is given as a function of the solid angle $\mathrm{d}\Omega = 2\pi\sin\alpha\,\mathrm{d}\alpha$

$$\frac{\mathrm{d}\sigma}{\mathrm{d}\Omega} = \frac{\rho_\mathrm{p}(\alpha)}{\sin\alpha}\left|\frac{\mathrm{d}\rho_\mathrm{p}(\alpha)}{\mathrm{d}\alpha}\right| d\Omega\ . \tag{6.38}$$

The transition from the CM-system to the L-system is trivial, if $M_2 \gg M_1$, since then, according to (6.16), $\theta_1 \approx \alpha$, $\mu \approx M_1$, and therefore $\mathrm{d}\sigma_\mathrm{CM} = \mathrm{d}\sigma_\mathrm{L}$. In general

$$\left(\frac{\mathrm{d}\sigma}{d\Omega}\right)_{\mathrm{L}1,2} \mathrm{d}\Omega_{\mathrm{L}1,2} = \left(\frac{\mathrm{d}\sigma}{d\Omega}\right)_{\mathrm{CM}} \mathrm{d}\Omega_\mathrm{CM} \tag{6.39}$$

with $\mathrm{d}\Omega_S = \sin\alpha\,\mathrm{d}\alpha\mathrm{d}\varphi$, $\mathrm{d}\Omega_{\mathrm{L}1} = \sin\theta_1\,\mathrm{d}\theta_1\mathrm{d}\varphi$, and $\mathrm{d}\Omega_{\mathrm{L}2} = \sin\theta_2\,\mathrm{d}\theta_2\mathrm{d}\varphi$. Since the collision takes place on one plane, the azimuthal angle φ is the same. Applying (6.16) and (6.17), we have

$$\begin{aligned}\cos\theta_1 &= \pm\frac{1}{\sqrt{1+\tan^2\alpha}} = \frac{\cos\alpha + M_1/M_2}{\sqrt{1+2(M_1/M_2)\,\cos\alpha + (M_1/M_2)^2}}\\ \cos\theta_2 &= \cos(\pi/2-\alpha/2) = \sin(\alpha/2)\end{aligned} \tag{6.40}$$

and after some transformations

$$\begin{aligned}\frac{\mathrm{d}\sigma(\theta_1,\varphi)}{\mathrm{d}\Omega_{\mathrm{L}1}} &= \frac{\mathrm{d}\sigma(\alpha,\varphi)}{\mathrm{d}\Omega_\mathrm{S}}\frac{(1+2(M_1/M_2)\,\cos\alpha+(M_1/M_2)^2)^{3/2}}{|1+(M_1/M_2)\cos\alpha|}\\ \frac{\mathrm{d}\sigma(\theta_2,\varphi)}{\mathrm{d}\Omega_{\mathrm{L}2}} &= \frac{\mathrm{d}\sigma(\alpha,\varphi)}{\mathrm{d}\Omega_\mathrm{S}}\,4\,\sin\frac{\alpha}{2}\ .\end{aligned} \tag{6.41}$$

Substituting for ρ_p from (6.31) together with its derivative

$$\rho_\mathrm{p} = \frac{C}{2E_\mathrm{r}}\cot\left(\frac{\alpha}{2}\right) \quad \text{and} \quad \frac{\mathrm{d}\rho_\mathrm{p}}{\mathrm{d}\alpha} = -\frac{C}{4E_\mathrm{r}}\frac{1}{\sin^2\frac{\alpha}{2}} \tag{6.42}$$

in (6.38), we find the cross-section for the Coulomb interaction (using $\cot'(x) = -1/(\sin x)^2$, $\cot(x/2) = \sin x/(1-\cos x)$ and $1 - \cos x = 2\sin^2(x/2)$)

$$\frac{\mathrm{d}\sigma}{\mathrm{d}\Omega} = \left(\frac{C}{4E_\mathrm{r}}\right)^2 \frac{1}{\sin^4 \frac{\alpha}{2}} = \left(\frac{Z_1 Z_2 e^2}{8\pi\epsilon_\mathrm{o} E_\mathrm{r}}\right)^2 \frac{1}{(1-\cos\alpha)^2} \,. \tag{6.43}$$

This is the Rutherford cross-section.

Summarizing, a detector which is positioned at angles (θ, φ) with respect to the beam axis and which covers a certain solid angle interval $\mathrm{d}\Omega$ (given by the effective size of the detector divided by the square of the distance between the target and the detector) counts

$$\mathrm{d}N = \Gamma N_\mathrm{t} \frac{\mathrm{d}\sigma}{\mathrm{d}\Omega}(\theta,\varphi)\,\mathrm{d}\Omega \left[\frac{\text{part.}}{s}\right] \tag{6.44}$$

events per second, where N_t is the number of scattering centers in the beam–target interaction volume. Usually, thin foils are applied in the experiments to reduce the effect of multiple scattering.

6.1.4 Interaction Potential $U(r)$

The collision process is mainly determined by the interaction potential, which must be chosen according to the situation under consideration. In plasma physics, usually, the Coulomb potential is applied and set to zero at distances larger than the Debye length (Sect. 5.4). Whether the calculations are performed with a "truncated" Coulomb potential or using the Debye length in an additional screening factor does not make much difference. The right choice of the interaction potential in solid state physics is more challenging since the screening effect of the electrons becomes important. Knowing the wave functions of all electrons would, in principle, allow for the determination of the exact interaction potential. However, the solution of the Schrödinger equation is quite complicated even for simple atoms. Hartree suggested in 1927 a simplified method, which has been developed further by Fock and Slater [94–96]. Based on this approach, a number of potentials with different screening factors have been proposed in the past. The used screening lengths have the general form

$$a_\mathrm{s} = a_\mathrm{o} \left(\frac{9\pi^2}{128}\right)^{1/3} Z_{12}^{-1/3} \tag{6.45}$$

with the Bohr radius $a_\mathrm{o} = 0.529 \times 10^{-10}$ m and effective atomic number

$$Z_{12} = (Z_1^x + Z_2^x)^y \tag{6.46}$$

where x and y are chosen. In other approximations, the dependence on the distance is changed, for example $U(r) \propto 1/r^2$. Such a functional dependence can only be valid in a certain, limited region away from the target atom. Thus, the average distance of approach should fall into this region. Such an adjustment has to be made for each ion–atom pair of interest as a function of the ion energy. A list of potentials more frequently used is given below.

1. Potential of "hard spheres"

$$U(r) = \begin{cases} \infty & \text{for} \quad r < r_1 + r_2 \\ 0 & \text{for} \quad r > r_1 + r_2 \end{cases} \tag{6.47}$$

where r_1 and r_2 are the radii of the colliding spheres. The corresponding cross-section is

$$\frac{\mathrm{d}\sigma}{\mathrm{d}\Omega} = (r_1 + r_2)^2/4\,. \tag{6.48}$$

2. Coulomb potential

$$U(r) = \frac{Z_1 Z_2 e^2}{4\pi\epsilon_o\, r} \tag{6.49}$$

with the atomic numbers Z_1 and Z_2. The cross section is (see (6.43))

$$\frac{\mathrm{d}\sigma}{\mathrm{d}\Omega_S} = \left(\frac{a_s}{4\,\varepsilon}\right)^2 \frac{1}{\sin^4\frac{\alpha}{2}} = \left(\frac{a_s}{2\,\varepsilon}\right)^2 \frac{1}{(1-\cos\alpha)^2} \tag{6.50}$$

where the screening length a_s and the "reduced" energy ε

$$\varepsilon = \frac{a_s E_o M_2 4\pi\epsilon_o}{e^2 Z_1 Z_2 (M_1 + M_2)} = E_r \bigg/ \left(\frac{Z_1 Z_2 e^2}{4\pi\epsilon_o\, a_s}\right) \tag{6.51}$$

have been introduced. It is worth noting that at higher energies and therefore closer distances of approach the Coulomb interaction potential becomes more and more realistic, since the incoming (high-energy) particle penetrates the electron shells deeply and *see* the "naked" atom core. The screening effect of the electrons is therefore reduced.

3. Firsov potential

$$U(r) = \frac{0.415 Z_1 Z_2 e^2 a_L}{4\pi\epsilon_o\, r^2} \tag{6.52}$$

where a_L is the Lindhard screening length

$$a_L = 0.8853\, a_o \left(Z_1^{2/3} + Z_2^{2/3}\right)^{-1/2} \tag{6.53}$$

with the Bohr radius a_o. This potential finds its application in the energy interval $0.1 \le \varepsilon_L \le 1$ with $\varepsilon_L = a_L E_o M_2 4\pi\epsilon_o/(e^2 Z_1 Z_2 (M_1 + M_2))$ (Lindhard reduced energy). The corresponding cross-section is

$$\begin{aligned} \frac{\mathrm{d}\sigma}{\mathrm{d}\Omega} &= \frac{0.415\pi^2 Z_1 Z_2 e^2 a_L (M_1 + M_2)}{4\pi\epsilon_o E_o M_2} \frac{\pi - \alpha}{\alpha^2 (2\pi - \alpha)^2 \sin\alpha} \\ &= \frac{0.415\, a_L^2 \pi^2 (\pi - \alpha)}{\varepsilon_L\, \alpha^2 (2\pi - \alpha)^2 \sin\alpha}\,. \end{aligned} \tag{6.54}$$

4. Screened Coulomb potentials

$$U(r) = \frac{Z_1 Z_2 e^2}{4\pi\epsilon_o r}\,\Phi\left(\frac{r}{a_s}\right) \tag{6.55}$$

where the function $\Phi(r/a)$ is represented as

$$\Phi\left(\frac{r}{a_s}\right) = \sum_{i=1}^{n} c_i \exp\left(-d_i\frac{r}{a_s}\right)\,, \quad \sum_{i=1}^{n} c_i = \Phi(0) = 1\,. \tag{6.56}$$

To this group belong the Bohr potential

$$\Phi(r) = \exp(-r/a_s) \tag{6.57}$$

with $a_s = a_o\left(Z_1^{2/3} + Z_2^{2/3}\right)^{-1/2}$, the Moliere potential

$$\Phi(r) = 0.35\exp\left(-0.3\frac{r}{a_s}\right) + 0.55\exp\left(-1.2\frac{r}{a_s}\right) + 0.1\exp\left(-6.0\frac{r}{a_s}\right) \tag{6.58}$$

with $a_s = 0.8853a_o\left(\sqrt{Z_1} + \sqrt{Z_2}\right)^{-2/3}$, the Kr-C potential

$$\begin{aligned}\Phi(r) = {} & 0.190945\exp(-0.278544\,r/a_s) + 0.473674\exp(-0.637174\,r/a_s) + \\ & 0.335381\exp(-1.919249\,r/a_s)\end{aligned} \tag{6.59}$$

with $a_s = 0.8853a_o\left(\sqrt{Z_1} + \sqrt{Z_2}\right)^{-2/3}$, and the ZBL potential

$$\begin{aligned}\Phi(r) = {} & 0.028171\exp(-0.20162\,r/a_s) + 0.28022\exp(-0.40290\,r/a_s) + \\ & 0.50986\exp(-0.94229\,r/a_s) + 0.18175\exp(-3.1998\,r/a_s)\end{aligned} \tag{6.60}$$

with $a_s = 0.8853a_o/(z_1^{0.23} + Z_2^{0.23})$. Unfortunately, the cross-section as well as the relation of the scattering angle as a function of the impact parameter cannot be given in terms of elementary functions for these interaction potentials.

5. Morse potential

$$U(r) = U_D\exp[-2C(r - r_o)] - 2U_D\exp[-C(r - r_o)] \tag{6.61}$$

where U_D is the energy minimum at distance r_o. The third parameter C can be used as well as the other two for fitting this analytical relation to a more realistic potential, for example, obtained by numerical calculations.

6.1.5 Binary Collision: General Case

In the general case of binary collisions, both colliding particles move. According to the conservation of energy and momentum, we have

$$\begin{aligned}\frac{1}{2}M_\mathrm{a}(\boldsymbol{v}_\mathrm{a})^2 + \frac{1}{2}M_\mathrm{b}(\boldsymbol{v}_\mathrm{b})^2 &= \frac{1}{2}M_\mathrm{a}(\boldsymbol{v}_\mathrm{a}^*)^2 + \frac{1}{2}M_\mathrm{b}(\boldsymbol{v}_\mathrm{b}^*)^2 \\ M_\mathrm{a}\boldsymbol{v}_\mathrm{a} + M_\mathrm{b}\boldsymbol{v}_\mathrm{b} &= M_\mathrm{a}\boldsymbol{v}_\mathrm{a}^* + M_\mathrm{b}\boldsymbol{v}_\mathrm{b}^* \end{aligned} \tag{6.62}$$

with the masses M_a and M_b of the particles a and b, respectively; $\boldsymbol{v}_\mathrm{a}$, $\boldsymbol{v}_\mathrm{b}$ are the velocity of both particles before the collision and $\boldsymbol{v}_\mathrm{a}^*$, $\boldsymbol{v}_\mathrm{b}^*$ are the velocities after the collision. A possible change of the internal energy of particles is not considered here. The relative velocities are $\boldsymbol{v}_\mathrm{r} = \boldsymbol{v}_\mathrm{a} - \boldsymbol{v}_\mathrm{b}$ and $\boldsymbol{v}_\mathrm{r}^* = \boldsymbol{v}_\mathrm{a}^* - \boldsymbol{v}_\mathrm{b}^*$; their absolute values are given by

$$\begin{aligned}|\boldsymbol{v}_\mathrm{r}| = v_\mathrm{r} &= \sqrt{(v_{ax} - v_{bx})^2 + (v_{ay} - v_{by})^2 + (v_{az} - v_{bz})^2} \\ |\boldsymbol{v}_\mathrm{r}^*| = v_\mathrm{r}^* &= \sqrt{(v_{ax}^* - v_{bx}^*)^2 + (v_{ay}^* - v_{by}^*)^2 + (v_{az}^* - v_{bz}^*)^2}\,. \end{aligned} \tag{6.63}$$

The velocities before and after the collision can be expressed using the relative velocities and the velocity of the center of mass $\boldsymbol{v}_\mathrm{c}$. By definition of the CM-system, the equation

$$M_\mathrm{a}(\boldsymbol{r}_\mathrm{a} - \boldsymbol{r}_\mathrm{c}) + M_\mathrm{b}(\boldsymbol{r}_\mathrm{b} - \boldsymbol{r}_\mathrm{c}) = 0 \tag{6.64}$$

relates the radius vectors of both particles $\boldsymbol{r}_\mathrm{a}$, $\boldsymbol{r}_\mathrm{b}$ to the vector $\boldsymbol{r}_\mathrm{c} = \boldsymbol{r}_\mathrm{a} - \boldsymbol{r}_\mathrm{b}$. Taking the derivative of (6.64) yields the velocity of the center of mass $\boldsymbol{v}_\mathrm{c} = \mathrm{d}\boldsymbol{r}_\mathrm{c}/\mathrm{d}t$

$$M_\mathrm{a}\boldsymbol{v}_\mathrm{a} + M_\mathrm{b}\boldsymbol{v}_\mathrm{b} = (M_\mathrm{a} + M_\mathrm{b})\boldsymbol{v}_\mathrm{c} \tag{6.65}$$

with respect to the laboratory system (compare with (6.11)). Since the total momentum is conserved, the velocity $\boldsymbol{v}_\mathrm{c}$ does not change. Performing several transformations leads to

$$\begin{aligned}\boldsymbol{v}_\mathrm{a} &= \boldsymbol{v}_\mathrm{s} + \frac{M_\mathrm{b}}{M_\mathrm{a} + M_\mathrm{b}}\boldsymbol{v}_\mathrm{r} = \boldsymbol{v}_\mathrm{s} + \boldsymbol{u}_\mathrm{a} \\ \boldsymbol{v}_\mathrm{b} &= \boldsymbol{v}_\mathrm{s} - \frac{M_\mathrm{a}}{M_\mathrm{a} + M_\mathrm{b}}\boldsymbol{v}_\mathrm{r} = \boldsymbol{v}_\mathrm{s} + \boldsymbol{u}_\mathrm{b} \\ \boldsymbol{v}_\mathrm{a}^* &= \boldsymbol{v}_\mathrm{s} + \frac{M_\mathrm{b}}{M_\mathrm{a} + M_\mathrm{b}}\boldsymbol{v}_\mathrm{r}^* = \boldsymbol{v}_\mathrm{s} + \boldsymbol{u}_\mathrm{a}^* \\ \boldsymbol{v}_\mathrm{b}^* &= \boldsymbol{v}_\mathrm{s} - \frac{M_\mathrm{a}}{M_\mathrm{a} + M_\mathrm{b}}\boldsymbol{v}_\mathrm{r}^* = \boldsymbol{v}_\mathrm{s} + \boldsymbol{u}_\mathrm{b}^* \end{aligned} \tag{6.66}$$

with $\boldsymbol{v}_\mathrm{c} = (M_\mathrm{a}\boldsymbol{v}_\mathrm{a} + M_\mathrm{b}\boldsymbol{v}_\mathrm{b})/(M_\mathrm{a} + M_\mathrm{b})$ and the velocities in the CM-system $\boldsymbol{u}_\mathrm{a}$, $\boldsymbol{u}_\mathrm{b}$, $\boldsymbol{u}_\mathrm{a}^*$, and $\boldsymbol{u}_\mathrm{b}^*$. Using these relations in (6.62) shows that the absolute values of the relative velocities as well as the absolute values of the velocities in the CM-system remain unchanged, i.e., $v_\mathrm{r} = v_\mathrm{r}^*$, $u_\mathrm{a} = u_\mathrm{a}^*$, and $u_\mathrm{b} = u_\mathrm{b}^*$. In the CM-system, the velocity vectors are turned only by an angle α (Fig. 6.5)

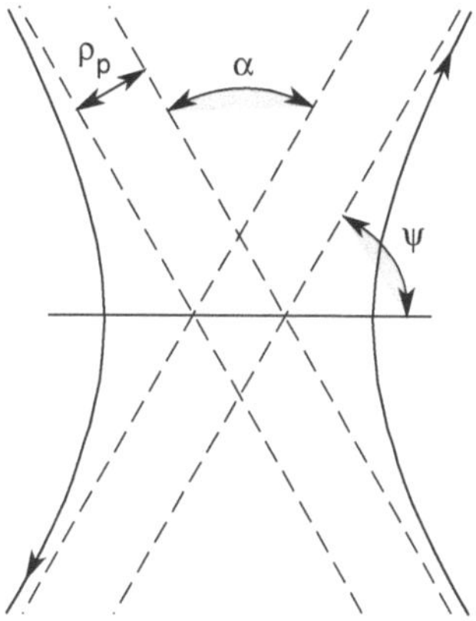

Fig. 6.5. Collision of two moving particles in the CM-system

$$\alpha = \pi - 2 \int_{r_{\min}}^{\infty} \frac{\rho_\mathrm{p}\, dr}{r^2 \sqrt{1 - \frac{\rho_\mathrm{p}^2}{r^2} - \frac{2U(r)}{\mu v_\mathrm{r}^2}}}\,, \tag{6.67}$$

which is determined by the interaction potential. In comparison to (6.30), the velocity v_o is replaced here by the relative velocity v_r.

Again, the problem of two colliding particles is reduced to the motion of one particle with reduced mass μ in the central force field $U(r)$. In the case of Coulomb interaction, we have the familiar solution of (6.67)

$$\tan\left(\frac{\alpha}{2}\right) = \frac{C}{2\rho_\mathrm{p} E_\mathrm{r}} = \frac{Z_1 Z_2 e^2}{4\pi\epsilon_\mathrm{o} \rho_\mathrm{p} \mu v_\mathrm{r}^2} \tag{6.68}$$

since $E_\mathrm{r} = \mu v_\mathrm{r}^2/2$ (see (6.31)). It is convenient to describe the rotation of the relative velocity vector in a coordinate system where the z-axis is directed along the relative velocity before the collision $\boldsymbol{v}_\mathrm{r}$ (Fig. 6.6). Then, we have for

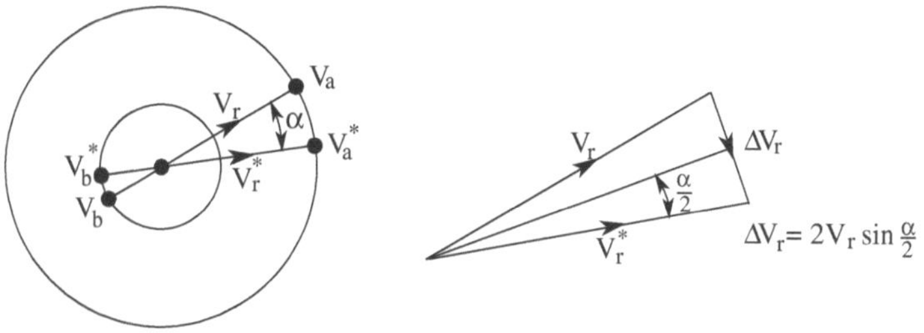

Fig. 6.6. Change of the relative velocity vector during collision. Note that $2v_\mathrm{r}\sin(\mathrm{d}\alpha/2) = v_\mathrm{r}\mathrm{d}\alpha$

the velocity change

$$\begin{aligned}\Delta v_{\mathrm{rx}} &= 2v_{\mathrm{r}}\sin(\alpha/2)\cos(\alpha/2)\cos\varphi = v_{\mathrm{r}}\sin\alpha\cos\varphi \\ \Delta v_{\mathrm{ry}} &= 2v_{\mathrm{r}}\sin(\alpha/2)\cos(\alpha/2)\sin\varphi = v_{\mathrm{r}}\sin\alpha\sin\varphi \\ \Delta v_{\mathrm{rz}} &= -2v_{\mathrm{r}}\sin(\alpha/2)\sin(\alpha/2) = -v_{\mathrm{r}}(1-\cos\alpha)\end{aligned} \tag{6.69}$$

and obtain, together with (6.66), the change of velocity in that coordinate system (note $\boldsymbol{v}_{\mathrm{c}}$ =const.) for the particle “a”

$$\begin{aligned}\Delta v_{\mathrm{ax}} &= [M_{\mathrm{b}}/(M_{\mathrm{a}}+M_{\mathrm{b}})]\, v_{\mathrm{r}}\sin\alpha\cos\varphi \\ \Delta v_{\mathrm{ay}} &= [M_{\mathrm{b}}/(M_{\mathrm{a}}+M_{\mathrm{b}})]\, v_{\mathrm{r}}\sin\alpha\sin\varphi \\ \Delta v_{\mathrm{az}} &= -[M_{\mathrm{b}}/(M_{\mathrm{a}}+M_{\mathrm{b}})]\, v_{\mathrm{r}}(1-\cos\alpha)\end{aligned} \tag{6.70}$$

and correspondingly for the particle “b”

$$\begin{aligned}\Delta v_{\mathrm{bx}} &= -[M_{\mathrm{a}}/(M_{\mathrm{a}}+M_{\mathrm{b}})]\, v_{\mathrm{r}}\sin\alpha\cos\varphi \\ \Delta v_{\mathrm{by}} &= -[M_{\mathrm{a}}/(M_{\mathrm{a}}+M_{\mathrm{b}})]\, v_{\mathrm{r}}\sin\alpha\sin\varphi \\ \Delta v_{\mathrm{bz}} &= [M_{\mathrm{a}}/(M_{\mathrm{a}}+M_{\mathrm{b}})]\, v_{\mathrm{r}}(1-\cos\alpha)\,.\end{aligned} \tag{6.71}$$

The angle φ is the polar angle in the (x,y)-plane and varies between 0 and 2π. The velocities after the collision are given by

$$\begin{aligned}\boldsymbol{v}_{\mathrm{a}}^{*} &= \boldsymbol{v}_{\mathrm{a}} + \frac{M_{\mathrm{b}}}{M_{\mathrm{a}}+M_{\mathrm{b}}}\,\Delta\boldsymbol{v}_{\mathrm{r}} \\ \boldsymbol{v}_{\mathrm{b}}^{*} &= \boldsymbol{v}_{\mathrm{b}} - \frac{M_{\mathrm{a}}}{M_{\mathrm{a}}+M_{\mathrm{b}}}\,\Delta\boldsymbol{v}_{\mathrm{r}}\,.\end{aligned} \tag{6.72}$$

In the global coordinate system, which is fixed for all collisions, the different components of $\Delta\boldsymbol{v}_{\mathrm{r}}$ are expressed as follows:

$$\begin{aligned}\Delta v_{\mathrm{rx}} &= (v_{\mathrm{rx}}/v_{\mathrm{r}\perp})v_{\mathrm{rz}}\sin\alpha\cos\varphi - (v_{\mathrm{ry}}/v_{\mathrm{r}\perp})v_{\mathrm{r}}\sin\alpha\sin\varphi - v_{\mathrm{rx}}(1-\cos\alpha) \\ \Delta v_{\mathrm{ry}} &= (v_{\mathrm{ry}}/v_{\mathrm{r}\perp})v_{\mathrm{rz}}\sin\alpha\cos\varphi + (v_{\mathrm{rx}}/v_{\mathrm{r}\perp})v_{\mathrm{r}}\sin\alpha\sin\varphi - v_{\mathrm{ry}}(1-\cos\alpha) \\ \Delta v_{\mathrm{rz}} &= -v_{\mathrm{r}\perp}\sin\alpha\cos\varphi - v_{\mathrm{rz}}(1-\cos\alpha)\end{aligned} \tag{6.73}$$

where $v_{\mathrm{r}\perp} = \sqrt{v_{\mathrm{rx}}^2 + v_{\mathrm{ry}}^2}$ and $v_{\mathrm{r}} = \sqrt{v_{\mathrm{rx}}^2 + v_{\mathrm{ry}}^2 + v_{\mathrm{rz}}^2}$. In the case of $v_{\mathrm{r}\perp} = 0$, we obtain, instead of (6.73), again (compare to (6.69))

$$\begin{aligned}\Delta v_{\mathrm{rx}} &= v_{\mathrm{r}}\sin\alpha\cos\varphi \\ \Delta v_{\mathrm{ry}} &= v_{\mathrm{r}}\sin\alpha\sin\varphi \\ \Delta v_{\mathrm{rz}} &= -v_{\mathrm{r}}(1-\cos\alpha)\,.\end{aligned} \tag{6.74}$$

The equations of (6.72) and (6.73) can be directly used in particle simulation codes, when the motion and collisions of each particle are followed. General energy and angular relations for two-particle collisions together with the evaluation of the corresponding cross-sections are given in [97–101].

6.2 Particle Transport in Matter

The penetration of a particle, ion, or atom through solid material is accompanied by a partial or full loss of its electrons. At high velocities, the particle is fully stripped of its electrons. An overlapping of electron shells occurs resulting in a quantum exclusion of possible states. Target electrons can be captured by the particle. The effective charge of the moving particle is a function of its velocity and the kind of material.

Conversely, the moving particle affects the target by polarizing the electron gas around it. Energy is transferred from the particle to the atomic positive nuclei, a process called nuclear stopping or elastic energy losses (Sect. 6.2.2), and to the target electrons called electronic or inelastic stopping (Sect. 6.2.3). Bohr concluded in his first papers that the energy loss of ions in matter can be divided into these two components. He estimated by applying simple recoil kinematics that the inelastic loss channel is the more important one. His reviews [102,103] provide up to now a very profound and clear physical description of the main processes and parameters. By losing its energy, the particle also changes its effective charge and, consequently, the interaction intensity with the target medium is influenced as well. This mutual dependence makes the analysis of particle penetration through matter difficult. Ziegler formulated in his comprehensive review of experimental data and fitting functions the main questions [104]:

1. How does an energetic charged particle lose energy to the quantized electron plasma of a solid?
2. How do you incorporate into this interaction the simultaneous distortion of the electron plasma caused by the particle?
3. How do you estimate the degree of ionization of the moving atom and describe its electrons when it is both ionized and within an electron plasma?
4. How do you calculate the screened Coulomb scattering of the moving atom with each heavy target nucleus it passes?
5. How do you include relativistic corrections to all of the above?

It was hoped that by introducing the so-called effective charge of the projectile that the stopping power theories developed for proton penetration could be used for partially-stripped heavy ions. In the model of Bohr, all electrons of the particle with velocities lower than the instantaneous particle velocity are assumed to be stripped off. Lamb suggested a similar effective charge approximation, but based on the energy rather than on the velocity of the ion's electrons [105]. Brandt and Kitagawa revised the Bohr suggestion of the ionization degree of ions traveling within solids and considered the velocity of the particle's electrons with respect to the Fermi velocity of the solid [106]. Electrons with velocities smaller than the relative velocity of the ions of the Fermi velocity are stripped off. Their new concept has proved to be quite accurate. The Fermi energy of the medium can be measured in experiments of electron transmission through thin foils.

The first unified approach to stopping and range theory was given by Lindhard, Scharff, and Schiott [107,108]. The range of ions could be predicted within a factor of 2 for all species and velocities up to the velocity of the stopping power maximum. The best agreement could be achieved for ions, which are neither fully stripped nor nearly neutral [104]. Further improvements of the theory are based on numerical methods, which allowed for the consideration of more realistic interaction potentials. Stopping powers can now be calculated with an average accuracy of several percents overall. Range distributions for amorphous elemental targets have about the same accuracy. The stopping power of ions in compounds has been analyzed in detail in [91]. The described treatment is based on the Bragg rule, i.e., a linear superposition of the inelastic stopping powers of the constituents weighted by their concentrations is applied.

Due to the complex bond structure in materials such as organic compounds, the determination of the correct excitation and mean ionization potentials remains a difficult task.

6.2.1 Definitions and Main Parameters

Starting with an energy E_o, the total range of particles penetrating a medium is given by

$$R_o = \int_{E_o}^{0} \left(\frac{\mathrm{d}E}{\mathrm{d}x} \right)^{-1} \mathrm{d}E \ . \tag{6.75}$$

Although, rather than being a stopping force, the rate of energy loss $\mathrm{d}E/\mathrm{d}x$, defined as the energy loss per unit distance x, is commonly called stopping power for historical reasons. The stopping power depends on the particle's velocity as described in Sects. 6.2.2 and 6.2.3. While slowing down, a particle may change its initial direction of motion by elastic encounters with target nuclei. Three types of possible trajectories can be distinguished (Fig. 6.7). The penetration of particles through matter can be described by several parameters. The path length is the total length of the trajectory, the depth is the distance from the surface to the position where the particle is stopped, and the spread is the distance from the point of impact to the surface point

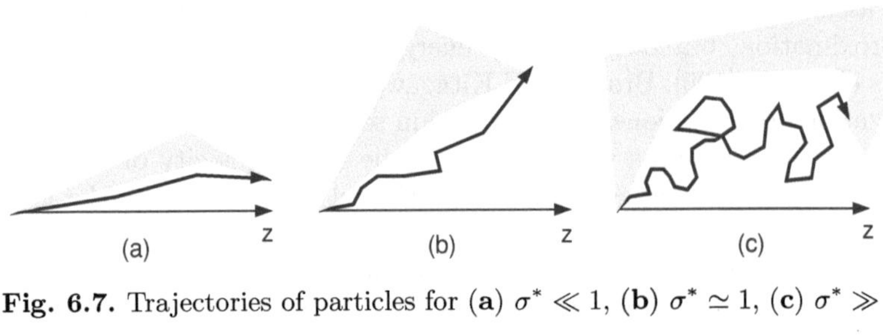

Fig. 6.7. Trajectories of particles for (**a**) $\sigma^* \ll 1$, (**b**) $\sigma^* \simeq 1$, (**c**) $\sigma^* \gg 1$

determined by projecting the point of rest onto the surface. In addition, we have the radial range, which denotes the distance from the point of impact to the point of rest in the material (Fig. 6.8). Furthermore, transverse and longitudinal projected ranges can be defined (see Chap. 10 in [109]).

Introducing the ratio of the total range to the transport length as a new parameter allows for the classification of the processes of stopping and scattering

$$\sigma^* = \frac{R_o}{l_{tr}} = R_o n_o \sigma_{tr} \tag{6.76}$$

where the transport length $l_{tr} = 1/(n_o \sigma_{tr})$ is determined by the atomic density n_o and the transport cross-section

$$\sigma_{tr} = 2\pi \int_0^{\pi} (1 - \cos\theta)\, \frac{d\sigma}{d\Omega}\, \sin\theta\, d\theta\ . \tag{6.77}$$

In the case of $\sigma^* \ll 1$ (Fig. 6.7a), the particle moves on a rather straight trajectory, since no deflection by an angle of near unity occurs. This situation changes if $\sigma^* \simeq 1$ (Fig. 6.7b), then at least, one strong deflection disturbs the straight motion. For $\sigma^* \gg 1$ (Fig. 6.7c), a more or less chaotic motion can be observed, which is described in the terminology of diffusion (random-walk) processes. The scattering parameter σ_{tr} gives the number of strong deflections along the total path. For non-relativistic electrons, we have approximately [110]

$$\sigma^* = \left[1 - \exp\left(-\frac{5.6}{Z_2 + 1}\right)\right]^{-1} \tag{6.78}$$

whereas for light ions in heavy media the parameter is about [110]

$$\sigma^* = \frac{6.33 M_1^{1/2}}{Z_1^{2/3} \varepsilon_o^{3/2}}\, \ln(1 + 0.7\varepsilon_o) \tag{6.79}$$

with the reduced energy ε_o (see (6.88)). The atomic number of the target material Z_2 is the key factor and largely defines the scattering behavior, i.e., the higher the ratio Z_2/Z_1, the more pronounced the scattering.

Distributions are characterized by their moments. By taking only the first two moments into account, a good agreement with experimental results can often be achieved. Thus, the Gaussian normal distribution

$$f(R_d) = \frac{1}{\sqrt{2\pi \overline{\Delta R_d^2}}} \exp\left(-\frac{(R_d - \overline{R_d})^2}{2\overline{\Delta R_d^2}}\right)\ , \tag{6.80}$$

here written for the depth distribution function, can also be applied for the energy distribution as well as for the distribution of scattering angles.

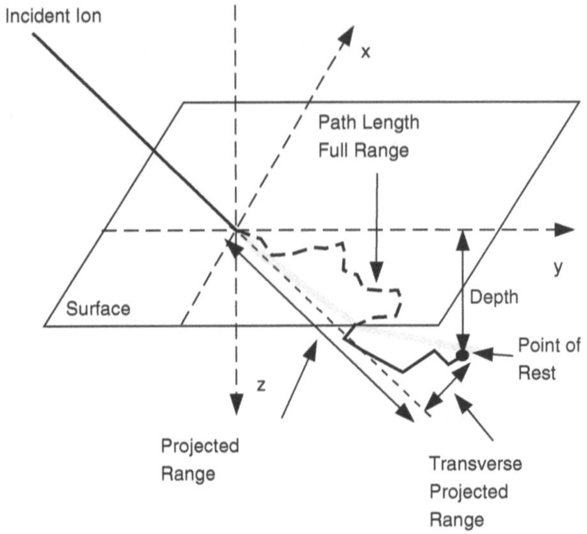

Fig. 6.8. Schematic drawing for the definitions of depth and of the different ranges as indicated

The depth distribution function (6.80) gives the fraction of particles that have reached a depth of R_d where $\overline{R_\mathrm{d}}$ is the mean depth and $\overline{\Delta R_\mathrm{d}^2} = \overline{(R - \overline{R_\mathrm{d}})^2}$ is the depth variance—the straggling parameter (Fig. 6.8). More sophisticated analytical distributions and theories of particle transport can be found in [107, 108, 111–113]. Today's theoretical investigations are usually assisted by computer simulation techniques (see Sect. 6.10). As an example, the ranges and standard deviations are given for certain ion–target combinations and energies in Table 6.1.

6.2.2 Elastic Energy Loss

During penetration, a particle has a large number of interactions with the nuclei in the target. In each collision, a part of its kinetic energy E_Δ is transferred. These losses are called elastic energy losses and are defined as the loss of energy per unit distance

$$-\frac{\mathrm{d}E}{\mathrm{d}x} = n_\mathrm{o} S_\mathrm{n} = n_\mathrm{o} \int\limits_{E_\Delta^\mathrm{min}}^{E_\Delta^\mathrm{max}} E_\Delta \frac{\mathrm{d}\sigma}{\mathrm{d}E_\Delta} \mathrm{d}E_\Delta \tag{6.81}$$

where n_o is the atom density in the target and

$$\frac{E_\Delta^\mathrm{max}}{E_\mathrm{o}} = \gamma_\mathrm{k} = \frac{4M_1M_2}{(M_1+M_2)^2} \tag{6.82}$$

gives the maximum energy. The maximum energy can be transferred in a elastic collision, namely during a head-on collision with $\alpha = \pi$ according to

Table 6.1. Mean depth and standard deviation for normal incidence [109, 114]

Ion	Target	Energy (eV)	$\overline{R_\mathrm{d}}$ (nm)	$\sqrt{\overline{\Delta R_\mathrm{d}^2}}$ (nm)
D	C	10	0.3	0.2
D	C	10^2	2	1.2
D	C	10^3	20	9
D	C	10^4	160	40
D	Si	10	0.7	0.4
D	Si	10^2	4	2
D	Si	10^3	25	13
D	Si	10^4	200	65
^{4}He	Si	10^2	2	2.5
^{4}He	Si	10^3	14	10
^{4}He	Si	10^4	100	60
^{4}He	Ni	10^2	2	0.8
^{4}He	Ni	10^3	7	5
^{4}He	Ni	10^4	45	20

(6.8); S_n is the stopping cross-section and $\mathrm{d}E/\mathrm{d}x$ the stopping power. For Coulomb interactions ($U(r) = Z_1 Z_2 e^2/(4\pi\epsilon_\mathrm{o} r) = C/r$), (6.81) can be readily calculated. For that purpose, we give the Rutherford cross-section (6.43) as a function of the transferred energy E_Δ instead of the scattering angle α (see (6.8) and (6.17))

$$E_\Delta = E_2 = 4E_\mathrm{o}\frac{M_1 M_2}{(M_1+M_2)^2}\cos^2\theta_2 = E_\mathrm{o}\gamma_\mathrm{k}\sin^2\frac{\alpha}{2} \tag{6.83}$$

since $\cos(\theta_2) = \cos(\pi/2 - \alpha/2) = \sin(\alpha/2)$ and

$$\mathrm{d}E_\Delta = E_\mathrm{o}\gamma_\mathrm{k}\sin\frac{\alpha}{2}\,\cos\frac{\alpha}{2}\,\mathrm{d}\alpha \tag{6.84}$$

with the kinematic factor γ_k given in (6.82), and $E_\mathrm{o} = M_1 v_\mathrm{o}^2/2$ is the kinetic energy of the incoming particle. Substituting now $\sin(\alpha/2)$ in terms of E_Δ yields the desired result

$$\frac{\mathrm{d}\sigma}{\mathrm{d}E_\Delta} = \frac{\pi C^2 M_1}{E_\mathrm{o} M_2}\,\frac{1}{E_\Delta^2} \tag{6.85}$$

since $E_\mathrm{r} = E_\mathrm{o} M_2/(M_1+M_2)$, $\sin\alpha = 2\sin(\alpha/2)\cos(\alpha/2)$, and $\mathrm{d}\Omega = 2\pi\,\sin\alpha\,\mathrm{d}\alpha$. Using (6.85) in (6.81) leads to

$$-\frac{\mathrm{d}E}{\mathrm{d}x} = n_\mathrm{o}\,S_\mathrm{n}(E_\mathrm{o}) = n_\mathrm{o}\int_{E_\Delta^\mathrm{min}}^{E_\Delta^\mathrm{max}} E_\Delta\,\frac{\mathrm{d}\sigma}{\mathrm{d}E_\Delta}\,\mathrm{d}E_\Delta = \frac{\pi C^2 n_\mathrm{o} M_1}{E_\mathrm{o} M_2}\ln\left[\frac{E_\Delta^\mathrm{max}}{E_\Delta^\mathrm{min}}\right]. \tag{6.86}$$

It is obvious from (6.86) that the minimum energy to be transferred $E_\Delta^{\min}$ cannot be set to zero. The problem to define the limits always occurs when integrations are performed based on the Rutherford cross-section. The cutoff at $E_\Delta^{\min}$ corresponds to a maximum impact parameter, which can be chosen to equal to the interatomic distance. Because $E_\Delta^{\min}$ appears in the logarithm, this parameter is of marginal importance despite the uncertainty in determining a suitable maximum impact parameter.

In practice, (6.81) does not allow an analytical presentation if realistic interaction potentials (Sect. 6.1.4) are considered. Often approximations fitted to numerical calculations are applied. It is helpful to use the normalized stopping cross-section

$$s_\mathrm{n}(\varepsilon) = \frac{(M_1 + M_2)\epsilon_\mathrm{o}}{M_1 Z_1 Z_2 e^2 a_\mathrm{L}} \, S_\mathrm{n}(E) \tag{6.87}$$

as a function of the reduced energy

$$\varepsilon = \frac{E}{E_\mathrm{L}} = E \, \frac{a_\mathrm{L} M_2 4\pi\epsilon_\mathrm{o}}{e^2 Z_1 Z_2 (M_1 + M_2)} \tag{6.88}$$

where a_L is the Lindhard screening length (6.53). Together with the normalized range r_ρ

$$r_\rho = \frac{x}{R_\mathrm{L}} = x \, \frac{4\pi M_1 M_2 n_\mathrm{o} a_\mathrm{L}^2}{(M_1 + M_2)^2} \, , \tag{6.89}$$

we have the relations

$$-\frac{\mathrm{d}E}{\mathrm{d}x} = n_\mathrm{o} S_\mathrm{n}(E) = -\frac{E_\mathrm{L}}{R_\mathrm{L}} \frac{\mathrm{d}\varepsilon}{\mathrm{d}r_\rho} = \frac{E_\mathrm{L}}{R_\mathrm{L}} \, s_\mathrm{n}(\varepsilon) = -\frac{n_\mathrm{o} a_\mathrm{L} Z_1 Z_2 e^2 M_1}{\epsilon_\mathrm{o}(M_1 + M_2)} \frac{\mathrm{d}\varepsilon}{\mathrm{d}r_\rho} \tag{6.90}$$

between the introduced parameters where $s_\mathrm{n}(\varepsilon) = \mathrm{d}\varepsilon/\mathrm{d}r_\rho$. Frequently used approximations are

$$s_\mathrm{n}^{Kr-C} = \frac{0.5\ln(1 + 1.2288\varepsilon)}{\varepsilon + 0.1728\sqrt{\varepsilon} + 0.008\varepsilon^{0.1504}} \tag{6.91}$$

$$s_\mathrm{n}^{U} = \frac{0.5\ln(1 + 1.1383\varepsilon)}{\varepsilon + 0.0132\,\varepsilon^{0.21226} + 0.19593\sqrt{\varepsilon}} \, . \tag{6.92}$$

6.2.3 Inelastic Energy Loss

The energy losses due to collisions with the target electrons are called inelastic energy losses. Especially at high energies of the particles, these losses dominate, since the light electrons are able to pick up considerable amounts of energy from the incident particle. The collisions with the nuclei cause mainly scattering and the energy losses (Sect. 6.2.2) are smaller. In accordance with the kind of model applied, the inelastic energy losses are called local, if the interaction with the bounded electrons in the clouds around the nuclei is considered [115,116]. In other (non-local) models, the electrons are considered as homogeneously distributed in the target by forming an electron gas of constant density as a good approximation to the conduction electrons in metals. The perturbation caused by the charged particles are assumed to be small.

The main parameter in the analysis is the ratio of the particle velocity to the Bohr velocity v_o/v_B. The latter value is equal to the velocity of the electrons in the first orbit of Bohr's atomic model

$$v_B = \frac{e^2}{4\pi\epsilon_o\hbar} = \frac{\hbar}{a_o m_e} = 2.18 \times 10^6\,\text{m/s} \tag{6.93}$$

with the Bohr radius $a_o = 4\pi\epsilon_o\hbar^2/(m_e e^2) = 5.29 \times 10^{-11}\,\text{m} = 0.529\,\text{Å}$. A proton with that velocity has an energy of about 25 keV. Three velocity regions can be distinguished. At small velocities of the incident particle, $v_o \ll v_B$, the energy losses are negligible. With increasing velocity, the energy losses increase as well and have their maximum at roughly $v_o = 3v_B Z_1^{2/3}$ where Z_1 is the atomic number of the incident ion. At higher velocities, the inelastic energy losses decrease, since the interaction time during the collisions becomes shorter (Fig. 6.9).

1. $v_o \ll v_B$

The positive charge of an ion slowly penetrating an electron gas causes a redistribution of the electrons within the Debye sphere (Sect. 5.2). Due to inertia, the center of this negatively-charged cloud falls a little bit behind the

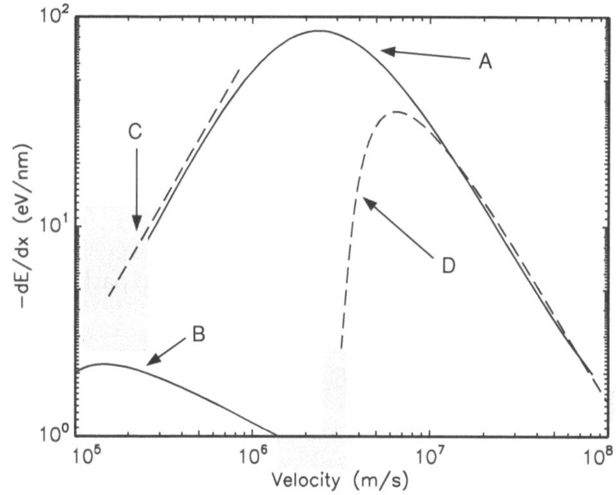

Fig. 6.9. Schematic presentation of energy losses as a function of the particle energy. (A) inelastic energy loss (B) elastic energy loss (C) Lindhard–Scharff formula (6.95) (D) Bethe–Bloch formula (6.100))

moving ion. This charge separation generates an electric field, which tends to slow down the ion. The energy loss of the ion as a consequence of the polarization action was found to be directly proportional to the ion velocity [117]. This is an analogy to the friction force acting on a moving sphere with radius r in a viscous fluid

$$-\frac{dE}{dx} \propto 6\pi\eta_{\mathrm{v}} v_{\mathrm{o}} r \tag{6.94}$$

according to the Stokes relation where η_{v} is the viscosity. The effects caused by a charged particle penetrating a quantized electron plasma have been analyzed in [118]. Lindhard provided a full treatment in the case of non-relativistic interaction and suggested in [92] together with Scharff the relation

$$-\frac{\mathrm{d}E}{\mathrm{d}x} = n_{\mathrm{o}} S_{\mathrm{LS}} = 2Z_1^{1/6} e^2 a_{\mathrm{o}} n_{\mathrm{o}} \frac{Z_1 Z_2}{\epsilon_{\mathrm{o}}(Z_1^{2/3} + Z_2^{2/3})^{3/2}} \frac{v}{v_{\mathrm{B}}} = k_{\mathrm{L}} \sqrt{E_{\mathrm{o}}} \tag{6.95}$$

or

$$-\frac{\mathrm{d}E}{\mathrm{d}x}\left[\frac{eV}{\text{Å}}\right] = 1.21\, n_{\mathrm{o}} \left[\frac{1}{\text{Å}^3}\right] \frac{Z_1^{7/6} Z_2}{(Z_1^{2/3} + Z_2^{2/3})^{3/2}} \sqrt{\frac{E_{\mathrm{o}}[\mathrm{eV}]}{M_1[\mathrm{amu}]}}\,, \tag{6.96}$$

which well describes the experimental data of inelastic energy losses in the velocity region $v_{\mathrm{o}} < v_{\mathrm{B}} Z_1^{2/3}$.

2. $v_{\mathrm{o}} \gg v_{\mathrm{B}}$

At high velocities of the incident ion, i.e., $v_{\mathrm{o}} \gg Z_2 v_{\mathrm{B}}$, the binding energies of the electrons and their own velocities can be neglected, i.e., the electrons of the target are assumed to be at rest. The corresponding cross-section for energy loss in binary collisions has already been derived in Sect. 6.2.2 and is (see 6.85)

$$\frac{\mathrm{d}\sigma}{\mathrm{d}E_{\Delta}} = \frac{2\pi Z_1^2 e^4}{(4\pi\epsilon_{\mathrm{o}})^2 m_{\mathrm{e}} v_{\mathrm{o}}^2 E_{\Delta}^2} \tag{6.97}$$

with $E_{\mathrm{o}} = M_1 v_{\mathrm{o}}^2/2$, $M_2 = m_{\mathrm{e}}$, $Z_2 = -1$, and $v_\infty = v_{\mathrm{o}1} - v_{\mathrm{o}2} = v_{\mathrm{o}} - v_{\mathrm{o}2} \simeq v_{\mathrm{o}}$. According to (6.81), for the energy losses along the path

$$-\frac{\mathrm{d}E}{\mathrm{d}x} = n_{\mathrm{o}} Z_2 \int_{E_{\Delta}^{\mathrm{min}}}^{E_{\Delta}^{\mathrm{max}}} E_{\Delta} \frac{\mathrm{d}\sigma}{\mathrm{d}E_{\Delta}} \mathrm{d}E_{\Delta} = n_{\mathrm{o}} \frac{2\pi Z_1^2 e^4 Z_2}{(4\pi\epsilon_{\mathrm{o}})^2 m_{\mathrm{e}} v_{\mathrm{o}}^2} \ln\left(\frac{E_{\Delta}^{\mathrm{max}}}{E_{\Delta}^{\mathrm{min}}}\right) \tag{6.98}$$

where $Z_2 n_{\mathrm{o}}$ gives the number of electrons per unit volume. The maximum energy to be transferred in ion–electron collisions is well-defined as

$$E_{\Delta}^{\mathrm{max}} = 4E_{\mathrm{o}} \frac{M_1 M_2}{(M_1 + M_2)^2} \simeq \frac{m_{\mathrm{e}}}{M_1} = 2m_{\mathrm{e}} v_{\mathrm{o}}^2\,. \tag{6.99}$$

However, the right choice of $E_{\Delta}^{\min}$ is questionable. It has been proposed to take the excitation energy for $E_{\Delta}^{\min}$, but the resulting energy losses would be too small. One could consider only collisions with an impact parameter smaller than the average distance between the electrons [119]. Darwin restricted his analysis to collisions with deep electron penetration [120].

Two decades later, Bethe [121] and Bloch [122, 123] derived their fundamental relation based on quantum mechanics for the stopping power of fast particles under the assumption that the oscillator strength can be represented by the ionization energy I_{ion}. In addition, the Born approximation has been applied, i.e., the amplitude of the wave scattered by the atomic electron field has to be small, when compared to the one of the incident (undisturbed) wave. The problem with the factor of 2 comparing the results of classical and quantum mechanics calculations was resolved by Bloch by a proper treatment of the impact parameter in both models. It is worth noting that the wavelength of the particle is rather small in comparison to the interatomic distance. Thus, the classical treatment might be justified for most of the velocities of interest. In the non-relativistic case, the Bethe–Bloch relations states

$$\begin{aligned} -\frac{\mathrm{d}E}{\mathrm{d}x} &= n_{\mathrm{o}} S_{\mathrm{BB}} = n_{\mathrm{o}} \frac{4\pi Z_1^2 Z_2 e^4}{(4\pi\epsilon_{\mathrm{o}})^2 m_{\mathrm{e}} v_{\mathrm{o}}^2} \ln\left(\frac{2 m_{\mathrm{e}} v_{\mathrm{o}}^2}{I_{\mathrm{ion}}}\right) \\ &= n_{\mathrm{o}} \frac{Z_1^2 Z_2 e^4}{4\pi\epsilon_{\mathrm{o}}^2 m_{\mathrm{e}} v_{\mathrm{o}}^2} \ln\left(\frac{2 m_{\mathrm{e}} v_{\mathrm{o}}^2}{I_{\mathrm{ion}}}\right) . \end{aligned} \tag{6.100}$$

For the mean ionization energy, several approximations are proposed such that $I_{\text{ion}} = 13 Z_2$ or $I_{\text{ion}} = 10.3 Z_2 (1 - 0.793 Z_2^{-2/3})$, thereby trying to include the contribution of excitation as well. Inherent to this treatment is that energies smaller than the ionization energy cannot be transferred.

The Bethe–Bloch formula remains the principal tool to evaluate the energy loss of particles with velocities of 1 MeV/amu to 2 GeV/amu, of course taking relativistic corrections at higher velocities into account [104]. Several other coefficients such as shell corrections have been added to improve the formula [124]. Excellent reviews of relativistic particle (> 10 MeV/amu) stopping powers are given in [125] and [126]. For velocities below 1 Mev/amu, this relation fails because the ion projectile may not be fully stripped of its electrons as assumed in the theory.

3. $v_{\mathrm{o}} \simeq v_{\mathrm{B}}$

At the maximum of energy losses, the particle velocity is of the same order as the velocities of the electrons in the target. Partially-stripped particles, overlapping electron clouds, and the necessity of dealing with binary collisions in their general form are some of the aspects to consider. Despite the theoretical effort [118, 127–131], no satisfactory solution could be found. Up to now, approximations of the type [132]

$$S = \frac{S_{\mathrm{LS}}\, S_{\mathrm{BB}}}{S_{\mathrm{LS}} + S_{\mathrm{BB}}} \tag{6.101}$$

where S_{LS} is the stopping power as given by the Lindhard–Scharff formula (6.95) at low velocities and S_{BB} is the Bethe–Bloch stopping power (6.100) at high velocities, are used in the intermediate energy region [104].

6.3 Material Modification by Ion Beams

Comprehensive reviews on material modifications due to ion beam impact are given in [133–138]. The main processes are defect production and agglomeration, compositional changes of alloys by preferential sputtering, Gibbsian segregation, displacement mixing, bombardment-induced decomposition, radiation-enhanced diffusion, and phase transformation. All these processes are governed by a number of parameters, such as ion type and energy, target temperature, structure and composition, and ion flux density. Thus, a large variety of structures and composition distributions can result from ion bombardment of solid compounds.

Preferential sputtering is triggered by differences of mass and surface binding energy. Atoms of different elements are emitted at different rates. The resulting compositional changes in the top atom layer are accompanied by segregation. This effect, often called Gibbsian adsorption, tends to minimize the surface free energy by increasing the concentration of one alloy constituent in the outermost atom layer relative to that in the bulk. Usually, segregation requires elevated temperatures, where thermally activated diffusion can occur but can also be bombardment-induced. During irradiation, a number of trajectories are terminated near the target surface and are followed by one or more low-energy, chemically-guided steps. The spatial separation between defect production and annihilation establishes defect fluxes, usually directed toward the surface. Since the motion of defects requires the motion of atoms, the fluxes of alloying elements couple to the defect fluxes. Ion bombardment provides sufficient energy leading to recoil generation, cascade evolution, atom mixing, and redistribution of constituents. Amorphization as a process of low-temperature mixing is often observed.

Defects are produced during collisions of ions and displaced atoms with other atoms with the result that bulk atoms receive an energy of $E_{\text{d}} \approx 25\,\text{eV}$, which is necessary to make an irreversible displacement out of a crystal lattice. Note that the displacement energy is not a well-defined quantity and ranges from 10–40 eV. It depends on the direction of the pushed target atoms, and how large its energy should be to form a stable Frenkel pair, i.e., a vacancy and an interstitial displaced atom. These are point defects. Based on the model proposed in [139], the number of produced Frenkel pairs N_{F} can be estimated as

$$N_{\text{F}} \simeq 0.8\,\frac{(1-R_{\text{E}})E_{\text{o}}}{2E_{\text{d}}} \tag{6.102}$$

where R_{E} is the energy reflection coefficient. This estimation can be improved by including the elastic energy loss in the cascade and the energy loss due to transmission and sputtering [109]. These additional losses obviously reduce the energy at disposal for damage production.

As a result of secondary processes such as diffusion and joining of point defects into clusters, more complex defects are formed: dislocation loops, vacancy pores, or, in the presence of a gas, bubbles. By ion irradiation, the material properties can be modified: (1) the yield stress of metals and alloys increases considerably at temperatures below about 0.35 of the (absolute) melting temperature leading to hardening and embrittlement, (2) swelling caused by the nucleation and growth of cavities occurs, and (3) plastic deformation of materials caused by the simultaneous action of irradiation and mechanical stress (irradiation creep) occurs as well as high-temperature embrittlement due to radiation-induced impurity segregation to grain boundaries [134,138,140,141]. Helium-containing materials often fail by intergranular fracture.

Since thermal diffusion in metals occurs predominantly by the position exchanges between atoms and vacancies, the presence of excess defects during ion bombardment at elevated temperatures results in radiation-enhanced diffusion. The analysis of temperature-controlled processes such as diffusion and segregation is complicated owing to the dependence of the corresponding coefficients of temperature, irradiation dose, ion energy, the structure of the material, as well as on the particle flux density in the case of radiation-stimulated effects. The theoretical study of the behavior of gases in materials is based on a set of transport equations that contain a large number of uncertain parameters such as trap binding energies, mobility, diffusivity, solubility, recombination coefficients, defect microstructure [142], and can, in the best of cases, explain the observed behavior. Especially in the case of graphite with its complex C–H bonding structure, even computer simulations are hampered due to the lack of reliable basic data [143–146].

Many of the listed processes, for example displacement mixing, can be described by the diffusion formalism by using effective diffusion coefficients, which are expressed in terms of jump rates and jump lengths (see (6.109)). The diffusion equation for constant, but anisotropic diffusion is given by

$$\frac{\partial n}{\partial t} = \sum_{i=1}^{d_\mathrm{m}} D_i \frac{\partial^2 n}{\partial x_i^2} \tag{6.103}$$

where $n(r,t)$ is the particle density for time t at the position $r = (x_1, ..., x_{d_\mathrm{m}})$, and d_m is the dimension. The change of the particle density in time is, therefore, defined by the difference of the incoming flux into a volume element and the outgoing particle flux. The convective part of the particle flux ($\Gamma = n\, v_\mathrm{convection}$) is neglected in (6.103). If all particles N started to diffuse at $t = 0$ from the same position $r = r_\mathrm{o}$, then the solution of (6.103) yields

$$n(x_\mathrm{i}, t) = N \left(\prod_{i=1}^{d_\mathrm{m}} \frac{1}{\sqrt{4\pi D_i\, t}} \right) \exp\left[-\sum_{i=1}^{d_\mathrm{m}} \left(\frac{x_i^2}{4D_i\, t} \right) \right] . \tag{6.104}$$

Hence, the surfaces of same particle density are spheres in the case of isotropic three-dimensional diffusion and ellipsoids

$$\frac{x^2}{2\nu_x\lambda_x^2 t} + \frac{y^2}{2\nu_y\lambda_y^2 t} + \frac{z^2}{2\nu_z\lambda_z^2 t} = \frac{x^2}{4D_x t} + \frac{y^2}{4D_y t} + \frac{z^2}{4D_z t} = \text{const.} \tag{6.105}$$

if the diffusion is different in different directions. The distance between the position reached after time t and the initial position $\Delta x_i = |x_i - x_{io}|$ is determined by the diffusion coefficient D_i (if $x_{io} = 0$)

$$\langle(\Delta x_i)^2\rangle = \frac{1}{N}\int_{-\infty}^{\infty} n(x_i,t)\, x_i^2 \, \mathrm{d}x_i = 2D_i t \tag{6.106}$$

where for one-dimensional diffusion $D = \langle(\Delta x)^2\rangle/(2t)$. In the case of isotropic diffusion, i.e., $D = D_1 = D_2 = D_3 = D_{d_\mathrm{m}}$, we have

$$|\Delta r|^2 = \sum_{i=1}^{d_\mathrm{m}} (\Delta x_i)^2 \tag{6.107}$$

and

$$\langle(\Delta r)^2\rangle = \sum_{i=1}^{d_\mathrm{m}} \langle(\Delta x_\mathrm{i})^2\rangle = 2d_\mathrm{m}\, Dt \tag{6.108}$$

with

$$D = \frac{1}{2d_\mathrm{m}} \frac{\langle(\Delta r)^2\rangle}{t} = \frac{1}{2d_\mathrm{m}}\, \nu\lambda^2 \,. \tag{6.109}$$

Using (6.109), the diffusion coefficient can be readily determined in particle simulation codes. For each particle, the distance from a starting point has to be obtained after a certain time t, which should be much larger than the mean collision time ($t \gg \tau$), and then their squares have to be averaged for all particles simulated. Another method to obtain the diffusion coefficient is via

$$D = \frac{1}{2d_\mathrm{m}} \lim_{t\to\infty} \frac{\langle(\Delta r)^2\rangle}{t} \tag{6.110}$$

by calculating the square of the distance between initial and current positions over the simulation time, and averaging this for all particles. This curve shows for larger time a linear asymptotic behavior; its slope is directly proportional to the diffusion coefficient according to (6.110).

In fusion experiments, erosion and powerful destruction events as well as deposition of material dominate the change of surface composition and surface structure. Not to forget, the rather large roughness of the used plasma-facing materials makes the discussion in terms of atom layers somewhat academic.

6.4 Retention and Tritium Inventory Control

In principle, four mechanisms can result in tritium retention: (1) direct implantation of tritium ions, (2) diffusion into the bulk, (3) production of tritium by transmutation nuclear reactions, and (4) codeposition of tritium with eroded material (Fig. 6.10).

The amount of implanted fuel atoms in ITER can be roughly estimated at

$$M_{\mathrm{T}} = m_{\mathrm{T}}\, c_{\mathrm{T/C}}\, n_{\mathrm{C}}\, S_{\mathrm{wall}}\, d_{\mathrm{impl}} = 2.3\,\mathrm{g} \tag{6.111}$$

where $m_{\mathrm{T}} = 5 \times 10^{-27}$ kg is the tritium atom mass, $c_{\mathrm{T/C}}$ is the concentration of tritium in carbon (here set to unity), $n_{\mathrm{C}} = 11.3 \times 10^{28}$ atoms/m^3 is the atomic density of graphite, $S_{\mathrm{wall}} \simeq 800\,\mathrm{m}^2$ is the area of all wall components in ITER, and $d_{\mathrm{impl}} \simeq 5 \times 10^{-9}$ m [114] is the average implantation depth according to an energy of about 100 eV of charge-exchange neutrals. With lower impact energies and grazing incidence, the implantation depth becomes much smaller and the implanted amount is far below the ITER limit of 350 g. Diffusion into the depth of the bulk material, e.g., along grain boundaries, could lead to a higher inventory, but the resulting amount of trapped tritium does not reach critical values [147, 148].

On the other hand, codeposited layers with a thickness of several hundred micrometers are often observed in fusion experiments using carbon-based wall components. The ratio of tritium to carbon in such deposited layers depends on the surface temperatures, but can reach values up to 0.4, and in so-called soft layers up to unity. At temperatures of 1000 degrees centigrade, this ratio goes down to 0.1. No saturation effect is observed, i.e., as long as new carbon is sticking to the surfaces tritium will be codeposited. This codeposition mechanism is clearly the dominant one with respect to tritium retention.

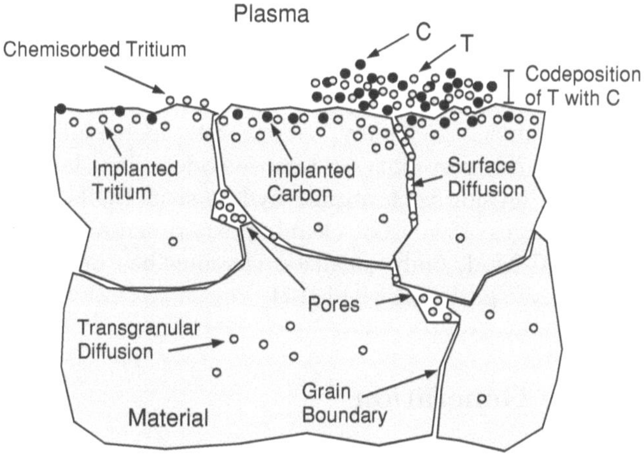

Fig. 6.10. Schematic of different retention and diffusion channels of tritium [146]

Unfortunately, not only carbon materials but also materials such as beryllium and tungsten show significant ability to be codeposited with fuel ions [149].

Prevention of film deposition or ensuring high re-erosion of deposited films is a necessity. Experience from plasma technology research, despite its opposite aim of achieving high deposition rates, could be useful. Recently, experiments in ASDEX-Upgrade have shown that injection of nitrogen into the divertor plasma led to a strong decrease in the amount of deposited material without deteriorating the performance of the main plasma [150]. Which mechanisms are involved is a question of ongoing research. Chemical sputtering, surface chemistry effects, and gas-phase reactions are considered. Nitrogen bondings with itself or with carbon atoms are highly stable and lead to the formation of volatile products such as N_2, CN, HCN, C_2N_2. This process of chemical erosion is effective only at higher energies ($E > 20\,\mathrm{eV}$). It has been found that the evolving surface state influences the deposition efficiency. The higher the nitrogen concentration at the surface, the lower the growth rate. In addition, the bond structure at the surface affects the sticking of impinging species. The radical molecule as well as the ion chemistry in the plasma is significantly modified under the presence of nitrogenated species [150]. All these effects could help to inhibit film formation and, therefore, have the potential to reduce the tritium inventory due to codeposition.

While oxygen is not a favored gas as far as tokamak operation is concerned, it is the basis for most techniques suggested to remove codeposited films [151].

Based on laboratory experiments on the oxidation of carbon films, which show a significant reaction of oxygen with the layers in the range between 500–750 K, ventilation of TEXTOR with oxygen has been investigated to remove redeposited carbon material and to release the incorporated hydrogen. A significant part of the injected oxygen adsorbs on the walls due to the formation of stable oxygen compounds. Partly, oxygen reacts with carbon to form CO and CO_2 [152].

The effect of thermo-oxidative removal of hydrogen from codeposited layers is not well understood. It is suggested that metal contamination could act as a catalyst, thus explaining the high erosion rates (a few μm/h) observed for the films, as compared to the graphite substrate [151].

While oxygen shows the highest removal rate, increased wall temperatures (e.g., 600 K) offer the possibility to remove codeposited layers by thermally-induced chemical erosion with atomic hydrogen in ECR (electron cyclotron resonance) discharges [153]. ECR cleaning experiments have been performed also in Alcator C-Mod. Only localized cleaning has been observed with a rather low effective sputtering yield [154].

6.5 Impurity Generation

Large energy and particle fluxes are deposited onto wall elements in fusion experiments. Owing to this intense plasma–wall interaction, material is eroded by several mechanisms. The emission of atoms consequently may lead to plasma contamination. The main erosion processes are physical and chemical sputtering, thermal and radiation-enhanced sublimation, erosion by arcs, and blistering. Furthermore, there are exotic erosion effects such as potential sputtering and Coulomb explosion events (Sect. 6.6). Almost all eroded particles leave the surface as neutrals in the ground state.

6.5.1 Physical Sputtering

Physical sputtering is the emission of surface atoms due to the impact of energetic particles and is described by momentum transfer. Target atoms involved in the developing collision cascade leave the surface if their received energy exceeds the surface binding energy E_s. The energy of the sputtered atoms is, therefore, reduced by E_s before escaping. Usually, the heat of sublimation as measured for real surfaces is used for the surface binding energy. Since the energy distribution of the sputtered particles has a maximum, the assumption of a planar potential at the surface is justified. Physical sputtering may be influenced by the target temperature due to a modification of the heat of sublimation. The surface binding energy decreases with increasing target temperature [155].

The most advanced analytical description of physical sputtering described in terms of collision cascade formation is presented by Sigmund and co-workers [156–158]. A comprehensive review of physical sputtering with respect to both experimental and theoretical investigations is given in [159].

The crucial quantity is the sputtering yield

$$Y = \frac{\text{number of emitted target atoms}}{\text{number of incident particles}} . \tag{6.112}$$

In contrast to the reflection coefficient for example, the sputtering yield cannot be seen as a probability since it can be larger than unity. The sputtering yield depends on the ion–target combination, the energy of the incident particle, and its incident angle. By using the differential sputtering yield $\mathrm{d}Y^2/\mathrm{d}E\mathrm{d}\Omega$, the angle and energy distribution of the sputtered atoms is characterized. Sputtering is a rather short event. Two depths can be distinguished. Most of the sputtered atoms come out of the first two atomic layers ($\approx 5\,\text{Å}$), whereas a layer with a thickness of 25–50% of the mean depth is involved in cascade formation and transfer of the energy from the projectile to the target atoms. Three regimes of sputtering can be separated: (1) the regime of few collisions, (2) the cascade regime, and (3) the thermal-spike regime. In the case of light ions and low energies, cascade formation does not occur, since only few target atoms are involved, and their trajectories reveal a strong anisotropy. In this case, the eight different mechanisms proposed by Behrisch [160] can be reduced to four main processes illustrated in Fig. 6.11 [109]. It is sufficient to consider only recoils of the first generation (primary knock-on atoms (PKA)) and of the second generation (secondary knock-on atoms (SKA)). These recoils are

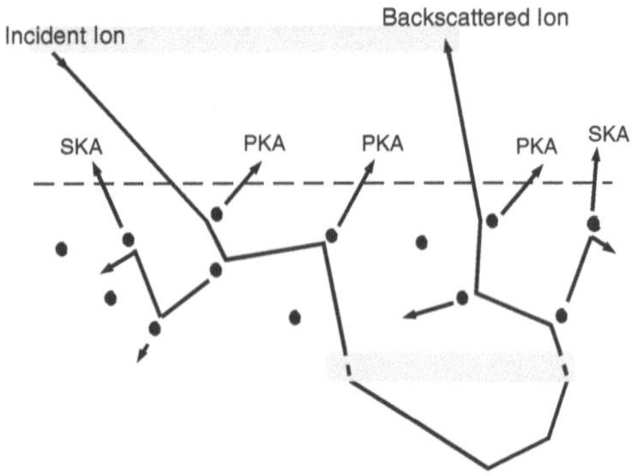

Fig. 6.11. Sputtering mechanisms for low-energy light ions. PKA: primary knock-on atom; SKA: secondary knock-on atom (see Fig. 12.1 of [109])

produced in the top atom layers by the ions crossing the surface during the impact and after being backscattered. For lower ion energies, the backscattering channel is more important, for higher energies most of the sputtered atoms are released during the impact of the ion.

The higher the energy and the mass of the incident ions, the more atoms are affected and recoils of several generations occur. The developing cascade can reach the surface and target atoms are emitted. The thermal spikes are, in fact, high-density cascades originated by high-energy heavy ions leading to an energy deposition on a very short time scale in a rather small volume. High temperatures are the results of these spikes accompanied by thermal sublimation and shock waves. Crater formation at the surface has been observed in the experiments.

As a rough estimation, the sputtering yield can be expressed as the ratio of energy E_{dep} deposited in the outermost layer of thickness d to the surface binding energy E_{s}

$$Y \simeq \frac{E_{\mathrm{dep}}}{E_{\mathrm{s}}} \tag{6.113}$$

since the target atoms that have received the energy E_{dep} have to overcome the surface potential. The energy is transferred in elastic collisions. According to (6.86)

$$E_{\mathrm{dep}} = n_{\mathrm{o}}\, S_{\mathrm{n}}(E_{\mathrm{o}})\, d \tag{6.114}$$

with the stopping cross-section S_{n} as a function of the ion energy E_{o}. Thus, the sputtering yield exhibits the same energy dependence as the stopping cross-section—low sputtering yield at low and high energies and a maximum at reduced energies of $\varepsilon = 0.1 - 1$. Taking the values $n_{\mathrm{o}}\, S_{\mathrm{n}}(E_{\mathrm{o}}) = 0.1\,\mathrm{eV/Å}$ $d = 5\,\mathrm{Å}$, and $E_{\mathrm{s}} = 5\,\mathrm{eV}$, we obtain a sputtering yield of $Y = 0.1$.

According to [161–163], the yield of physical sputtering can be calculated for normal incidence by

$$Y(E, \alpha = 0^{\mathrm{o}}) = Q_{\mathrm{y}}\, s_{\mathrm{n}}(\varepsilon) \left(1 - \left(\frac{E_{\mathrm{th}}}{E}\right)^{2/3}\right) \left(1 - \frac{E_{\mathrm{th}}}{E}\right)^{2} \tag{6.115}$$

where E is the particle energy, and $s_n(\varepsilon)$, the nuclear stopping cross-section, is to be taken as [164]

$$s_n(\varepsilon) = \frac{0.5\ln(1 + 1.2288\varepsilon)}{\varepsilon + 0.1728\sqrt{\varepsilon} + 0.008\varepsilon^{0.1504}} \tag{6.116}$$

based on the Kr-C interaction potential with $\varepsilon = EM_2a_L4\pi\epsilon_o/[Z_1Z_2e^2(M_1 + M_2)] = E/E_{TF}$ and the Lindhard screening length $a_L = 0.04685/(Z_1^{2/3} + Z_2^{2/3})^{1/2}$ nm. Z_1 and Z_2 are the nuclear charges, and M_1 and M_2 are the masses in atomic mass units of the incident particle and the target atom, respectively. There are some analytical expressions for the parameter Q_y and the threshold energy E_{th} [165–167], but usually these values are obtained by fitting the relation (6.115) to experimental and/or simulation data (see Table 6.2). For energies below the threshold energy E_{th}, the sputtering yield is zero. This energy threshold naturally appears as a result of the surface binding energy. E_{th} should be larger than E_s, at least. However, the energy has to be transferred from the incident ion to a target atom and the target atom itself should change its initial direction after collision and move to the surface. This requires at least one additional collision. Bohdansky suggested the following relations [161]

$$E_{th} = \begin{cases} E_s/(\gamma_k(1-\gamma_k)) & \text{for } M_1/M_2 \leq 0.2 \\ 8E_s\,(M_1/M_2)^{2/5} & \text{for } M_1/M_2 > 0.2 \end{cases} \tag{6.117}$$

Table 6.2. Parameters needed to calculate the sputtering yield from (6.115) [168]

Target – particle	H	D	T	He	Self-sputtering
Be ($E_s = 3.38$ eV)					
E_{th} (eV)	13	13	15	16	24
E_{TF} (eV)	256	282	308	720	2208
Q_y	0.07	0.11	0.14	0.28	0.67
C ($E_s = 7.42$ eV)					
E_{th} (eV)	31	28	30	32	53
E_{TF} (eV)	415	447	479	1087	5688
Q_y	0.05	0.08	0.10	0.2	0.75
Fe (E_s =4.34 eV)					
E_{th} (eV)	61	32	23	20	31
E_{TF} (eV)	2544	2590	2635	5517	174122
Q_y	0.07	0.12	0.16	0.33	10.44
Mo (E_s =6.83 eV)					
E_{th} (eV)	172	83	56	44	49
E_{TF} (eV)	4719	4768	4817	9945	533127
Q_y	0.05	0.09	0.12	0.24	16.27
W (E_s =8.68 eV)					
E_{th} (eV)	447	209	136	102	62
E_{TF} (eV)	9871	9925	9978	20376	1998893
Q_y	0.04	0.07	0.1	0.2	33.47

with the kinematic factor $\gamma_k = 4M_1M_2/(M_1+M_2)^2$ (the power of (5/2) in the relation given in [161] is a misprint). Fitting experimental and calculated data for many ion–target combinations, the following approximations have been proposed [164]

$$\frac{E_{\rm th}}{E_{\rm s}} = 7\left(\frac{M_2}{M_1}\right)^{-0.54} + 0.15\left(\frac{M_2}{M_1}\right)^{1.12} . \tag{6.118}$$

The dependence upon the angle of incidence α with respect to the surface normal can be calculated as the following [169]:

$$Y(E,\alpha) = \frac{Y(E,\alpha=0)}{(\cos\alpha)^{f_{\rm y}}} \exp\left(f_{\rm y}\left[1-\frac{1}{\cos\alpha}\right]\cos\alpha_{\rm max}\right) . \tag{6.119}$$

The values $f_{\rm y}$ and $\alpha_{\rm max}$ are often used also as fitting parameters or can be estimated with the relations [168] $f_{\rm y} = \sqrt{E_{\rm s}}(0.94 - 0.00133\,M_2/M_1)$ and $\alpha_{\rm max} = \pi/2 - a_{\rm L}n^{1/3}/\sqrt{2\varepsilon\sqrt{E_{\rm s}/(\gamma_{\rm k}E)}}$ where $E_{\rm s}$ is the surface binding energy (heat of sublimation) in eV, n is the density of the target material in atoms/m^3, and $\gamma_{\rm k} = 4M_1M_2/(M_1+M_2)^2$. The angle $\alpha_{\rm max}$ corresponds to the maximum of the sputtering yield.

The energy distribution of sputtered particles, which leave the surface predominantly as neutrals, is given by [157]

$$\frac{dY}{dE} \propto \frac{E}{(E+E_{\rm s})^{3-2m_{\rm p}}} \tag{6.120}$$

where E is the energy of the sputtered atoms. The maximum of the energy distribution is at $E_{\rm s}/(2-2m_{\rm p})$. Often, the value $m_{\rm p}=0$ for a hard sphere interaction potential, which is reasonable at low energies, is assumed [170]. However, better agreement with experimental and simulated data is found, if for the exponent $m_{\rm p}$ of the power potential a value of 1/6 is taken [157,170]. A cosine distribution for the angular distribution of sputtered atoms is the consequence of an isotropic flux of recoils in the solid.

In order to calculate the sputtering yields due to the bombardment of the plasma ions onto the material surface, the energy and angular distribution of the impinging particles have to be considered. A twofold averaged sputtering yield, defined as the yield averaged over the energy and angle of the incident ions is obtained by double integration [171,172]. In addition, the effect of sheath acceleration should be included, since the incoming ions, which are assumed to have a Maxwellian velocity distribution at the sheath entrance, change their directions and energies quite considerably when crossing the electric sheath above the surface. In addition to thermal energy flux, ($\propto 2k_{\rm B}T_{\rm i}$) the impurity ions of charge state q obtain an energy of about $q|e\phi_{\rm w}| \simeq q\,3k_{\rm B}T_{\rm e}$ in the sheath. The energy of (singly-charged) ions is then roughly five times the electron temperature for $T_{\rm i} = T_{\rm e}$ (see Sect. 8). The resulting average yield is then a function of the plasma temperature and the charge state of the ions

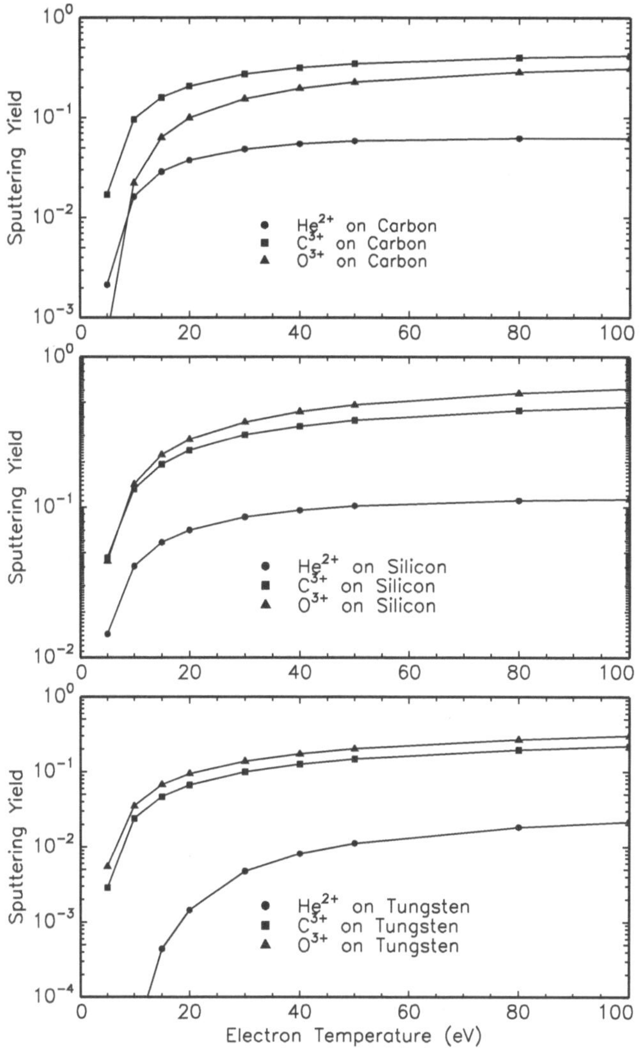

Fig. 6.12. Sputtering yields as a function of the plasma temperature for different ion–target combinations as indicated in the figure [173]

(Fig. 6.12). In contrast to (6.115), the threshold character disappears, as shown in Fig. 6.12, since there are always particles in the tail of the Maxwellian distribution with energies exceeding the threshold energy for sputtering as small as the plasma temperature might be.

Sputtering yields obtained with a Monte Carlo simulation program are given in Table 6.3.

A useful approximation of sputtering yields for fusion application can be found in [174]. Of course, the transformation of the velocity distribution in

Table 6.3. Calculated sputtering yields for various ion–target combinations obtained with the simulation program TRVMC, a version of the TRIM code [172] where a shifted Maxwellian distribution was used as the velocity distribution of the incident ions

Yield	$T_e = 5\,eV$	$T_e = 10\,eV$	$T_e = 20\,eV$	$T_e = 40\,eV$
$Y_{D^+ \to C}$	2.4×10^{-4}	3.7×10^{-3}	1.3×10^{-2}	2.1×10^{-2}
$Y_{D^+ \to Si}$	2.5×10^{-4}	4.0×10^{-3}	1.5×10^{-2}	2.6×10^{-2}
$Y_{D^+ \to Mo}$	–	1.0×10^{-6}	1.9×10^{-4}	3.1×10^{-3}
$Y_{D^+ \to W}$	–	–	7.0×10^{-8}	5.4×10^{-5}
$Y_{C^{3+} \to C}$	6.1×10^{-3}	3.7×10^{-2}	9.9×10^{-2}	1.9×10^{-1}
$Y_{C^{3+} \to Si}$	4.8×10^{-2}	1.4×10^{-1}	2.7×10^{-1}	4.1×10^{-1}
$Y_{C^{3+} \to Mo}$	1.2×10^{-2}	7.1×10^{-2}	1.8×10^{-1}	3.2×10^{-1}
$Y_{C^{3+} \to W}$	2.45×10^{-4}	1.6×10^{-2}	7.2×10^{-2}	1.6×10^{-1}

the sheath can only be tackled to its full extent within the scope of kinetic simulations.

The erosion of material exposed to a plasma, containing impurities, such as carbon shows a complicated non-linear behavior. This was demonstrated using computer simulation for the ion bombardment of various target materials [175]. Transitions from an erosion phase to a deposition phase and also reversed situations have been observed [172, 176]. Even under steady state plasma conditions, the erosion of an exposed target changes during the bombardment due to the change in surface composition. In the case of thin layers, wall conditioning layers for example, the lifetime depends on the kind of substrate material [177]. Owing to the higher reflection probability of plasma ions for high-Z materials, substrates made of such materials show higher erosion of the top layers. The general complexity arising in multi-component systems will be the subject of further investigation by experiments and numerical simulations in the near future.

While analytic approaches are able to show dependencies and main trends more clearly, computer simulations provide detailed information based only on a very limited set of assumptions regardless of complexity of the geometry, material structure, and composition (Sect. 6.10). With respect to ion–solid interaction, two methods using particles can be distinguished: (1) the binary collision approximation, and (2) the molecular dynamics method. Both techniques are discussed in [109] where also a large bibliography of simulation studies is given.

The sputtered particles leave the surface predominantly as neutrals. The small fraction of ions emitted depends strongly, however, on the electronic surface conditions via the work function. Oxygen tends to increase the positive fraction, and alkaline metals the negative fraction. The positive fraction increases with increasing velocity of the sputtered atoms. A similar statement applies to the excitation state of emitted target atoms, since ionization can be regarded as the highest excitation state.

6.5.2 Chemical Erosion

Chemical erosion is characterized by the formation of volatile molecules during chemical reactions of the incident particles with the target atoms. Chemical erosion (or sputtering) is important in case of carbon-based materials under the bombardment with hydrogen and oxygen leading to an emission of a wide spectrum of hydrocarbon molecules and carbon oxides [178–180]. Since oxygen atoms react chemically with carbon and form CO and CO_2 at a total yield of almost unity, a reduction of the oxygen level in the plasma will thus decrease the carbon contamination.

The complex chemistry of the C–H system has been partly resolved by Küppers and co-workers [181–186] regarding the thermal chemical reactivity. During the bombardment with hydrogen isotopes, the carbon atoms in the implantation zone are hydrogenated and form a complex C–H bond structure there. With increasing surface temperature, radicals such as CH_3 are released, while at temperatures above 600 K the recombination of hydrogen to H_2 prevails leading to a reduction of the erosion yield. These temperature-controlled processes are supported by radiation damage owing to the energy transfer from the incident ions to the atoms in the lattice. Thus, open bonds for hydrogen attachment are provided, if the ion energy exceeds a certain threshold. At low surface temperature, thermal release of hydrocarbons is not possible. However, the binding energies of the radicals ($\approx 1\,\mathrm{eV}$) at the surface is much smaller than the sublimation energy of carbon (7.4 eV) and they can be released by a bond breaking mechanism induced by ion impact. This process is called ion-induced desorption of hydrocarbon radicals [187,188] and was responsible of a still significant erosion yield at room temperature and low-impact energies as observed in the experiments.

By combining thermal and ion-induced effects as described above, a semi-empirical relation has been deduced and adjusted to available experimental data ([189] and personal communication with J. Roth (2000))

$$Y_{\mathrm{chem}}(E) = (Y_1 + Y_2 + Y_3)/4 + (Y_4 + Y_5)/8 \tag{6.121}$$

with (i=1–5)

$$Y_i = Y_i^{\mathrm{surf}} + Y_i^{\mathrm{therm}}(1 + C_{\mathrm{d}} \cdot Y^{\mathrm{damage}}) \tag{6.122}$$

$$Y^{\mathrm{damage}} = Q_{\mathrm{y}} s_{\mathrm{n}}(\varepsilon)[1 - (E_{\mathrm{thd}}/E)^{2/3}](1 - E_{\mathrm{thd}}/E)^2$$

$$Y_i^{\mathrm{surf}} = \frac{s_i Q_{\mathrm{y}} s_{\mathrm{n}}(\varepsilon)[1 - (E_{\mathrm{th}}/E)^{2/3}](1 - E_{\mathrm{th}}/E)^2}{1 + \exp[(E - 65)/40]}$$

$$Y_i^{\mathrm{therm}} = \frac{0.0439\, s_i \, \exp(-c_i/T_{\mathrm{s}})}{2 \cdot 10^{-32} \Gamma + \exp(-c_i/T_{\mathrm{s}})} \tag{6.123}$$

with

$$s_i = \frac{1}{1 + 3 \cdot 10^7 \exp(-1.4/T_{\mathrm{s}})} \times \left[\frac{2 \cdot 10^{-32} \Gamma + \exp(-c_i/T_{\mathrm{s}})}{2 \cdot 10^{-32} \Gamma + [1 + 2 \cdot 10^{29} \exp(-1.8/T_{\mathrm{s}})/\Gamma] \exp(-c_i/T_{\mathrm{s}})} \right] \tag{6.124}$$

Table 6.4. Parameters for $Y_{\text{chem}}(E)$

Parameter – particle	H	D	T
Q_{y}	0.035	0.1	0.12
C_{d}	250	125	83
E_{th} (eV)	31	27	29
E_{ths} (eV)	2	1	1
E_{thd} (eV)	15	15	15

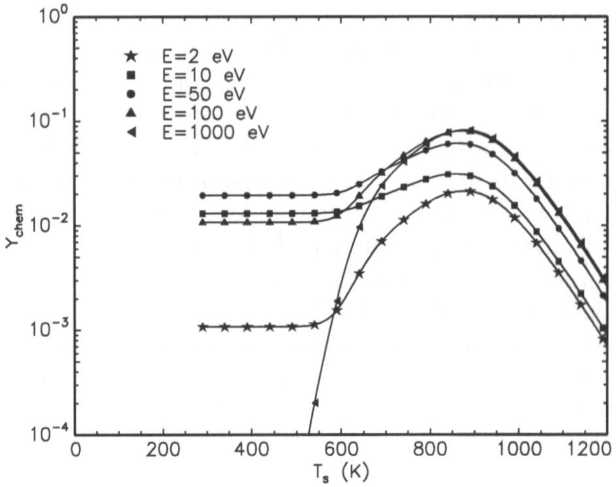

Fig. 6.13. Yields of chemical sputtering of graphite as a function of the target temperature and energy according to (6.121) for an ion flux density of $\Gamma = 10^{22}\ 1/(\text{m}^2\text{s})$

and $c_1 = 1.865$, $c_2 = 1.7$, $c_3 = 1.535$, $c_4 = 1.38$, $c_5 = 1.26$. The ion flux density Γ is given in ions/(m^{2}s); the energy of the incident ions E and the surface temperature T_{s} should be inserted in electron volts. If the energy E is smaller than the threshold energy E_{ths}, then the contribution of the surface process is zero ($Y_i^{\text{surf}} = 0$). The yield Y^{damage} should be omitted if $E < E_{\text{thd}}$. The parameters C_{d}, Q_{y}, E_{th}, E_{ths}, and E_{thd} are given in Table 6.4. The nuclear stopping cross-section $s_{\text{n}}(\varepsilon)$ is given by (6.116). Using (6.121), yields of chemical erosion by deuterium ions with different energies have been calculated (Fig. 6.13).

To call chemical erosion of carbon-based materials a complex process is an understatement. It depends not only on the ion energy, surface temperature, and ion flux, but also on the structure and composition of the particular graphite material. Small additives (dopants) such as boron or metals (Si, Ti) change the erosion yield considerably [190–192] owing to their long-range effect on the electronic structure in the C–H network.

The observed reduction of chemical erosion at higher flux densities [193–197] is a favorable effect but still not completely understood. It has been suggested that time-consuming rearrangements of chemical bonds in the hydrogenation process of carbon atoms to hydrocarbon radicals as the precursor for sputtering might be the reason for the flux dependence [187]. A dependence of Γ^{-1}, observed in some experiments, results in a constant amount of eroded material in spite of a growing flux of incident ions. Using Bayesian probability analysis, a number of experimental data provided by several fusion experiments has been fitted and the following dependence has been suggested [198]

$$Y_{\mathrm{chem}}(E, T_{\mathrm{s}}, \Gamma) = \frac{Y_{\mathrm{chem,low\,flux}}(E, T_{\mathrm{s}})}{1 + \{\Gamma/(6 \times 10^{21}[1/(\mathrm{m}^2\mathrm{s})])\}^{0.54}} \tag{6.125}$$

yielding a decrease in the erosion yield with $\Gamma^{-0.54}$ at high ion fluxes. The erosion yield at $10^{24}\,1/(\mathrm{m}^2\mathrm{s})$ is expected to be only $Y_{\mathrm{chem}} \leq 0.005$ [199].

Synergistic effects naturally arising in a plasma that consists (besides hydrogen) of different ions such as carbon, oxygen, and noble gases (such as argon, neon, and helium) play a significant role [200, 201]. These heavier impurity ions, often injected for radiation enhancement (see Sect. 9.1.5), increase the chemical erosion of carbon by an order of magnitude owing to efficient C–H bond breaking in the material [202].

The released hydrocarbons cover a wide spectrum of different C_xH_y molecules, the composition of which depends sensitively on the ion energy and surface temperature. The type of released hydrocarbon is decisive for further transport and reactions in the plasma.

Despite the progress achieved over the last years in describing basic processes of chemical erosion by means of molecular dynamics (MD) simulations [203–207], this method suffers from still insufficient computer capabilities that restrict the calculations to very short time scales, far too short for a realistic description. Because of the lack of knowledge, constant chemical erosion yield of about $Y_{\mathrm{chem}} = 0.01 - 0.02$ and a fixed spectrum of released hydrocarbons are usually assumed in impurity codes such as the WBC [208] or ERO code [209, 210] for prediction of erosion and redeposition in fusion experiments.

Reviews on chemical erosion can be found in [4, 6].

6.5.3 Radiation-Enhanced Sublimation

Enhanced erosion yields have been observed for ion bombardment of carbon-based materials at target temperatures above 1200 K. This effect is called radiation-enhanced sublimation (RES) and is related to diffusion and sublimation of interstitials created by the ion impact. This occurs when the transferred energy exceeds a threshold (displacement energy) which is about 25–35 eV for carbon. The mobility of interstitials in graphite is higher (migration energy 0.3–0.8 eV) as the vacancy mobility increases (migration energy 3.5–4.5 eV). The migrating interstitials may recombine immediately at the end of the so-called collision chain, some of them migrate freely until they either recombine with vacancies, dislocations, grain boundaries or they agglomerate into interstitial clusters. After arriving at the surface, the interstitials are assumed to be thermally desorbed due to the low binding energies [211]. The yield of RES is determined by the number of interstitials produced in the surface layer with the thickness of the average diffusion length of interstitials. Interstitials originated in deeper layers do not have a chance to reach the surface and are annihilated at internal defect structures.

The RES is particularly high at low flux density [212, 213]. It could be shown that the energy dependence of physical sputtering according to (6.115) can also be applied for RES, using the same threshold energy but a modified value for Q_{y} [6]

$$Q_{\mathrm{RES}} = Q_{\mathrm{y}} + 54 M_1^{1.18} \exp[-E_{\mathrm{RES}}/k_{\mathrm{B}} T_{\mathrm{s}}] \left(\frac{\Gamma}{10^{20}}\right)^{-0.1} \tag{6.126}$$

with the surface temperature T_{s}, the ion flux density Γ in ions/(m^2 s), and E_{RES}=0.75–0.85 eV. The decrease of RES with increasing ion flux, i.e., increasing interstitial production, is predicted to go as $\Gamma^{-0.25}$, but in experiments a weaker dependence on $\Gamma^{-0.1}$ has been found. The carbon atoms are emitted according to the Maxwellian energy distribution characterized by the surface temperature and a cosine angle distribution. The emission of clusters (C_2, C_3) is significantly reduced compared to thermal sublimation.

Below 600 K, the erosion is dominated by physical sputtering. In the case of hydrogenic impact, chemical hydrocarbon formation occurs between about 600 and 1200 K. RES dominates above 1200 K where the erosion yield increases monotonically until it exceeds at 2000 K the physical sputtering yield by more than a factor of 10 [211]. Above 2800 K, thermal sublimation sets in.

Radiation-enhanced sublimation by neutron bombardment has been found to be one or two orders of magnitude larger than the physical sputtering yield of neutrons due to momentum transfer in collision cascades. The summarized erosion by neutrons is about three orders of magnitude smaller than initiated by ions and charge-exchange neutrals [214].

6.5.4 Thermal Evaporation

With rising surface temperatures, more and more atoms are able to leave the target. The resulting evaporation flux density is determined by the vapor pressure p as a function of surface temperature T_{s}

$$\Gamma_{\mathrm{subl}} = s\frac{n\bar{v}}{4} = \frac{s\,p}{\sqrt{2\pi M_2 k_{\mathrm{B}} T_{\mathrm{s}}}} = s\;2.6 \times 10^{24} \frac{p(T_{\mathrm{s}})[Pa]}{\sqrt{M_2[amu] T_{\mathrm{s}}[K]}} \frac{\mathrm{atoms}}{\mathrm{m^2 s}} \tag{6.127}$$

using $n = p/k_{\mathrm{B}}T_{\mathrm{s}}$ and $\bar{v} = \sqrt{8k_{\mathrm{B}}T_{\mathrm{s}}/(\pi M_2)}$. The sticking coefficients for metals are about s =0.6–0.9 for graphite, which emits C, C_2, and C_3 particles. The coefficient s is in the range of 0.05 for a C_3 cluster and 0.4 for C atoms [166].

The vapor pressure $p(T_s)$ can be described in terms of the heat of vaporization (or sublimation) E_s

$$p(T_s) = p_o \exp\left(-\frac{E_s}{k_B T_s}\right) \tag{6.128}$$

where k_B is Boltzmann's constant. Under steady state conditions the vapor pressure is determined by the equilibrium between sublimation and recondensation at the surface. A comparison of different materials with respect to sublimation can be found in [215].

For carbon, the following relation is useful

$$\Gamma^C_{subl} = C_{subl} \cdot (T_s)^g \exp(-E_s/k_B T_s) \tag{6.129}$$

with $g = 3.25$, $C_{subl} = 2.5 \times 10^{20}\,K^{-3.25}/s$, and $E_s = 7.42\,eV$ [216].

Thermal sublimation and melting are practically unavoidable in fusion experiments at so-called "leading" edges of divertor and limiter plates, places where the magnetic field lines strike normally on the surface. The geometrical effect of heat load reduction due to grazing incidence is therefore nullified.

6.5.5 Blistering

Blistering is the appearance of bubbles on the surface as a result of the formation of poorly dissolved gases. During bombardment with He and hydrogen ions, the formation of blisters at the surface has been observed at high fluences. This temperature-controlled effect occurs, when the He or hydrogen concentration in the material reaches values at which accumulation occurs, e.g., at grain boundaries it is energetically favored over interstitial positioning. The accumulation causes stress in the thin surface layer and can lead finally to flaking. From experiments with 1–15 keV He it is seen that the depth of maximum helium concentration is usually about a factor of 3 smaller than the cap thickness. Hence, the assumption of high gas pressure, which finally overcomes the mechanical strength, has to be modified. The effect of exfoliation can be explained by the formation of stress between the implanted surface layers and in the bulk material due to swelling. The stress finally leads to ruptures near the interface of the implanted and unimplanted region [217].

The critical fluence at which blisters start to appear has been found experimentally to be in the range of 10^{21} to 10^{22} He/m^2. Thickness of blister skins and size of blisters increase with energy. At a He energy of 3 MeV, the critical fluence for tungsten is about 0.5×10^{22} He/m^2 with an average blister diameter of about 130 μm [218]. For hydrogen and deuterium the fluences are an order of magnitude higher.

When the irradiation dose is increased the shells of the bubbles break and new blisters can form on the exposed surface. For low surface temperatures (0.1–0.2 T_m), where T_m is the melting temperature, blistering occurs. For $T < 0.4$–$0.5\ T_m$ flaking prevails, and for $T < 0.6\ T_m$ blisters are formed with dimensions that decrease as T is raised. Finally, for $T > 0.6\ T_m$ a porous surface appears, which is subjected to neither blistering nor flaking. The erosion rate is the highest for flaking. Fortunately, surface roughness and broad energy and angular distributions of incident ions can mitigate or even suppress the blister formation.

Recently, bubble formation has been observed on powder metallurgy tungsten irradiated by hydrogen [219], deuterium [220] and helium ions [221] at surprisingly low energy ($< 100\,\mathrm{eV}$) but with a high flux density ($> 10^{22}\ 1/(\mathrm{m}^2\ \mathrm{s})$). A threshold energy for bubble formation has been found to be about 15 eV for helium bombardment. Hydrogen blister formation has been detected only at surface temperatures below 950 K. The mechanism of bubble formation is still under discussion since simple models fail to reproduce the observations. The projected range of He ions, for example, is estimated to be less than 10 Å. The He contained in the bubbles is suggested to be supplied from the surface by diffusion processes.

6.6 Charge Effects

Already by approaching the surface, at a distance of about 0.5–1.5 nm, the incident ions extract electrons from the first atomic layers and recombine before penetrating into the material [222]. The reversed process, called contact ionization, occurs if the gas atoms have an ionization potential less than the work function of the wall material. Then the gas atoms will be ionized upon contact with the surface. Thus cesium vapor (ionization potential 3.9 eV) ionizes when getting in contact with a tungsten surface (work function 4.5 eV). If the metal surface is heated at the same time to thermionic emission temperature, electrons will be emitted and a plasma is formed with the positive ions. This technique, used in Q-machines, establishes a highly-ionized, low-energy plasma free of instabilities, i.e., a quiescent plasma.

Multiply-charged ions carry a rather large amount of potential energy and relax within times of about 10 fs via dielectronic processes in a transient "hollow" atom [223]. The electronic potential energy is transferred rapidly onto very small surface areas and associated with intense electronic excitation and emission of electrons, photons, and bulk atoms. The conversion of electronic excitations into motion of atoms, a process still under discussion, could lead to significant erosion yield and surface modification especially in the case of insulator materials. In contrast to the kinetically-induced sputtering process, this phenomenon is called potential sputtering [224,225]. Although of less concern in fusion experiments, potential sputtering holds great promise as a tool for more gentle nanostructuring of surfaces, i.e., modification of the topmost layers. Applications range from information storage via material processing to biotechnology [226].

An atomic cluster at speeds in excess of the Bohr velocity loses its binding valence electrons after transversing a few atomic layers, while retaining its initial geometry. In addition to the accompanied electronic excitation of the target, the strong Coulomb repulsion among the ions of the fast cluster can result in a destructive event: the Coulomb explosion. The associated massive energy deposition finds applications in solid state detectors [227], material modification [228], and inertial confinement fusion [229].

6.7 Diffusion-Controlled Sputtering

When an alloy is sputtered, the surface composition changes as a result of preferential sputtering of one of the constituent elements. In addition, diffusion processes affect the composition profile, the altered layer extending several atomic layers below the surface [230, 231].

In the simplest model of a binary alloy the equations of one-dimensional diffusion in the solid interior normal to the surface have to be considered. Assuming constant diffusion coefficients one has

$$\frac{\partial c_{\mathrm{A}}}{\partial t} = D_{\mathrm{A}} \frac{\partial^2 c_{\mathrm{A}}}{\partial x^2} \quad \text{and} \quad \frac{\partial c_{\mathrm{B}}}{\partial t} = D_{\mathrm{B}} \frac{\partial^2 c_{\mathrm{B}}}{\partial x^2} \tag{6.130}$$

where c_{A} and c_{B} are the number fractions of the "A" atoms and "B" atoms, respectively. By definition we have $c_{\mathrm{A}} + c_{\mathrm{B}} = 1$. Therefore, both equations are only consistent with $D_{\mathrm{A}} = D_{\mathrm{B}} = D$. Processes such as segregation, radiation-enhanced diffusion, and ion mixing can be roughly described by introducing an effective diffusion coefficient.

Experimentally, distances are measured relative to the actual surface. Let z be the depth of a point below this surface, which initially for $t = 0$ is at $x = 0$ (see Fig. 6.14). In general, the recession or deposition velocity $v^{\mathrm{s}}(t)$ will not be constant, and we have

$$z = x - \int_0^t v^{\mathrm{s}}(t')\, dt' \tag{6.131}$$

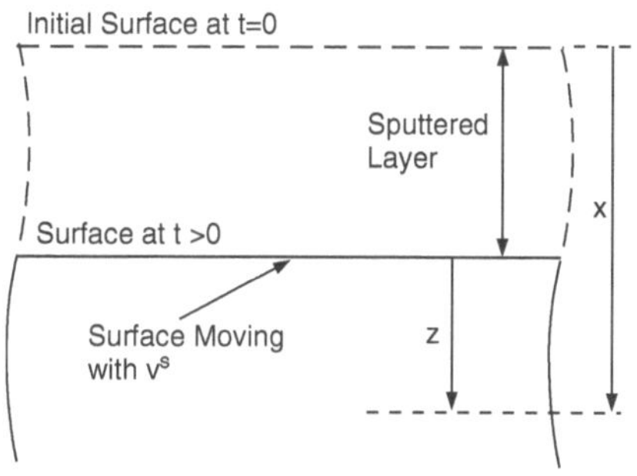

Fig. 6.14. Definition of the coordinates x and z

where $v^{\mathrm{s}}(t) > 0$ if the surface is receding. The diffusion equation for component A takes on the form

$$\frac{\partial c_{\mathrm{A}}}{\partial t} = D\frac{\partial^2 c_{\mathrm{A}}}{\partial z^2} + v^{\mathrm{s}}(t)\frac{\partial c_{\mathrm{A}}}{\partial z} \tag{6.132}$$

assuming the atomic density of the alloy remaining constant.

Suppose that effective volumes $1/n_{\mathrm{A}}^{\mathrm{o}}$ and $1/n_{\mathrm{B}}^{\mathrm{o}}$ can be assigned to single atoms of A and B, which remain essentially constant through a range of alloy compositions. Then $n_{\mathrm{A}}^{\mathrm{o}}$ and $n_{\mathrm{B}}^{\mathrm{o}}$ denote the densities in atoms per unit volume of pure material A and B, respectively. Then, for N_{A} and N_{B} atoms in a total volume V of alloy one has

$$V = N_{\mathrm{A}}\frac{1}{n_{\mathrm{A}}^{\mathrm{o}}} + N_{\mathrm{B}}\frac{1}{n_{\mathrm{B}}^{\mathrm{o}}} . \tag{6.133}$$

If $n_{\mathrm{A}} = c_{\mathrm{A}}n$ and $n_{\mathrm{B}} = c_{\mathrm{B}}n$ are the local densities (in atoms per unit volume) of the alloy constituents A and B, one obtains

$$1 = n_{\mathrm{A}}\frac{1}{n_{\mathrm{A}}^{\mathrm{o}}} + n_{\mathrm{B}}\frac{1}{n_{\mathrm{B}}^{\mathrm{o}}} = \frac{c_{\mathrm{A}}n}{n_{\mathrm{A}}^{\mathrm{o}}} + \frac{(1-c_{\mathrm{A}})n}{n_{\mathrm{B}}^{\mathrm{o}}} \tag{6.134}$$

with n being the density of the alloy

$$n = \frac{n_{\mathrm{A}}^{\mathrm{o}}n_{\mathrm{B}}^{\mathrm{o}}}{c_{\mathrm{A}}n_{\mathrm{B}}^{\mathrm{o}} + (1-c_{\mathrm{A}})n_{\mathrm{A}}^{\mathrm{o}}} . \tag{6.135}$$

Balancing all fluxes caused by erosion, reflection, deposition, and redeposition, the change of surface concentrations $\Delta n_{\mathrm{s}}^{\mathrm{A}}$ and $\Delta n_{\mathrm{s}}^{\mathrm{B}}$ (note, n_{s} are given in particles/m^2) during a time interval Δt is given by

$$\begin{aligned}\frac{\Delta n_{\mathrm{s}}^{\mathrm{A}}}{\Delta t} &= \Gamma_{\mathrm{A}}^{ero} - \Gamma_{\mathrm{A}}^{\mathrm{dep}} - \Gamma_{\mathrm{A}}^{\mathrm{redep}} + \Gamma_{\mathrm{A}}^{refl} \\ &= \Gamma_{\mathrm{A}}^{ero}[1 - f_{\mathrm{A}}^{\mathrm{redep}}(1 - R_{\mathrm{A}}^{\mathrm{redep}})] - \Gamma_{\mathrm{A}}^{\mathrm{dep}}(1 - R_{\mathrm{A}}^{\mathrm{dep}})\end{aligned} \tag{6.136}$$

and

$$\frac{\Delta n_{\mathrm{s}}^{\mathrm{B}}}{\Delta t} = \Gamma_{\mathrm{B}}^{ero}[1 - f_{\mathrm{B}}^{\mathrm{redep}}(1 - R_{\mathrm{B}}^{\mathrm{redep}})] - \Gamma_{\mathrm{B}}^{\mathrm{dep}}(1 - R_{\mathrm{B}}^{\mathrm{dep}}) \tag{6.137}$$

where $n_{\mathrm{s}}^{A,B}$ is the number of atoms per unit surface element, f^{redep} is the redeposition probability, and R^{dep} and R^{redep} are the reflection coefficients for deposited and redeposited particles, respectively. Using (6.136) and (6.137), the recession or deposition velocity $v^{\mathrm{s}}(t)$ can be derived

$$v^{\mathrm{s}}(t) = \frac{\Delta n_{\mathrm{s}}^{\mathrm{A}}}{\Delta t}\frac{1}{n_{\mathrm{A}}^{\mathrm{o}}} + \frac{\Delta n_{\mathrm{s}}^{\mathrm{B}}}{\Delta t}\frac{1}{n_{\mathrm{B}}^{\mathrm{o}}} = v_{\mathrm{A}}^{\mathrm{s}}(t) + v_{\mathrm{B}}^{\mathrm{s}}(t) . \tag{6.138}$$

It is worth noting that the sputtering coefficients depend, of course, on the actual composition in the surface layer $c_{\mathrm{A}}^{\mathrm{surface}}$ and $c_{\mathrm{B}}^{\mathrm{surface}}$. Firstly, the surface binding energy is changed with composition [159, 232, 233]. Furthermore, the sputtering yield depends on the composition, since the momentum transfer is changed. Particularly, very light elements show up with large escape depths, larger than the usually assumed two or three atomic layers. In a first approximation, the assumption of linear correlation between the partial sputtering yield and surface concentration can be made, i.e., $Y_{i\to j}^{\text{partial}} = c_j^{\text{surface}} Y_{i\to j}^{\text{c}}$.

Since the sputtering and deposition processes are mainly localized in the first atomic layers of the alloy, they can be used to define the boundary condition for the diffusion equation, which describes the evolution of the composition of much deeper layers. At the surface, we have

$$\begin{aligned} v_{\text{A}}^{\text{s}}(t)n_{\text{A}}^{\text{o}} &= \left. \left(Dn\frac{\partial c_{\text{A}}}{\partial z} + v^{\text{s}}(t)c_{\text{A}}n \right) \right|_{\text{surface}} \\ v_{\text{B}}^{\text{s}}(t)n_{\text{B}}^{\text{o}} &= \left. \left(Dn\frac{\partial c_{\text{B}}}{\partial z} + v^{\text{s}}(t)c_{\text{B}}n \right) \right|_{\text{surface}} \end{aligned} \qquad (6.139)$$

The first term on the right side (if taken as negative) is the diffusion flux entering the thin surface layer. The second term is due to the receding/growing surface. Changes of the alloy density due to diffusion can be neglected.

The relations (6.132), (6.138), and (6.139) together with the conservation relation $c_{\text{B}} = 1 - c_{\text{A}}$ constitute a self-consistent set of highly non-linear equations [230, 231] with the initial condition of a given composition profile at $t = 0$.

A coupling of this mixing model with a plasma simulation code is required to analyze consistently the non-linear behavior of the plasma–wall interaction. The coefficients $f_{\text{A,B}}^{\text{redep}}$ and the fluxes of ions to the surface can be obtained using the plasma code and serve as input parameters in the mixing model. In turn, the impurity concentration in the plasma is mainly determined by the eroded amount, while the "effective" sputtering yields $Y_{\text{A,B}\to\text{A,B}}$ depend sensitively on the current surface composition.

The relative importance of the main processes involved (erosion/deposition on one hand, diffusion on the other) can be estimated by means of a dimensionless number, the Peclet number used in fluid mechanics: $P_{\text{e}} = v\Delta z/D$. The term Δz is some characteristic length, which is in our case the thickness of several atomic layers, since most of the sputtered atoms originate from a very thin region at the surface. The "convective" velocity v is here the recession/deposition speed v^{s}.

6.8 Backscattering

Some particles can eventually leave the surface after traveling in the material. They are backscattered in a certain range of energy and angle with a probability given by the backscattering (or reflection) coefficient. In the experiments, the reflection of impurities can often not be distinguished from sputtering of the same impurity at the surface. Almost all reflected particles are neutrals. Backscattering is characterized by a broad energy spectrum of the reflected particles—energies from nearly zero to values close to the initial energy are found.

The probability that a particle impinging with energy E_o under a direction (θ_o, φ_o) is reflected with an energy E $(E \leq E_o)$ into a solid angle $\Omega = (\theta, \varphi)$ is given by the differential reflection coefficient $\mathrm{d}^2R(E_o, \theta_o, \varphi_o; E, \theta, \varphi)/\mathrm{d}E\mathrm{d}\Omega$. This coefficient gives the probability for a particle to make the transition from a state (E_o, Ω_o) before incidence to the state (E, Ω) after backscattering. The total reflection coefficient of particles is given by

$$R_\mathrm{N} = \int_\Omega \int_0^{E_o} \frac{\mathrm{d}^2R(E_o, \theta_o; E, \theta, \varphi)}{\mathrm{d}E\mathrm{d}\Omega} \, \mathrm{d}E \, \mathrm{d}\Omega \, . \tag{6.140}$$

The energy reflection coefficient gives the fraction of energy carried away by the reflected particles

$$R_\mathrm{E} = \frac{1}{E_o} \int_\Omega \int_0^{E_o} E \, \frac{\mathrm{d}^2R(E_o, \theta_o; E, \theta, \varphi)}{\mathrm{d}E\mathrm{d}\Omega} \, \mathrm{d}E \, \mathrm{d}\Omega \, . \tag{6.141}$$

The theoretical description of backscattering started with models of one strong deflection, e.g., for electrons [234–236] and for ions [237, 238], which are applicable at higher incident energies (see Sect. 6.8.1). In fact, one of few quantitative surface analysis techniques, the Rutherford backscattering method (RBS), is based on this simple one-collision model. The two-collision models available [239–241] help to refine the RBS spectra analysis by reducing the background signal.

For lower incident energies of the particles, the assumption of diffusive transport due to many collisions with small deflection can be used [242, 243] (see Sect. 6.8.2).

As usual, the region of intermediate energies causes problems in the theoretical analysis. The most straightforward (and most intricate) way is the solution of Boltzmann's transport equation by selecting appropriate collision cross-sections for the energy region of interest [112, 244–256]. Often, the particle backscattering coefficient R_N is obtained based on the calculated implantation distribution $R(z)$ in the material [257, 258]

$$R_\mathrm{N} = \int_{-\infty}^{0} R(z) \, \mathrm{d}z \, . \tag{6.142}$$

Cutting simply the depth profile calculated for an infinite material is of course only a very crude approximation of the boundary condition.

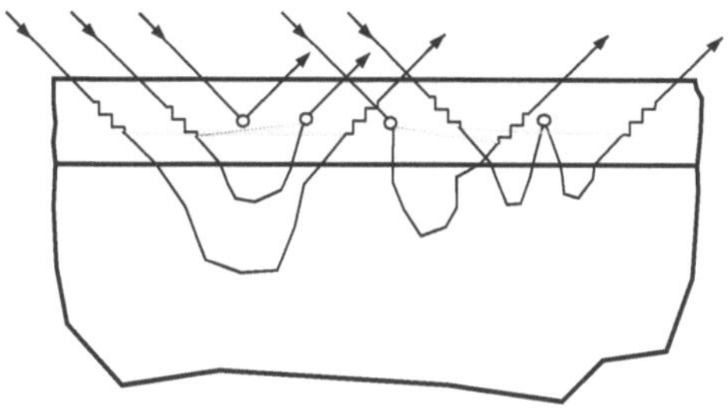

Fig. 6.15. Additional trajectories that appear after a thin layer is added on top of a half-infinite material. Elastic collisions are indicated with small circles; energy loss due to collisions with target electrons is represented by springs [260]

The correct boundary condition with respect to backscattering can be considered using the so-called imbedding method [259]. The trick here is to add (in mind) a thin layer on top of a half-infinite material of the same kind (Fig. 6.15).

Since the reflection probability for such a system should be the same before and after this operation, an equation can be deduced by notifying that the sum of all additional processes in this thin layer is zero. The result is an equation for the differential reflection coefficient $R(E,\Omega)$ that can be solved analytically in certain cases (for electrons [261], for ions [260]). Chandrasekhar invented this method by studying the transmission and reflection of light in the earth atmosphere.

6.8.1 One-Collision Model

Fast particles ($\sigma^* \ll 1$) can be assumed to be backscattered in one strong deflection with a target atom. Along the way down to the position of collision, sometimes deep in the material, as well as on the path back to the surface (after the collision), the particle moves on a rather straight trajectory. Along this path of length l it loses its energy in collisions with electrons. The probability that a particle scatters in a certain depth into a solid angle interval $\mathrm{d}\Omega$ is then given by

$$\mathrm{d}^2 R(E_\mathrm{o},\Omega_\mathrm{o};E,\Omega) = \frac{\mathrm{d}\sigma_\mathrm{el}}{\mathrm{d}\Omega}(E_\mathrm{t},\alpha)\, n_\mathrm{o}\, \mathrm{d}z_\mathrm{t}\, \mathrm{d}\Omega \tag{6.143}$$

where α is the scattering angle, and $n_\mathrm{o}\mathrm{d}z$ is the number of target atoms per unit square. The initial energy of the particle E_o is simply related to the energy of the particle E after leaving the surface by

$$E_\mathrm{t} = E_\mathrm{o} - \int\limits_0^{z_\mathrm{t}} \left|\frac{\mathrm{d}E}{\mathrm{d}l}\right|_\mathrm{in} \mathrm{d}l\,, \quad E = E_\mathrm{t} - \Delta E_\mathrm{el} - \int\limits_0^{z_\mathrm{t}/\cos\theta} \left|\frac{\mathrm{d}E}{\mathrm{d}l}\right|_\mathrm{in} \mathrm{d}l \tag{6.144}$$

where E_t is the particle energy in depth z_t just before collision and θ is the reflection angle with respect to the surface normal (Fig. 6.16). According to

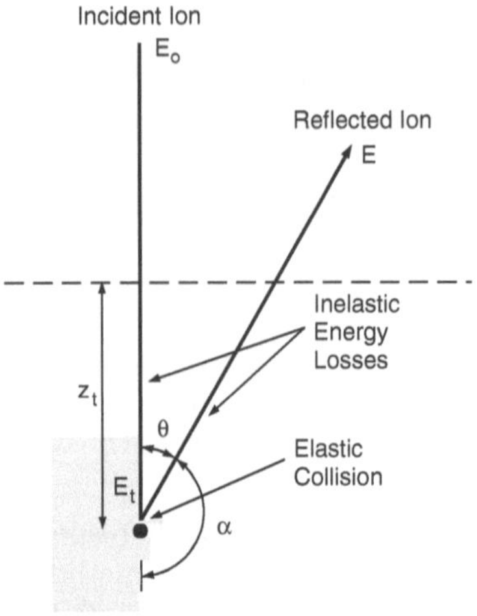

Fig. 6.16. Graphical depiction of the one collision model for normal incidence

(6.83), the elastic energy loss is $\Delta E_{\mathrm{el}} = E_{\mathrm{t}}\gamma_{\mathrm{k}}\sin^2(\alpha/2)$. With (6.144), the energy E is definitely linked to a certain depth z_{t}. Specifying the law of inelastic energy loss, $(\mathrm{d}E/\mathrm{d}l)_{\mathrm{in}}$, z_{t} and $\mathrm{d}z_{\mathrm{t}}$ can be expressed in terms of E and $\mathrm{d}E$, respectively. Thus substituting E_{t}, z_{t}, and $\mathrm{d}z_{\mathrm{t}}$ in (6.143) yields the desired relation for the reflection coefficient $\mathrm{d}^2R(E_{\mathrm{o}},\theta_{\mathrm{o}};E,\theta,\varphi)/(\mathrm{d}E\mathrm{d}\Omega)$. Analytically, this operation can be performed only for a limited class of presentations of $(\mathrm{d}E/\mathrm{d}l)_{\mathrm{in}}$ (Sect. 6.2). For simplification, equal scattering angles in the laboratory system and in the center-of-mass system have been assumed, valid for $M_1/M_2 \ll 1$.

6.8.2 The Diffusion Model

In the case of $\sigma^* \gg 1$, the motion of the particle in the solid becomes fully isotropic after traveling the distance of the transport length l_{tr}. Reaching the depth $z_{\mathrm{t}} = l_{\mathrm{tr}}\cos\theta_{\mathrm{o}}$ (θ_{o} is the incident angle), the particles start to move in all directions until they are stopped in the material or have left the surface. The affected volume has the shape of a sphere and is defined by the total range R_{o} minus the transport length as its radius (Fig. 6.17). In this model [242], the particle reflection coefficient R_{N} is simply given by the ratio of the spherical segment

$$V_{\mathrm{R}} = 2\pi(R_{\mathrm{o}} - z_{\mathrm{t}})^2(R_{\mathrm{o}} - 2z_{\mathrm{t}})/3 \tag{6.145}$$

to that of the total volume of the sphere $V = 4\pi(R_{\mathrm{o}} - z_{\mathrm{t}})^3/3$, i.e.,

$$\begin{aligned} R_{\mathrm{N}} &= \frac{V_{\mathrm{R}}}{V} = \frac{1}{2}\,\frac{R_{\mathrm{o}} - 2z_{\mathrm{t}}}{R_{\mathrm{o}} - z_{\mathrm{t}}} = \frac{1}{2}\,\frac{1 - 2l_{\mathrm{tr}}\cos\theta_{\mathrm{o}}/R_{\mathrm{o}}}{1 - l_{\mathrm{tr}}\cos\theta_{\mathrm{o}}/R_{\mathrm{o}}} \\ &= \frac{1}{2}\left(\frac{1 - 2\cos\theta_{\mathrm{o}}/\sigma^*}{1 - \cos\theta_{\mathrm{o}}/\sigma^*}\right) \end{aligned} \tag{6.146}$$

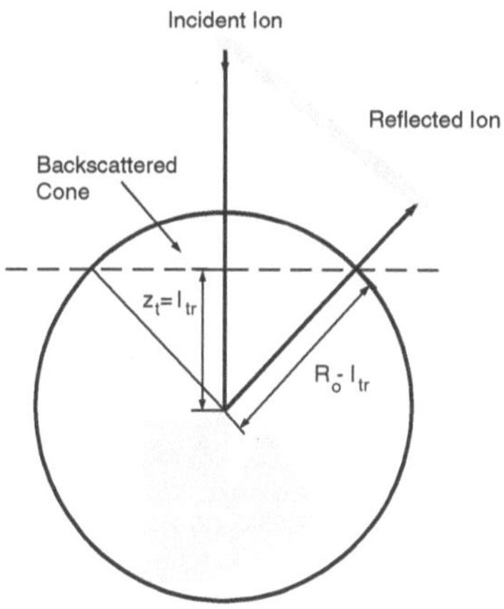

Fig. 6.17. Geometry of the diffusion model for the case $\theta_o = 0$

under the condition that

$$R_o > l_{tr}(1 + \cos\theta_o)\ . \tag{6.147}$$

The particles should have enough energy to leave the surface. Interestingly, the ratio of the surface area section $2\pi(R_o - 2z_t)$ to the total surface area of the sphere $4\pi(R_o - z_t)^2$ as well as the ratio of the characteristic lengths $(R_o - 2z_t)/(R_o - z_t)$ give the same result (6.146). The prefactor 1/2 has to be considered in the latter case, since half of the particles are moving in directions away from the surface.

This rather coarse approximation works surprisingly well for keV electrons [242,243]. However, its applicability is linked to condition (6.147); the model is only valid for $\sigma^* > 2\cos\theta_o$. For large values of σ^*, (6.146) predicts a maximum reflection probability of 1/2.

6.8.3 Approximations

The main parameter determining backscattering is the dimensionless quantity [260, 261] σ^* defined in (6.76). For half-infinite materials a somewhat better adjusted parameter

$$\sigma_R^* = \frac{R_o}{l_{tr}[1 - \cos(\pi/2 - \theta_o)]} \simeq \frac{E_o/|dE/dl|}{l_{tr}[1 - \cos(\pi/2 - \theta_o)]} \tag{6.148}$$

can be used where θ_o is the angle of incidence with respect to the surface normal. For normal incidence, i.e., $\theta_o = 0$, then $\sigma_R^* = \sigma^*$. The parameter σ_R^* gives, on average, the number of strong deflections required to leave the target. Knowing σ_R^*, the reflection probability can then be estimated. It increases with increasing σ_R^*.

Simple approximations of the total reflection coefficient can be constructed using the backscattering parameter σ_{R}^{*} in the form of $R_{\mathrm{N,E}} = a\,[1-\exp(-b\,\sigma_{\mathrm{R}}^{*})]$ or $R_{\mathrm{N,E}} = c\,\exp(-d/\sigma_{\mathrm{R}}^{*})$ where a, b, c, and d are fitting parameters. Analytical relations for R_{N} and R_{E} can be found in [256, 262, 263]. For light ions such as H^{+}, D^{+}, and He^{+} in the energy range from 0.01 to 100 keV, the approximation [256]

$$R_{\mathrm{N}} = 1 - \exp(-0.0513\,\sigma_{\mathrm{R}}^{*}) \tag{6.149}$$

can be used. Other fitting relations are [264]

$$R_{\mathrm{N}} = [(1 + 3.2\varepsilon^{0.34})^{1.5} + (1.33\varepsilon^{1.5})^{1.5}]^{-0.67} \tag{6.150}$$

$$R_{\mathrm{E}} = [(1 + 7.1\varepsilon^{0.35})^{1.5} + (5.3\varepsilon^{1.5})^{1.5}]^{-0.67} \tag{6.151}$$

with the reduced energy ε according to (6.88). Further relations can be found in the review on ion reflection by Eckstein [6].

In the energy range below 10 eV, any chemical affinity of the incident particle, for example hydrogen, to the target leads to a decrease of the reflection probability [265], since the energy of incidence becomes comparable to the surface binding energy which is in the eV range. This effect is usually not included in the approximations.

Despite the significant progress in the quantitative description of backscattering over the years, the powerful methods of computer simulation on fast computers available today have in fact terminated further theoretical development. Advanced programs such as TRIM [266–269] and MARLOWE [270,271] provide comprehensive information with a high degree of accuracy in all details. The quality of the results depends almost solely on the appropriate choice of the interaction potential. In the case of complex structures, geometries, and material mixture, computer simulations reveal their overwhelming power.

6.9 Electron Emission

Due to the interaction of the ionic crystal lattices with the free moving conduction electrons, an electric field arises near the surface confining the electrons inside the material. This field has to be overcome by those electrons leaving the surface. At room temperature, the electrons are captured in the material. Electrons can be emitted by providing them with sufficient energy to overcome the surface potential barrier. Alternatively, the barrier may be modified, thereby increasing the probability of escape of high-energy electrons. Electrons may be emitted on account of electron collisions (secondary electron emission (SEE)) (Sect. 6.9.1), or collisions with positive ions, neutrals, or metastables. They can also be emitted due to photon absorption, surface heating (Sect. 6.9.2), or the application of an electric field (field emission) (Sect. 6.9.3). The photoelectric ejection of electrons, where the energy of the incident photons must be larger than the surface work function, is a minor effect in fusion experiments. Electron-induced secondary electrons make up the major part of electron emission.

Fast positive ions induce electron emission from surfaces by intense local heating. Low-energy ions cause electron emission in a two-step process. First, they capture an electron from the surface thereby becoming neutralized in an excited state. Then, the excitation energy is transferred to a second surface electron, which can escape if its energy is larger than the surface work function. The electron emission coefficient for slow ions is considerably smaller than unity. No electron emission is expected due to neutral particles with thermal energy. However, fast neutrals can initiate electron ejection. This effect is used to detect neutral particles.

Excited metastable neutrals produce secondary electrons due to the same mechanisms as explained for the ions. The helium metastable level, for example, is at 19.7 eV, while the work function of magnesium is 3.01 eV. Thus, secondary electrons are emitted with a maximum energy of 16.7 eV.

Note that the total yield of electron emission might include a fraction of backscattered electrons.

6.9.1 Secondary Electron Emission (SEE)

True secondary electrons have energies in the eV range. Traditionally, emitted electrons with energy lower than 50 eV are considered to be secondary, while electrons with higher energies are attributed to reflected electrons. The coefficient δ_{SEE} can be expressed in the following form [272]

$$\delta_{\mathrm{SEE}} = \delta_{\max}(2.72)^2 \frac{E}{E_{\max}} \exp\left[-2\sqrt{\frac{E}{E_{\max}}}\right] \tag{6.152}$$

as a function of the electron energy E. Values of $\delta_{\max}$ and $E_{\max}$ for materials of interest in fusion devices are listed in Table 6.5.

The yield of secondary electron emission increases with the angle of incidence θ (taken with respect to the surface normal) according to [274]

$$\delta(\theta) = \delta(0)/\cos^{\gamma_{\mathrm{SEE}}}\theta \tag{6.153}$$

Table 6.5. Parameters for δ_{SEE} from [273]

Target	Z	$\delta_{\max}$	$E_{\max}$ (eV)
Be	4	0.5	200
C	6	1.0	300
Fe	26	1.3	400
Mo	42	1.25	375
W	74	1.4	650

with $\gamma_{SEE} = 1$ for all materials with nuclear charge $Z > 10$. For light elements such as Be, γ_{SEE} is close to 1.3.

6.9.2 Thermionic Electron Emission

The amount of electron emission in the case of heated surfaces depends on the number of electrons in the metal with thermal energies greater than the thermionic work function W_T and results in a flux of

$$j_s = \frac{4\pi m_e e k_B^2}{h^3} T_s^2 \exp\left(-\frac{W_T}{k_B T_s}\right) \tag{6.154}$$

according to the Richardson–Dushman law with the constant

$$A_{RD} = 4\pi m_e e k_B^2 / h^3 = 1.2 \times 10^6 \,\mathrm{A/(m^2K^2)} \,. \tag{6.155}$$

Often, the constant A is corrected by fitting (6.154) to experimental data. For carbon, $A_{RD}^C = 3 \times 10^{-5}\,\mathrm{A/(m^2K^2)}$ and $W_T^C = 4.5\,\mathrm{eV}$ [275], for tungsten the following values can be taken $A_{RD}^W \simeq A_{RD} = 1.2 \times 10^{-6}\,\mathrm{A/(m^2K^2)}$ and $W_T^W = 4.5\,\mathrm{eV}$. Small additives, for example a few percents of thorium in tungsten, can change the surface work function of certain materials significantly.

The thermal (or thermionic) emission of electrons can be considered as an analogy to the vaporization of a fluid. Let us calculate the flux of electrons out of the electron gas above the metal surface toward the surface. In thermal equilibrium, this rate of condensation is equal to the rate of vaporization, i.e., emission of electrons. The number of possible states of electrons in the electron gas (in the solid as well as above the surface) is

$$\mathrm{d}N_Z = 2\frac{\mathrm{d}p_x \mathrm{d}p_y \mathrm{d}p_z}{h^3} \tag{6.156}$$

where the factor of 2 owes to the fact that for each triad of momentum (p_x, p_y, p_z) two different directions of the electron spin are allowed. The probability $f_P(E, T_s)$ to be in a quantum state of energy E is defined by the Fermi–Dirac distribution

$$f_P(E, T_s) = \frac{1}{1 + \exp[(E - E_F)/k_B T_s]} \tag{6.157}$$

with the Fermi energy E_F. The electron density is determined by setting

$$\mathrm{d}n = f_P \,\mathrm{d}N_Z = \frac{2}{h^3} \frac{\mathrm{d}p_x \mathrm{d}p_y \mathrm{d}p_z}{1 + \exp[(E - E_F)/k_B T_s]} \,. \tag{6.158}$$

Since the energy of a motionless electron in a vacuum U_o is by several $k_B T_o$ larger than the Fermi energy, the unity in the denominator can be neglected (Fig. 6.18). The concentration of electrons with momentum in the interval $(\mathrm{d}p_x, \mathrm{d}p_y, \mathrm{d}p_z)$ is then

$$\mathrm{d}n = \frac{2}{h^3} \exp\left(-\frac{E - E_F}{k_B T_s}\right) \mathrm{d}p_x \mathrm{d}p_y \mathrm{d}p_z \,. \tag{6.159}$$

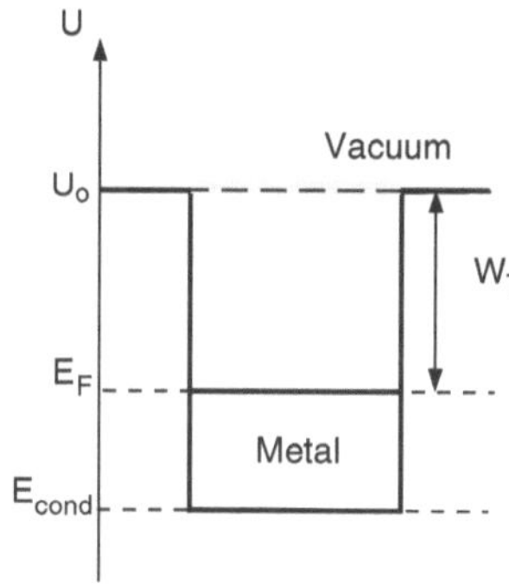

Fig. 6.18. Potential energy of an electron in a metal (W_{T}: thermionic work function, E_{F}: Fermi energy, E_{cond}: minimum energy of the conduction band, U_{o}: energy of a motionless electron in a vacuum)

Having the z-axis directed into the metal, the current density of electrons to the surface with momentum in $(\mathrm{d}p_x, \mathrm{d}p_y, \mathrm{d}p_z)$ is given by

$$\mathrm{d}j_{\mathrm{s}} = e\frac{p_z}{m_{\mathrm{e}}}\,\mathrm{d}n\,. \tag{6.160}$$

To obtain the total current density, (6.160) has to be integrated over positive p_z-values and over all p_x, p_y-values

$$j_{\mathrm{s}} = \frac{2e}{h^3 m_{\mathrm{e}}}\exp\left(-\frac{U_{\mathrm{o}} - E_{\mathrm{F}}}{k_{\mathrm{B}}T_{\mathrm{s}}}\right)\int\limits_0^{\infty}\int\limits_{-\infty}^{\infty}\int\limits_{-\infty}^{\infty}\exp\left(-\frac{p_x^2 + p_y^2 + p_z^2}{2m_{\mathrm{e}}k_{\mathrm{B}}T_{\mathrm{s}}}\right)p_z\,\mathrm{d}p_z\mathrm{d}p_x\mathrm{d}p_y\,. \tag{6.161}$$

Here, the energy as the sum of the potential and kinetic energy

$$E = U_{\mathrm{o}} + \frac{1}{2m_{\mathrm{e}}}(p_x^2 + p_y^2 + p_z^2) \tag{6.162}$$

in the vacuum is used. The integration over p_x and p_y yields $\sqrt{2\pi m_{\mathrm{e}} k_{\mathrm{B}} T_{\mathrm{s}}}$ since $\int_{-\infty}^{\infty}\exp(ax^2)\mathrm{d}x = \sqrt{\pi/a}$, while the integration over p_z gives $m_{\mathrm{e}}k_{\mathrm{B}}T_{\mathrm{s}}$. Thus,

$$j_{\mathrm{s}} = \frac{4\pi m_{\mathrm{e}} e k_{\mathrm{B}}^2}{h^3}\,T_{\mathrm{s}}^2\exp\left(-\frac{U_{\mathrm{o}} - E_{\mathrm{F}}}{k_{\mathrm{B}}T_{\mathrm{s}}}\right) \tag{6.163}$$

as already given above (see (6.154)). The thermionic work function $W_{\mathrm{T}} = U_{\mathrm{o}} - E_{\mathrm{F}}$ (Fig. 6.18) is equal to the work required to transfer an electron (with the largest kinetic energy in the metal) out of the metal into a vacuum state with no kinetic energy.

6.9.3 Electron Emission by the Application of an Electric Field

Schottky showed that a reduction of the effective work function by an applied electric field enhances the electron emission. This is a quantum tunneling effect and requires fields on the order of 10^7–10^8 V/cm. However, significant emission enhancement is found already at a field of about 10^6 V/cm. This is ascribed to surface roughness resulting in higher electric fields at the tips of small surface protrusions. Another explanation is the lowering of the effective work function due to surface contaminations.

6.10 Modeling of Particle–Solid Interaction

Starting in the 1950s and continuing today, computer simulation has been an indispensable tool in the analysis of particle–solid interaction processes [109]. The following sections are dedicated to the two main classes of simulation techniques in this field—the so-called molecular (or classical) dynamics model (MD) (Sect. 6.10.1) and the Monte Carlo methods (MC) (Sect. 6.10.2). Short descriptions of the basic ideas are given together with a list of steps to be executed in a computer program.

6.10.1 Molecular Dynamics

Wherever the assumption of binary collisions fails, the many-particle interaction should be adequately described. The motion of each particle of a system consisting of N particles is defined by the force F_i exerted on this particle by the rest of them, i.e., $N-1$. Self-forces are not considered, i.e., a particle does not act on itself. Whether the interatomic forces are calculated by an empirical potential or by ab initio techniques, the method for moving a particle in the MD simulation is the same, just integrating the Newtonian equations of motion

$$M_i \frac{\mathrm{d}^2 \boldsymbol{r}_i(t)}{\mathrm{d}t^2} = M_i \frac{\mathrm{d}\boldsymbol{v}_i(t)}{\mathrm{d}t} = \boldsymbol{F}_i = \sum_{j=1, i \neq j}^{N} \boldsymbol{F}_{ij} \tag{6.164}$$

in time for each particle i consecutively. The forces are usually derived from a potential energy $U(r)$, i.e., $\boldsymbol{F}_{ij} = -\nabla U(r_{ij})$, with the distance $r_{ij} = \sqrt{\boldsymbol{r}_{\mathrm{i}}^2 - \boldsymbol{r}_j^2}$ between the particles i and j. The advantage of tight-binding molecular dynamics over classical potential simulations is the explicit incorporation of the real electronic structure and bonding of the material. These data are evaluated by codes based on first principles.

MD simulations are very expensive. Especially, the calculation of the interatomic forces is much more time-consuming than the integration of (6.164). The analyzed systems consist, therefore, of no more than a hundred to a few thousand particles. Thus, the system size, for example, a crystal lattice, is rather small in real dimensions. Usually, periodic boundary conditions are applied or the atoms at the border are simply fixed. The simulation starts with some initial configuration, which is then relaxed to find the minimum energy structure. To introduce temperature into the system, the velocity of each particle is scaled at every time step in the way that the total kinetic energy of the system is given by $k_B T_s/2$ per each degree of freedom. The temperature of the system can also be coupled to an external heat reservoir allowing the temperature to become a dynamical variable.

The integration of (6.164) over small time steps can be performed using various numerical methods, for example, by the Runge–Kutta method of different orders. However, low-order algorithms have some advantages: no storage of high derivatives of positions, velocities, etc., is necessary, and larger time steps can be used without risking the violation of energy conservation. In addition, the number of force calculations in each time step should be reduced as much as possible, in the ideal case of one force calculation per time step. There is a trend to favor the family of Verlet algorithms [276], and, in particular, the velocity Verlet algorithm [277]

$$\boldsymbol{v}_i(t+\Delta t/2) = \boldsymbol{v}_i(t) + \frac{\Delta t}{2M_i}\boldsymbol{F}_i(t) \tag{6.165}$$

$$\boldsymbol{r}_i(t+\Delta t) = \boldsymbol{r}_i(t) + \Delta t\, \boldsymbol{v}_i(t+\Delta t/2) \tag{6.166}$$

$$\boldsymbol{v}_i(t+\Delta t) = \boldsymbol{v}_i(t+\Delta t/2) + \frac{\Delta t}{2M_i}\boldsymbol{F}_i(t+\Delta t)\,. \tag{6.167}$$

This scheme advances the velocities $\boldsymbol{v}$ and coordinates $\boldsymbol{r}$ over a time step Δt, which is on the order of femtoseconds. The time step should be an order of magnitude less than the period of the fastest oscillation, for example, bond stretching takes about 10 fs.

After step (6.166), the force $\boldsymbol{F}_i(t+\Delta t)$ is calculated, knowing the new positions, for step (6.167). The Verlet scheme has the advantage of high precision, while exhibiting low drift, i.e., the total energy fluctuates about some constant value. Furthermore, the Verlet method is symplectic, which means that the system can be traced back by reversing the momenta of all particles. Non-symplectic methods, such as the predictor–corrector schemes, usually have problems with energy conservation for longer simulation times.

In general, numerical methods cannot accurately follow the true trajectories for very long times. The ergodic behavior of classical trajectories, i.e., the fact that nearby trajectories diverge from each other exponentially fast due to the Lyapunov instability, sets a limit. However, the averaged values remain unaffected. Simulations run typically 10^3–10^6 Δt steps, corresponding to a few nanoseconds of real time, only in special cases extending to microseconds. It is thus important to check, whether an equilibrium is reached in the simulation during these rather short times. Unfortunately, many physical processes of interest would require an analysis on much larger time and spatial scales.

6.10.2 Monte Carlo Methods

This class of methods got its name simply because of the use of random numbers. In particle simulation, the trajectory of each single particle is followed in time. At each time step decisions have to be made, for example, whether a collision occurs or not. When applying random numbers (generated by a pseudo-random generator), the code becomes surprisingly short and efficient. Of course, one could generate a whole crystal structure and keep track of the positions of all target atoms and of the projectile to decide which certain target atom will be the next collision partner. The effort is significantly reduced, if a fixed mean free path is assumed and the impact parameter for the next collision is diced out. Since most real materials are amorphous or polycrystals, this approach gives, on average, a quite realistic description.

The final results are obtained by a statistical analysis of all trajectories.

To calculate total quantities such as the particle and energy reflection coefficient, several thousand trajectories are sufficient. However, millions of trajectories have to be calculated, if one wants to know, for example, the angular and energy distribution of reflection $\mathrm{d}^2R(E_o,\theta_o,\varphi_o;E,\theta,\varphi)/\mathrm{d}E\mathrm{d}\Omega$. Considering processes such as implantation, damage production, reflection, sputtering, and transmission requires a detailed knowledge of the penetration of charged particles through material. In contrast to analytical solutions of the Boltzmann transport equation, the MC methods have significant advantages:

- Realistic interaction potentials can be used instead of inaccurate approximations.
- Real surface geometries, inhomogeneities in structure such a cracks, pores, and fibers can be taken into account.
- Arbitrary ion–material combinations are possible to consider multi-element compounds and alloys of any desired composition.
- Information may be obtained, which is not accessible either to experimental investigations or to analytical studies.

Regardless of the many options and variations, let us discuss here the main steps in performing a MC particle simulation. Every particle starts from a defined position, usually at the surface, with a certain initial energy E_o into a certain initial direction given by (θ_o,φ_o). It changes its direction due to binary collisions with target atoms and loses its energy predominantly in collisions with electrons. The position of the next collision partner is determined either by saving all atomic positions of the crystal matrix (as done in the MARLOW code [270,271]) or by evaluating the impact parameter using a random number (as done in programs such as TRIM [266–269] or ACAT [278]). The latter approach is justified, when the surface affected by the particle beam is much larger than the grain size of the material. The following steps should be programmed:

1. Definition of the ion–material system (Z_1, Z_2, M_1, M_2, n_o).
2. Definition of the initial energy E_o, direction $\Omega_o = (\theta_o, \varphi_o)$, and position. In the case of crystal materials, the point of incidence at the surface can be varied by using two random numbers.
3. Each particle moves in steps of the mean free path $l_p = n_o^{-1/3}$. Thus, the distance between two collisions is just equal to the interatomic

distance. Moving along this path, the particle loses energy in collisions with electrons

$$\Delta E_{\text{in}} = \int_0^L \frac{\mathrm{d}E}{\mathrm{d}l}_{\text{in}} \mathrm{d}l \simeq \frac{\mathrm{d}E}{\mathrm{d}l}_{\text{in}} l_{\text{p}} = \frac{\mathrm{d}E}{\mathrm{d}l}_{\text{in}} n_{\text{o}}^{-1/3} \tag{6.168}$$

where $(\mathrm{d}E/\mathrm{d}l)_{\text{in}}$ is the inelastic energy loss per unit length (see Sect. 6.2.3). The simple model of constant mean free path turns out to be very effective, and accurate. In other models the distance to the next collision is evaluated at each step. For this, one has to define the minimum deflection angle and to consider in some way the energy dependence of scattering and energy loss. This can become a problem in the case of a very long mean free path.

4. At the end of l_{p}, a collision occurs. The impact parameter ρ_{p} is determined with a random number RND, which is distributed uniformly in the unit interval [0, 1]

$$\rho_{\text{p}} = \sqrt{\frac{\text{RND}}{\pi n_{\text{o}}^{2/3}}} . \tag{6.169}$$

5. The scattering angle α is calculated in the center-of-mass system (Sect. 6.1.1) and then transferred into the laboratory system

$$\tan\theta_1 = \frac{\sin\alpha}{\cos\alpha + M_1/M_2} \tag{6.170}$$

or (since $\sin^2(\alpha/2) = (1 - \cos\alpha)/2$)

$$\tan\theta_1 = \frac{2\sqrt{C(1-C)}}{1 - 2C + M_1/M_2} \tag{6.171}$$

with $C = \sin^2(\alpha/2)$. The new direction vector is given by the cosine law of a spherical triangle

$$\cos\theta_i = \cos\theta_{i-1}\cos\theta_1 + \sin\theta_{i-1}\sin\theta_1\cos(\psi) \tag{6.172}$$

where $\cos\theta_{i-1}$ is the angle with respect to the surface normal before and $\cos\theta_i$ the angle after the ith collision (Fig. 6.19). The azimuthal angle $\psi = 2\pi\text{RND}$ is calculated using another random number.

6. The energy loss in collision with the target atom is calculated by (6.83)

$$\Delta E_{\text{el}} = 4E\frac{M_1 M_2}{(M_1 + M_2)^2}\sin^2\frac{\alpha}{2} \tag{6.173}$$

and subtracted from the current particle energy, i.e., $E = E - \Delta E_{\text{el}}$.

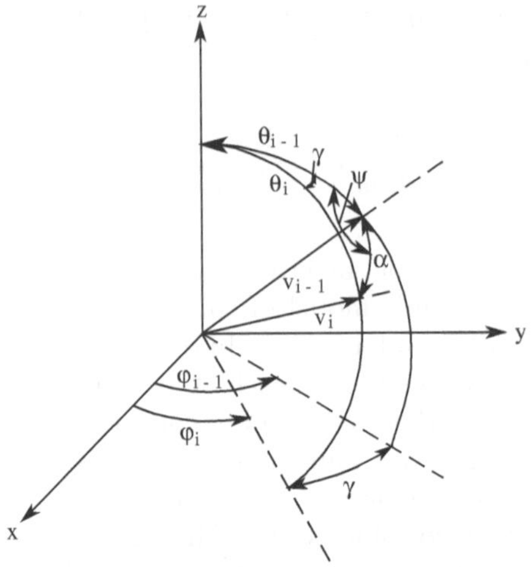

Fig. 6.19. Change of the velocity vector in the ith collision

7. A control of energy and position is performed. If the energy E becomes smaller than a so-called cutoff energy (usually on the order of the binding energy of several eVs), the particle is stopped. While staying in the region of modeling, i.e., as long as the particle has not left the surface, the motion of the particle is followed further by repeating steps 3 through 7.

The listed steps are repeated many times, for many particles. This is followed by a statistical analysis. The implantation profile, for example, is obtained simply by saving the end position of all particles. In analyzing sputtering phenomena, the knock-on target atoms, which obtained from collisions sufficient energy to leave their position in the lattice, have to be followed in the same way as done for the projectile. The simulation of cascade formation and development demands only a slight increase of the coding effort. At the surface, the target atom has to overcome the surface potential. This results in a change of its emission direction due to refraction in this planar potential (Sect. 6.2.4 in [109]).

In (6.169), a random number RND is used to model the distribution of all possible impact parameters $f(\rho_\mathrm{p})$. We have

$$\mathrm{RND} = \int\limits_0^{\rho_\mathrm{p}} f(\rho_\mathrm{p}')\,\mathrm{d}\rho_\mathrm{p}' \,, \quad \rho_\mathrm{p} = \rho_\mathrm{p}^{\max}\sqrt{\mathrm{RND}} \,, \tag{6.174}$$

since the differential cross-section is $\mathrm{d}\sigma = 2\pi\rho_\mathrm{p}\,\mathrm{d}\rho_\mathrm{p}$. We obtain

$$f(\rho_\mathrm{p}) = (\mathrm{d}\sigma/\mathrm{d}\rho_\mathrm{p})/\sigma_\mathrm{total} = 2\pi\rho_\mathrm{p}/(\pi\rho_\mathrm{p}^2) \,. \tag{6.175}$$

For a cylindrical volume of length l_p, the maximum impact parameter which corresponds to the smallest scattering angle is given by

$$\pi(\rho_\mathrm{p}^{\max})^2 l_\mathrm{p} = n_\mathrm{o}^{-1} \,. \tag{6.176}$$

Combining (6.174) with (6.176) leads to (6.169).

The usage of realistic interaction potentials requires a considerable effort in evaluating the scattering angle α, since the conversion (6.30) should be done numerically, an effort which is not fully reduced by applying the so-called "magic" scheme invented by Biersack and Haggmark [266]. In the range of reduced energies $0.1 < \varepsilon < 10$, the following approximation can be of use [279]

$$\sin^2\left(\frac{\alpha}{2}\right) = \frac{1}{1 + [C_1\,\varepsilon\,(\rho_\mathrm{p}/a_\mathrm{L})^{C_2}]^2} \tag{6.177}$$

with $C_1 = 4 - 2\log_{10}\varepsilon$ and $C_2 = 1.4 - 0.4\log_{10}\varepsilon$. For energies $\varepsilon > 10$, the analytical relation derived for the Coulomb potential (see (6.32)) is valid

$$\sin^2\left(\frac{\alpha}{2}\right) = \frac{1}{1 + (2pE_\mathrm{r}/A_\mathrm{c})^2} = \frac{1}{1 + (2\varepsilon\rho_\mathrm{p}/a_\mathrm{L})^2} \tag{6.178}$$

with the reduced impact parameter $\rho_\mathrm{p}/a_\mathrm{L}$ and $A_\mathrm{c} = Z_1 Z_2 e^2/(4\pi\epsilon_\mathrm{o})$.

The given algorithm (steps 1–7) can be readily translated into a computer language resulting in a code whose core routine consists of a few dozen lines only, but is able to produce results which are inaccessible to theoretical analysis, and that just by exploiting a common PC. The assumption of independent particle motion cannot be held in any case, since changes in composition and structure produced by one particle naturally affect the motion of the others. These highly non-linear effects are the subject of so-called dynamical simulations (for example see the description of the TRIDYN code in [280]).

The Monte Carlo method is much faster than molecular dynamics simulations and runs with several million particles are not unusual. However, its restriction is closely connected to the used assumption of binary collisions, which is not valid in the sub-eV energy range, as discussed already in Sect. 5.4.

7

Electrical Coupling

Beginning from Langmuirs initial and fundamental study [281,282] in 1929 concerning the contact of a neutral plasma with an uncharged material surface, the related questions have been the subject of intense investigation. Up to now, especially by using the so-called Langmuir probes to measure plasma density and temperature in plasmas, some points are still open [283, 284].

Under steady state conditions, the electron flux and the ion flux to a floating plate immersed into a plasma must be equal. Frequently, the electrons in a plasma have a larger temperature than the ions, i.e., $T_\mathrm{e} > T_\mathrm{i}$, but even for equal temperatures, the electrons, because of larger velocity, quickly charge a plate exposed in the plasma. A potential difference, called the Langmuir sheath potential, between the plate and the plasma is established, such that almost all electrons, except the fastest, are repelled, and the slower ions are attracted in order to satisfy the steady ambipolarity condition $\Gamma_\mathrm{i} = \Gamma_\mathrm{e}$. This sheath region is extended over only a few Debye lengths in the cases without magnetic field.

The flux density

$$\boldsymbol{\Gamma} = \int f(\boldsymbol{v})\, \boldsymbol{v}\, \mathrm{d}^3\boldsymbol{v} \tag{7.1}$$

is determined by the velocity distribution $f(\boldsymbol{v})$ of the particles. In thermalized plasmas with temperature T, a Maxwellian distribution (see Sect. A.1)

$$f(\boldsymbol{v})\mathrm{d}^3\boldsymbol{v} = n_\infty \left(\frac{m}{2\pi k_\mathrm{B}T}\right)^{(3/2)} \exp\left(-\frac{m\boldsymbol{v}^2}{2k_\mathrm{B}T}\right) \mathrm{d}^3\boldsymbol{v} \tag{7.2}$$

is established, where n_∞ is the density of the particles in the undisturbed region. However, in the sheath the velocity distribution of the particles is strongly distorted due the fact that in this thin region the electrostatic forces are stronger than the inertial forces. The trajectories of charged particles near the surface differ significantly from those in the undisturbed plasma.

7.1 Electron Flux Density

In general, all electrodes immersed in a plasma are negatively charged with respect to the surrounding plasma, i.e., $\phi_{\text{electrode}} < \phi_{\text{plasma}}$, irrespectively whether they act as an anode or cathode in the electric circuit. The reason for this is that the fast electrons (at the same temperature as the ions they have a much higher velocity due to their smaller mass) have to be repelled from the electrodes, otherwise the electron current out of the plasma would be too high and could not be compensated by an ion flux. A preferential loss of electrons would result, and the charge neutrality of the plasma and, therefore, the plasma itself would be destroyed.

The flux density of electrons along the x-axis, which is directed away from the probe surface into the plasma, is given by

$$\Gamma(x) = n_\infty \left(\frac{m_\mathrm{e}}{2\pi k_\mathrm{B} T}\right)^{(3/2)} \int\limits_{-\infty}^{\infty} \exp\left(-\frac{m_\mathrm{e} v_y^2}{2k_\mathrm{B} T_\mathrm{e}}\right) \mathrm{d}v_y$$
$$\times \int\limits_{-\infty}^{\infty} \exp\left(-\frac{m_\mathrm{e} v_z^2}{2k_\mathrm{B} T_\mathrm{e}}\right) \mathrm{d}v_z \int\limits_{-\infty}^{-v_\mathrm{min}} v_x \exp\left(-\frac{m_\mathrm{e} v_x^2}{2k_\mathrm{B} T_\mathrm{e}}\right) \mathrm{d}v_x \quad (7.3)$$

taking (7.2) as the velocity distribution near the sheath entrance. Electrons with $v_x > 0$ move away from the surface and cannot reach it. With a retarding potential, only electrons with sufficient velocity v_min in the direction toward the surface have the chance to reach the surface. Electrons in the velocity interval $v_x = [-v_\mathrm{min}, 0]$ are pushed away. Because of energy conservation, the sum of potential and kinetic energy is constant along x

$$E = \frac{m_\mathrm{e} v_x^2}{2} + Q\phi(x) = \frac{m_\mathrm{e} v_x^2}{2} - e\phi(x)\,. \quad (7.4)$$

In the undisturbed plasma far away from the probe, the potential is usually set to zero, i.e., $\phi(x = \infty) = 0$. An electron with a certain velocity $v_{x=\infty} = v_\infty$ in the undisturbed plasma can only reach a certain position x if the condition

$$E\Big|_\infty = E\Big|_x$$
$$E_\infty \geq \frac{m_\mathrm{e} v_x^2}{2}\Big|_{v_x=0} - e\phi(x) \quad (7.5)$$

is satisfied. In the point of return, $v_x = 0$. $E_\infty = m_\mathrm{e} v_\infty^2/2$ should be at least equal to $-e\phi(x)$ (note that $\phi(x) < 0$). Hence, the minimum velocity required to reach a position x yields

$$v_\mathrm{min} = \sqrt{\frac{-2e\phi(x)}{m_\mathrm{e}}}\,. \quad (7.6)$$

Note that $v_{\min}$ is always positive. The minus sign in the limit of the integral in (7.3) indicates that negative velocities in the x-direction are required to approach the probe surface due to the definition of the x-axis. Performing the integrations in (7.3) yields

$$\begin{aligned} \Gamma_\mathrm{e}(x) &= n_\infty \sqrt{\frac{m_\mathrm{e}}{2\pi k_\mathrm{B} T_\mathrm{e}}} \int\limits_{-\infty}^{-\sqrt{-2e\phi(x)/m_\mathrm{e}}} v_x \exp\left(-\frac{m_\mathrm{e} v_x^2}{2k_\mathrm{B}T_\mathrm{e}}\right) \mathrm{d}v_x \\ &= -n_\infty \sqrt{\frac{k_\mathrm{B}T_\mathrm{e}}{2\pi m_\mathrm{e}}} \exp\left(-\frac{-e\phi(x)}{k_\mathrm{B}T_\mathrm{e}}\right) \end{aligned} \tag{7.7}$$

or, as given in the familiar form with the mean electron velocity $\langle v_\mathrm{e}\rangle = \sqrt{8k_\mathrm{B}T_\mathrm{e}/(\pi m_\mathrm{e})}$

$$\Gamma_\mathrm{e}(x) = -\frac{n_\infty \langle v_\mathrm{e}\rangle}{4} \exp\left(\frac{e\phi(x)}{k_\mathrm{B}T_\mathrm{e}}\right) . \tag{7.8}$$

The integrals over v_y and v_z give each $\sqrt{2\pi k_\mathrm{B}T_\mathrm{e}/m_\mathrm{e}}$. For $\phi(x) = \phi_\mathrm{electrode}$, i.e., $x = 0$, the relation (7.8) yields the electron flux reaching the electrode. The prefactor in (7.8) is equal to the Boltzmann factor appearing by analyzing the transformation of the velocity distribution in a potential field. As known, the flux of particles with a Maxwellian velocity distribution through a arbitrary surface is $n_\infty\langle v\rangle/4$ (Sect. A.1).

7.2 Ion Flux Density

The analysis is considerably simplified if the ions are assumed to be cold, i.e., $T_\mathrm{i} = 0$. In fact, the exact results differ only marginally from the obtained results using the cold ion model [285]. This is due to the large ratio of the ion mass to the electron mass. Even at $T_\mathrm{i} \simeq T_\mathrm{e}$, the velocity ratio (and therefore the flux ratio, since $n_\mathrm{e} = n_\mathrm{i}$ in the undisturbed plasma) remains almost unchanged. With $T_\mathrm{i} = 0$, the ion velocity away from the surface is zero, but for $T_\mathrm{i} \simeq T_\mathrm{e}$ the ion velocity is still much smaller than the electron velocity. To achieve balance of fluxes, the ions have to be accelerated toward to surface. In the attracting potential of the so-called presheath, they obtain a certain velocity. The velocity at the sheath entrance v_s is determined by energy conservation

$$0 = E_\mathrm{s} + Q\,\phi_\mathrm{s} = \frac{m_\mathrm{i} v_\mathrm{s}^2}{2} + Q\,\phi_\mathrm{s} \tag{7.9}$$

leading to

$$v_\mathrm{s} = \sqrt{-\frac{2Q\,\phi_\mathrm{s}}{m_\mathrm{i}}} \tag{7.10}$$

where Q is the ion charge and ϕ_s the potential drop between $x = x_\infty$ and the position $x = x_\mathrm{s}$ at the sheath entrance. For a position x in the sheath

$(0 \leq x \leq x_\mathrm{s})$, energy and flux conservation give

$$E_\mathrm{s} = \frac{m_\mathrm{i} v_x^2}{2} + Q\,\phi(x) \tag{7.11}$$

and

$$n_\mathrm{s} v_\mathrm{s} = n_x v_x \tag{7.12}$$

since no sources or sinks are available inside the sheath. Combining (7.11) and (7.12) leads to an expression for the ion density in the sheath

$$n(x) = n_\mathrm{s} \frac{v_\mathrm{s}}{v_x} = n_\mathrm{s} \sqrt{\frac{2E_\mathrm{s}}{m_\mathrm{i}}} \frac{1}{v_x} = n_\mathrm{s} \sqrt{\frac{E_\mathrm{s}}{E_\mathrm{s} - Q\,\phi(x)}} \,. \tag{7.13}$$

The ion flux density

$$\Gamma_\mathrm{i} = -n_\mathrm{s} v_\mathrm{s} \tag{7.14}$$

is determined by the density at the sheath entrance n_s and the ion velocity there. Since we have not yet determined the potential ϕ_s, the value of v_s remains so far unknown. In the next section, a new derivation is presented. It is based on a criterion of minimum energy.

Considering a source free sheath and presheath, the question comes up, how the continuous loss of charged particles at the surface is compensated, since the plasma otherwise would be rapidly extinguished when a material surface is immersed into it. The ions are recombining with surface electrons and leave the surface either as fast backscattered neutrals or slow emitted molecules. Most of the incident electrons contribute to the surface charge of the material. In a first approximation, the material surface can be regarded as fully absorbing. In equilibrium, a plasma source is clearly required. However, the incorporation of a source into the theoretical models is difficult. Besides the transition from the presheath to the sheath, a new transition border separating the plasma source region from the presheath is then needed with the problem of defining appropriate boundary conditions. Usually, in order to simplify the analysis, a flux according to a half-Maxwellian distribution is assumed to be injected into the presheath.

While the dimension of the electric sheath is more or less well-defined (a few Debye lengths), the dimension of the presheath is still an open question. As described in the next section, the potential drop in the presheath is small enough to be sustained in the plasma on a much larger spatial scale length than the Debye length. In the experiments, the presheath region overlaps into the plasma source region. A clear distinction—as a theorist would desire to make—is not possible. In magnetized plasmas, most of the plasma particles are supplied into a certain flux tube by transport across the magnetic field lines. To formulate and to analyze a consistent model is a challenging task.

7.3 Bohm Criterion with the "=" Sign

Two regions can be distinguished: (1) the electric sheath, a thin layer with the thickness of a few Debye lengths where charge neutrality is violated, and (2)

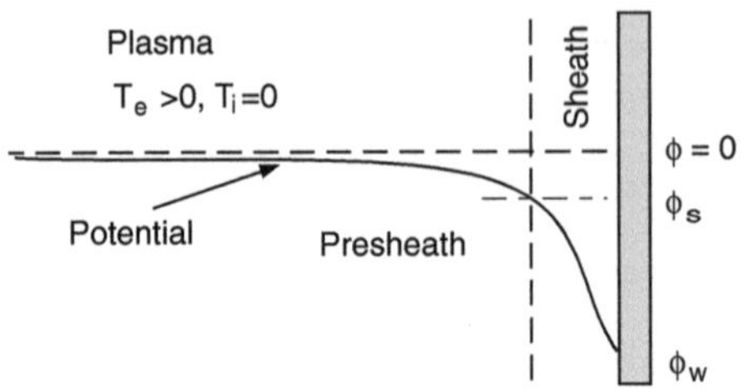

Fig. 7.1. Potential distribution in the plasma disturbed by an immersed material probe

the presheath, an extended region where quasineutrality prevails (Fig. 7.1). The potential at infinity, i.e., in the undisturbed region, is here set to zero ($\phi_\infty = 0$). The term ϕ_s denotes the potential at the sheath entrance and gives the potential drop in the presheath and ϕ_w is the potential at the wall (Fig. 7.1). Let us describe the ion flux toward the plate simply by (7.14)

$$\Gamma_{i,s} = n_{pl,s} v_{i,s} \tag{7.15}$$

where $v_{i,s}$ is the ion velocity and $n_{pl,s}$ is the plasma density at the sheath entrance. In the undisturbed plasma region, the ions usually do not have a notable flow velocity. In the small presheath potential drop, however, they are accelerated up to $v_{i,s}$ at the sheath edge. In addition to the assumption $T_i = 0$, it is assumed that the ions starting with $v_{i,\infty} = 0$ far away from the probe and fall then collisionlessly through the presheath potential drop. Using the equation of energy conservation gives (7.10)

$$v_{i,s} = \sqrt{\frac{-2e\phi_s}{m_i}} \tag{7.16}$$

for singly charged ions (Q = +e). The electron flux in the sheath is given by the relation (7.8), which serves as an excellent approximation

$$\Gamma_{e,w} = \frac{n_{pl,s}}{4} \langle v_e \rangle \exp\left\{ \frac{e(\phi_w - \phi_s)}{k_B T_e} \right\}. \tag{7.17}$$

Note that the fluxes of electrons and ions are spatially constant in the sheath. They are equal at each position. While the ions are simply accelerated toward the plate (their density is decreasing but their velocity increases), the electron flux consists of energetic electrons moving toward the plate and reaching it, and of less energetic electrons, which are repelled and turned back from the collector. Summarizing both contributions results in an electron flux directed toward the plate equal to the ion flux and that at each position in the sheath.

Substituting the relations (7.15) and (7.17) together with (7.16) into the ambipolarity condition $\Gamma_{i,w} = \Gamma_{e,w}$ yields, after some rearrangements,

$$\begin{aligned} \phi_w &= \phi_s + \frac{k_B T_e}{e} \ln\left(\frac{4 v_{i,s}}{\langle v_e \rangle} \right) \\ &= \phi_s + \frac{k_B T_e}{2e} \left[\ln\left(\frac{4\pi m_e}{m_i} \right) + \ln\left(\frac{-e\phi_s}{k_B T_e} \right) \right]. \end{aligned} \tag{7.18}$$

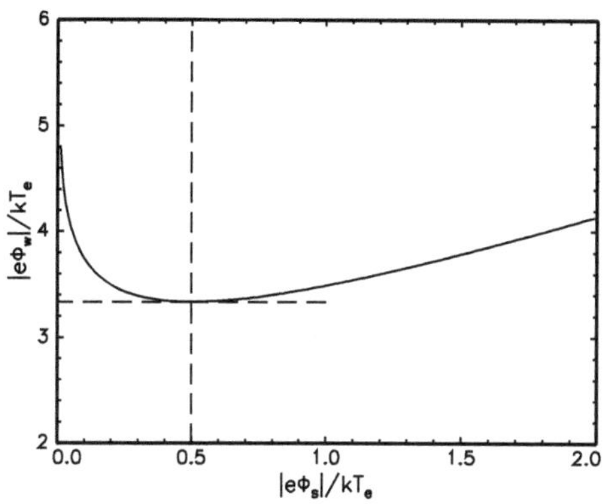

Fig. 7.2. The wall potential as a function of the presheath potential drop according to (7.18) with $m_e/m_i = 1/1836$. The wall potential ϕ_w attains a minimum for $\phi_s = -(0.5\, k_B T_e/e)$

It is worth noting that for a non-conducting surface the balance of electron and ion fluxes is established locally at each point of the surface by the arising potential in the electric sheath according to the electron temperature right above this position. In the case of a conductor, this criterion is replaced by

$$0 = \int (\Gamma_i - \Gamma_e)\, dS \tag{7.19}$$

where the integration has to be performed over the whole surface of the conductor, and the plasma parameters as well as the fluxes may differ with position.

For given T_e, m_i, and m_e, there is equation (7.18), but with two unknown potentials ϕ_w and ϕ_s. As the required additional relation, a minimum energy argument may be applied. Plotting ϕ_w (7.18) as a function of ϕ_s (see Fig. 7.2) we notice that there is a minimum. It is easily found by differentiating (7.18)

$$\frac{d\phi_w}{d\phi_s} = 0 = 1 + \frac{1}{2}\frac{k_B T_e}{e\phi_s} \tag{7.20}$$

leading to

$$\phi_s = -\frac{k_B T_e}{2e}\,. \tag{7.21}$$

Inserting (7.21) back into (7.18) defines the wall potential

$$\phi_w = \frac{k_B T_e}{2e}\left[\ln\left(\frac{2\pi m_e}{m_i}\right) - 1\right] \tag{7.22}$$

including the potential drop $(k_B T_e/2e)\ln[2\pi m_e/m_i]$ in the sheath [281] and the potential difference in the presheath of $-k_B T_e/(2e)$. The potential ϕ_w given in (7.22) is called floating potential, since no currents are drawn from the collector by an external circuit.

Not only does the potential at the wall ϕ_{w} but also the energy flux density arriving the surface

$$\boldsymbol{P}(\phi) = \frac{m}{2} \int v^2 \boldsymbol{v} f(\boldsymbol{v}, \phi) \mathrm{d}^3 \boldsymbol{v} \tag{7.23}$$

has a minimum for the wall potential given by (7.22). This is not surprising, due to the fact that besides thermal contribution, the energy flux is linked directly to the value of the wall potential as the potential difference between the surface and the undisturbed plasma region. As seen in (7.23), the particle distribution function depends on the potential. Hence, the principle of minimum energy flux could also have served as a criterion to derive (7.21). It was shown in experiments [286] that the energy flux reaches a minimum when attaining the floating potential (if ion reflection is not considered, i.e., $R_{\mathrm{E}} = 0$).

With the potential drop in the presheath (7.21), the ion velocity at the sheath entrance is given by the ion sound speed c_{s} (for $T_{\mathrm{i}} = 0$)

$$v_{\mathrm{i,s}} = \sqrt{\frac{k_{\mathrm{B}} T_{\mathrm{e}}}{m_{\mathrm{i}}}} = c_{\mathrm{s}} \tag{7.24}$$

according to (7.16).

This is, in fact, the Bohm criterion, but obtained here with the equal sign in contrast to other derivations stating $v_{\mathrm{i,s}} \geq \sqrt{k_{\mathrm{B}} T_{\mathrm{e}}/m_{\mathrm{i}}}$. The question, whether the equals sign holds or not, led to an extended discussion in the literature, see for example the review articles [287–291]. Most of the complications arise from the more or less artificial division of the plasma into sheath and presheath regions with the problem of defining correct boundary conditions. The ion flux density toward the surface is then given by

$$\Gamma_{\mathrm{i}} = n_{\mathrm{s}} c_{\mathrm{s}}. \tag{7.25}$$

Since the plasma is recombining at an exposed material surface, a source to replace the charged particles as well as a energy source is needed. The energy required for building-up the potential difference between the plasma and the material is delivered by the hot electrons and is supplied to ensure stationary conditions. In the case of a very localized plasma source, the presheath potential drop occurs just close to its location [285, 292]. If the plasma is generated by ionization and heated in a larger volume or is provided by cross-field transport, the presheath will be distributed over a larger region.

The necessity to describe the whole system including the plasma source, the presheath, considering also collisions, and the sheath renders analytical solutions impossible and code simulations are required. The most powerful simulation tool for analyzing the plasma–material transition is the so-called PIC (particle-in-cell) method (see Sect. 7.6). Considering the electric sheath, parameters such as the reflection and emission coefficients, ion temperature, collisionality, magnetic field geometry, and the impurity concentration are of importance.

With increasing ion temperature ratio $\tau = T_\mathrm{i}/T_\mathrm{e}$, the potential in the sheath and in the presheath drops, since less effort is required to accelerate the ions up to the sound speed at the sheath entrance. However, the affected change of the potential distribution is rather small [285, 292] and could be related to the remaining large discrepancy between electron and ion velocities even for $T_\mathrm{i} \geq T_\mathrm{e}$.

Riemann has shown [293] that the Bohm criterion is also satisfied in a collisional plasma, but the ion acceleration in the presheath is affected by collisional friction. The potential drop across the collisional zone can be larger in magnitude than the presheath voltage drop for a plasma with no collisions. The additional potential drop is required to accelerate the ions to the sound speed in the presence of ion momentum loss. With higher collisionality, the potential drop increases modestly with a logarithmic dependence.

Processes such as ion reflection and sputtering do not have a direct effect on the electric sheath, since incoming ions recombine with surface electrons and are reflected as neutrals; as well, almost all sputtered particles are neutral atoms. This is in contrast to electron emission. Each emitted electron which moves away from the surface is equivalent to a plasma ion moving toward the surface considering the current balance. In the (improbable) case of in-sheath ionization of reflected and emitted atoms, new charge carriers would appear and a plasma source would be established near the surface resulting in a change of the potential distribution [294].

7.4 Space Charge Limited Currents

Under certain conditions, a region of negative space charge is established in front of the surface and causes a limitation of electron emission, for example, of thermionic emission of electrons [295–298]. More accurately, the electron current into the plasma is limited, not the emission current out of the surface which, in the case of thermionic emission, is a function of the surface temperature only. A large part of emitted electrons, however, is immediately forced back to the surface by the electric field in that region of negative space charge. This effect starts to play a role when the electric field at the surface becomes zero.

The ion current density j_i and ion density n_i as well as j_e and n_e for the plasma electrons are given in the cold ion approximation ($T_\mathrm{i} = 0$) by the expressions [295, 299] (see also (7.15,7.17))

$$j_\mathrm{i} = e n_\mathrm{pl} \sqrt{\frac{2E_\mathrm{i}}{m_\mathrm{i}}} \,, \qquad n_\mathrm{i} = n_\mathrm{pl} \sqrt{\frac{E_\mathrm{i}}{E_\mathrm{i} - e\phi}}$$

$$j_\mathrm{e} = -e \left[n_\mathrm{pl} - \frac{j_\mathrm{ce}/e}{\sqrt{2(-e\phi_\mathrm{w})/m_\mathrm{e}}} \right] \frac{\langle v_\mathrm{e} \rangle}{4} \exp\left(\frac{e\phi_\mathrm{w}}{k_\mathrm{B} T_\mathrm{e}} \right)$$

$$n_e = \left[n_{pl} - \frac{j_{ce}/e}{\sqrt{2(-e\phi_w)/m_e}} \right] \exp\left(\frac{e\phi}{k_B T_e} \right) \tag{7.26}$$

where E_i denotes the ion energy at the sheath entrance. In the derivation of (7.26), the neutrality condition $n_{pl} = n_e + n_{ce} = n_i$ is applied, where n_{pl} is the plasma density in the undisturbed region. The current density of the electron emission j_{ce} is given by the Richardson–Dushman relation (6.154) in the case of thermionic electron emission, and their density in the sheath is defined by

$$n_{ce} = \frac{j_{ce}}{e v_{ce}} = \frac{j_{ce}/e}{\sqrt{2(e\phi - e\phi_w)/m_e}} \tag{7.27}$$

where v_{ce} is the velocity of the emitted electrons which fall freely through the sheath. The derivation of the expression (7.27) is similar to that of the ion density (7.13) in Sect. 7.2. The potential at the wall ϕ_w as deduced from the current balance $j_i + j_e + j_{ce} = 0$

$$e\phi_w = k_B T_e \ln \left[\left(\sqrt{\frac{2E_i}{m_i}} + j_{ce}/(e n_{pl}) \right) \Big/ \left(\frac{v_e^T}{4} \left(1 - \frac{j_{ce}}{e n_{pl}} \sqrt{m_e/(2(-e\phi_c))} \right) \right) \right] . \tag{7.28}$$

The ion energy at the sheath entrance E_i can be determined using the Poisson equation

$$\frac{d^2\phi}{dx^2} = \frac{e}{\epsilon_o}(n_e + n_{ce} - n_i) \tag{7.29}$$

from which a relation for the electric field E is easily obtained (multiplying by $d\phi/dx$ and integrating from ∞ to x)

$$\frac{1}{2}E^2 = \frac{1}{2}\left(\frac{d\phi}{dx} \right)^2 = \int_0^\phi \frac{e}{\epsilon_o}(n_e + n_{ce} - n_i)\, d\phi \geq 0. \tag{7.30}$$

Using in (7.30) the expressions of (7.26) for the densities and expanding $e\phi/T_e$ and $e\phi/E_i$ in a Taylor series yields a limit for the ion energy at the sheath entrance ($\phi(x = \infty) = 0$)

$$E_i \geq \frac{k_B T_e}{2} + \frac{E_i j_{ce}}{n_{pl} e} \sqrt{\frac{m_e}{2}} (-e\phi_w)^{-3/2} \left[\frac{k_B T_e}{2} - e\phi_w \right] . \tag{7.31}$$

For zero emission ($j_{ce} = 0$), this relation reduces to the Bohm criterion, i.e., $E_i \geq k_B T_e/2$ (Sect. 7.3).

Finally, postulating zero electric field at the cathode, a condition for critical emission current density $j_{\rm ce}^*$ can be defined by integrating (7.30)

$$0 = \left[1 - \frac{j_{\rm ce}^*}{n_{\rm pl}e}\sqrt{\frac{m_{\rm e}}{2(-e\phi_{\rm w})}}\right] kT_{\rm e}\left[\exp\left(\frac{e\phi_{\rm w}}{k_{\rm B}T_{\rm e}}\right) - 1\right]$$
$$- \frac{2j_{\rm ce}^*}{n_{\rm pl}e}\sqrt{\frac{m_{\rm e}}{2}(-e\phi_{\rm w})} + 2E_{\rm i}\left[\sqrt{1 - \frac{e\phi_{\rm w}}{E_{\rm i}}} - 1\right] . \qquad (7.32)$$

The emission current cannot exceed its critical value, $j_{\rm ce} \le j_{\rm ce}^*$, even in the case of increasing surface temperature.

7.5 Effect of Magnetic Field Geometry

Under oblique incidence, two parameters determine the behavior of the sheath: (1) the "magnetization" parameter ξ (11.2) that compares the gyro-radius ρ of electrons and ions with the Debye length $\lambda_{\rm D}$, and (2) the angle α between the magnetic field lines and the surface plane. In fusion relevant experiments, the electron gyro-radius is nearly equal to the Debye length, i.e., $\xi_{\rm e} \simeq 1$, but the ion gyro-radius is much larger, i.e., $\xi_{\rm i} \gg 1$. In order to reduce the heat load onto the divertor and limiter plates, the angle α is kept as small as possible, usually about 1 to 3°. Then, the presheath reveals a double structure. It is composed of a collisional presheath and a magnetic presheath. The thickness of the magnetic presheath is found to be approximately $(c_{\rm s}/\omega_{\rm i})\cos\alpha$ and varies with ion mass, electron temperature, magnetic field, and angle α [300–302]. For $\alpha < 3°$, the thickness is approximately equal to the ion gyro-radius $\rho_{\rm i} \simeq c_{\rm s}/\omega_{\rm i}$. The collisional presheath thicknesses vary with the ion-neutral collision mean free path and/or the extent of the plasma source region (Fig. 7.3).

As in the case without a magnetic field, the presheath has the function to accelerate the ion to sound speed. This value is reached at the entrance of the magnetic presheath along the magnetic field lines. In the magnetic presheath,

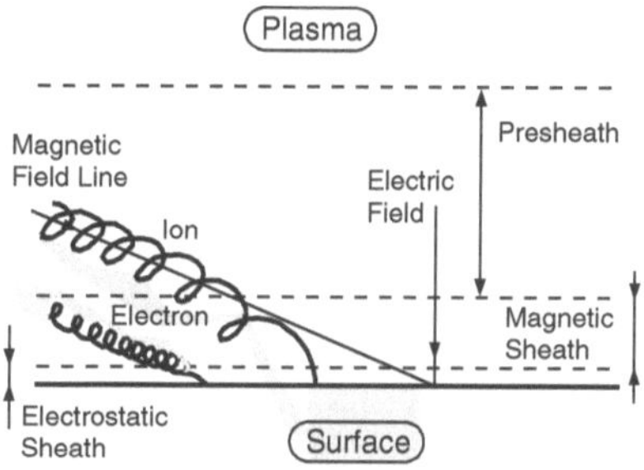

Fig. 7.3. Schematic of the different plasma zones in the near-surface region

the ion trajectories are bent over in a way that at the entrance of the electric sheath the sound speed is already reached with respect to the surface normal.

The electrostatic sheath with a thickness of about the Debye length is characterized by the break-down of quasineutrality, i.e., by the onset of space charge effects, causing strong electric fields, whereas quasineutrality still holds in the collisional as well as in the magnetic presheath. The potential difference (see (7.22) and (7.28)) between the surface and the undisturbed plasma remains nearly the same regardless of the angle of incidence α [300]. However, at oblique incidence, the floating potential decreases and may become positive with respect to the plasma potential as shown in experiments in the plasma generator PSI-2 [303, 304].

In the cold ion model, the ion flux density is equal to $\Gamma_{\rm i} = n_{\rm pl} v_{\rm i} \sin\alpha$ and the electron flux density is about $\Gamma_{\rm e} = (1/4) n_{\rm pl} v_{\rm e}^{\rm T} \exp(e\phi_{\rm w}/k_{\rm B}T_{\rm e}) \sin\alpha$, since the electrons are bounded to the magnetic field lines up to the entrance of the electric sheath. So that the flux balance $\Gamma_{\rm e} = \Gamma_{\rm i}$, results in a potential $\phi_{\rm w}$ are independent of α.

These results hold as long as the angle is larger than one degree. In the case of perfect parallel magnetic field lines ($\alpha = 0$), which in practice can hardly be achieved, the surface potential becomes positive, i.e., $\phi_{\rm w} > 0$ [305], since the ions with their larger gyro-radius should be repelled and the electrons must be attracted. For steady state, an ambipolar flux of ions and electrons toward the surface should be provided by collisions and/or anomalous transport.

The velocity distribution of ions, which are accelerated in the sheath, is distorted in a way that the average impact angle with respect to the surface normal is about 60 degrees [306, 307]. Sputtering of the surface by the ions is enhanced, owing not only to the ion energy increase but also due to the shallow angle of impact according to the angular dependence of physical sputtering (6.119).

Grazing incidence of the magnetic field lines affects also the effective emission of electrons. Some of emitted electrons might be led back to the surface by gyration. The crucial parameter here is the energy of the emitted electrons, which is, for example, different in the processes of thermionic emission and secondary electron emission. The latter process is characterized by emission energies of a few eV up to some tens of eV, while hot surfaces emit electrons with an energy corresponding to the surface temperature of about 0.1–0.2 eV. Electrons with small velocities have a larger probability to escape from the surface. An electron with nearly zero energy might be returned after one gyration, but its velocity component with respect to the surface normal will be as small as it was at the time of emission, while during the gyration the electron increases its velocity along the magnetic field lines owing to the electric field of the sheath. The Lorentz force has the tendency to return the electron to the wall while the electric field force pushes it away from the surface. If an electron is not absorbed during the first gyration, it will propagate into the plasma. With increasing energy of the emitted electrons and decreasing angle of the magnetic field lines with respect to the surface, the escape probability becomes smaller and, correspondingly, the effective emission into plasma is reduced [308]. In the case of magnetic field lines which are directed almost parallel to the surface, its roughness starts to affect the escape probability strongly, since for real materials the roughness is usually larger than the gyroradius of the emitted electrons.

7.6 Modeling of the Electric Sheath

An appropriate description of the electric sheath is not possible without a full kinetic treatment. Fluid approximations or even gyro-kinetic methods fail at considering the transformation of the velocity distribution in the sheath. To reduce the effort of MD calculations, which deals with $N \times (N-1)$ interactions in a system consisting of N particles, an averaged field can be introduced. This field is established by all the particles in the system. The transition from the description in terms of forces between the particles to a model in terms of a force field has been already mentioned in Sect. 5.7. A single particle moves in the field of other particles. Using the Poisson equation the potential distribution and, subsequently, the electric field $\boldsymbol{E} = -\nabla\phi$ can be determined just by knowing the positions of all charged particles at a certain time. The accuracy of the field description depends only on the spatial resolution, i.e., the cell size of the applied numerical mesh. For example, a spatial resolution on the order of the interparticle distance would result in a high accuracy but would mean a higher computational cost.

The particles are assigned according to their positions to grid points. By the way, this operation removes the uncertainty in describing the infinite potential at zero distance from the particle. The cell size should not be smaller than the Debye length (Sect. 5.2), otherwise the plasma cannot be correctly modeled. Such an inadequate choice would be promptly penalized by a diverging, strongly fluctuating numerical solution. On the other hand, collisions with an impact parameter smaller than the cell size are neglected (see the discussion in Sect. 5.4). Fluctuations of the potential at the grid points are alleviated by increasing the number of particles in one cell. In such simulations, the number of used particles are much smaller than the particle number in real plasma systems. The particles in the simulation act as representatives of a group of real particles having their averaged properties—not the cumulative ones. Comprehensive reviews on the so-called particle-in-cell (PIC) methods together with a detailed description of the required numerical methods are given in [309, 310].

7.6.1 Principles of PIC Simulations

The motion of "representative" particles, ions, and electrons are followed in time, while at each time step the density distribution of the particles as well as the corresponding potential distribution are evaluated on a mesh. Different interpolation techniques can be applied to assign the particle position to the grid points and to obtain the field values for a certain position. The magnetic field B can be assumed in many applications to be fully determined by an external, and constant magnetic field B_o. This assumption is valid as long as the internal currents are sufficiently small to generate significant internal magnetic fields. A PIC simulation can be structured as follows:

1. Definition of the initial conditions. At time $t = 0$, the position and velocities of all particles (ions and electrons), and thus the particle distribution function $f(\boldsymbol{v}, \boldsymbol{r}, t = 0)$, has to be defined.
2. Definition of time step Δt, cell size Δx and the number of particles N^{max}. The numerical model is scaled to the real system using the factor

$$f_{\text{norm}} = \frac{n_{\text{pl}} V_{\text{pl}}}{N^{\text{max}}} \tag{7.33}$$

   where V_{pl} is the considered (real) plasma volume and n_{pl} the plasma density.
3. The particles are assigned to the grid points to obtain the density distribution. Two of the most common methods of interpolation are shown in Fig. 7.4 and Fig. 7.5. The NGP method assigns the particles simply to the nearest grid points. The density at a grid point i is then given by $n_{\text{i}} = N_{\text{i}}/\Delta x$ where N_{i} is the number of particles at the grid point i. Using the CIC method (cloud-in-cell), both grid points in the neighborhood receive a "fraction" of the particle charge Q_{c}

$$Q_i = Q_{\text{c}} \left[\frac{\Delta x - (x - X_i)}{\Delta x} \right] \quad \text{and} \quad Q_{i+1} = Q_{\text{c}} \left[\frac{x - X_i}{\Delta x} \right] \tag{7.34}$$

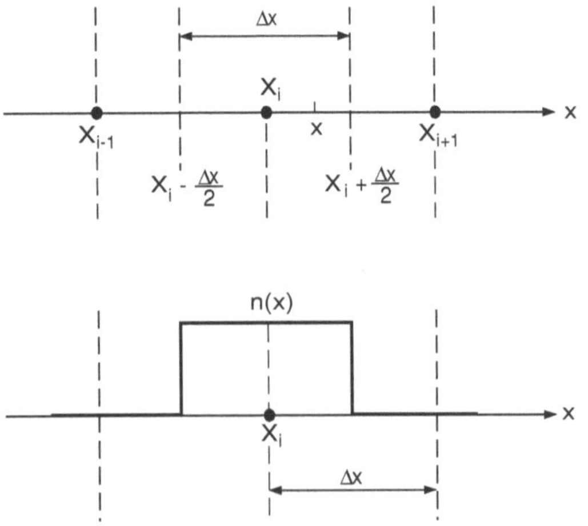

Fig. 7.4. NGP method

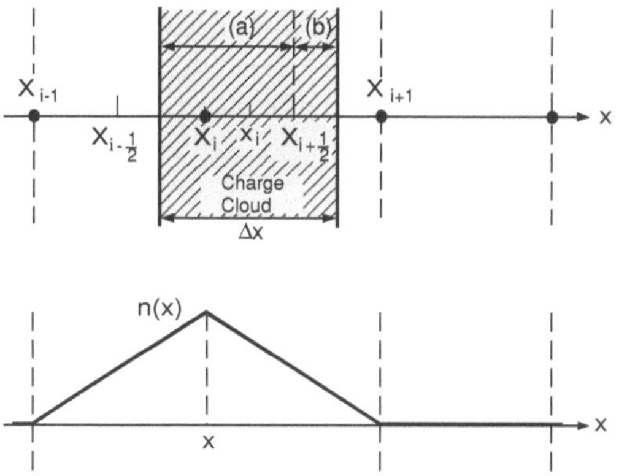

Fig. 7.5. CIC method

realizing a linear interpolation scheme. X_i, X_{i-1}, and X_{i+1} denote the grid point positions.

4. The distribution of the potential and the electric field on the grid points are obtained by numerical solution of the Poisson equation satisfying the given boundary conditions.
5. The value of the electric field at a certain position x is given by the value of the nearest grid point (NGP method) or by the values at the two grid points in the neighborhood

$$E(x) = \left[\frac{X_{i+1} - x}{\Delta x}\right] E_i + \left[\frac{x - X_i}{\Delta x}\right] E_{i+1}, \tag{7.35}$$

if $X_i < x < X_{i+1}$, according to the CIC method.
6. All particles are moved forward by one time step to new positions according to the equation of motion

$$m_{\mathrm{e,i}} \frac{\mathrm{d}\boldsymbol{v}_{\mathbf{e,i}}}{\mathrm{d}t} = \boldsymbol{F} = Q_{\mathrm{e,i}} \left(\boldsymbol{E} + [\boldsymbol{v} \times \boldsymbol{B}_{\mathrm{o}}]\right). \tag{7.36}$$

The numerical solution of (7.36) is given in Sect. A.2.
7. Steps 3–6 are repeated, until the desired calculation time $t^{\max} = N_{\mathrm{t}}\,\Delta t$ is reached, where N_{t} is the number of time steps.
8. Evaluation and output of physical quantities and distributions.

It is very important to use the same interpolation scheme for the density (step 3) and force (step 5) calculations, otherwise non-physical results such as the motion of a particle in its own field may occur. The extrapolation of the presented algorithm to two and three dimensions is straightforward. Usually, the equation of motion is in any case calculated in three dimensions, but a fast solver of the higher-dimensional Poisson equation should be chosen and the interpolation schemes should be extended to consider 4 neighborly grid points in the case of two-dimensional simulations and 8 grid points in the case of three-dimensional simulations.

7.6.2 Boundary Conditions

Due to the fact that in the simulation a certain plasma region is modeled, the definition of boundary conditions for the particles and the electric potential is essential. Numerical instabilities arise, for example, if a particle which has passed one border is put back into the simulation region at a position with a higher potential. This may happen if periodic boundary conditions for the particle but aperiodic conditions for the potential are defined. A periodic region is usually thought of as a fraction of an infinite extended plasma. Integrating the Poisson equation $\mathrm{d}E/\mathrm{d}x = \rho/\epsilon_o$ with the charge density $\rho = Q_i n_i - e\, n_e$ over the system length L

$$\int_x^{x+L} \frac{\mathrm{d}E}{\mathrm{d}x}\,\mathrm{d}x = E(x+L) - E(L) = \frac{1}{\epsilon_o}\int_x^{x+L} \rho\, dx = \frac{L}{\epsilon_o}\langle\rho\rangle \tag{7.37}$$

shows that the total charge of the system should be zero in the case of periodic boundary conditions, $E(x+L) = E(L)$. It is thus contradictory to demand periodic boundary conditions without ensuring zero total charge in the modeled system. In addition, the average electric field $\langle E\rangle$ in a region with periodic boundary conditions must also be zero since

$$\int_x^{x+L} \frac{\partial\phi}{\partial x}\,dx = \phi(x+L) - \phi(L) = 0 = -\int_x^{x+L} E\,dx = -L\,\langle E\rangle. \tag{7.38}$$

If a particle has left the region at one boundary, it is put into the system at the opposite boundary with the same velocity (in value and sign) in the case of periodic boundary conditions. Particles which belong to the two half-intervals $[0, \Delta x/2]$ and $[L - \Delta x/2, L]$ at the borders are counted up and assigned to the grid points at $x = 0$ and $x = L$ to ensure equal density there.

Having one or more material surfaces in the plasma, the condition of total zero charge must be given up, since at the surfaces charges can be collected. The resulting electric fields caused by a surface charge density σ are given by

$$E_o = \frac{\sigma_o}{\epsilon_o} \quad \text{and} \quad E_L = -\frac{\sigma_L}{\epsilon_o} \tag{7.39}$$

if one surface is located at $x = 0$ and the other at $x = L$. A thin plasma layer of thickness Δx and charge density ρ can always be considered as an electrode with surface charge density of $\sigma = \rho\Delta x$.

Particles (ion or electron) which strike the surface of electrodes are taken away from the list of "active" plasma particles and contribute to the corresponding surface charge. This models recombination at the surface. The simplest way to establish an equilibrium situation, i.e., particle conservation, is to feed an ion–electron pair into the plasma region when an electron and an ion are *lost* at the electrodes.

In the Poisson equation, densities have to be calculated. For the grid points at the electrodes, we have, for example, using the NGP scheme,

$$\rho(0) = \frac{\sigma_\mathrm{o}}{\Delta x} + \frac{\sum\limits_{[0,\Delta x/2]} \text{(all particles)}}{(\Delta x/2)\, S_\mathrm{pl}} \tag{7.40}$$

and

$$\rho(L) = \frac{\sigma_\mathrm{L}}{\Delta x} + \frac{\sum\limits_{[L-\Delta x/2,L]} \text{(all particles)}}{(\Delta x/2)\, S_\mathrm{pl}} \tag{7.41}$$

where S_{pl} is the plasma cross-section.

7.6.3 Choice of Time Step and Spatial Resolution

Whether a simulation is successful depends on the right choice of three parameters: (1) the cell size Δx for the calculation of the density distributions and the Poisson equation, (2) the time step Δt by following the motion of the particles, and (3) the number of particles N^max used in the simulation.

Due to numerical approximations, error fields can hardly be avoided. These non-physical electric fields lead to an increase of the kinetic energy of the particle in time—a process called stochastic heating. The electrons with their smaller mass are especially affected. Better interpolation schemes (preferring the CIC method to the NGP scheme) and smaller time steps help to reduce the problem. Fluctuating error fields affect the trajectories of the particles leading to a collision-like behavior. By increasing the number of particles in each cell, this effect of "numerical" collisions can be minimized. The number of particles should be at least $N^\mathrm{max}\lambda_\mathrm{D}/l = 20$, much better is a value of

$$N^\mathrm{max} = 200\frac{l}{\lambda_\mathrm{D}} \tag{7.42}$$

where l is the length of the considered plasma system. Only if the Debye sphere is properly filled with particles, can the occurrence of potential fluctuations be restricted.

The grid size should be smaller than the Debye length

$$\Delta x < \lambda_\mathrm{D} \quad \rightarrow \quad \Delta x \approx 0.2\lambda_\mathrm{D}. \tag{7.43}$$

Usually 5 grid points are sufficient to resolve the Debye length. The time step has to be chosen in accordance to the fastest process—plasma oscillation. The gyration of electrons is also characterized by a frequency of the same order of magnitude in many plasma experiments, i.e., $\omega_\mathrm{ce} \simeq \omega_\mathrm{pe}$ (see Sect. 11). Such a time step $\Delta t \ll 2\pi/\omega_\mathrm{ce}$ is much smaller than actually required for the much slower ions. Many simulations, especially in the past, have been performed by using another mass ratio m_i/m_e instead of the correct one with the aim to use larger time steps. A ratio of $m_i/m_e = 100$ is not unusual. This approximation is justified, since many processes show an asymptotic behavior with increasing mass ratio. The Courant–Friedrichs–Lewy criterion [311] connects the time step and the grid size to the maximum velocity $v_{e,i}^{max}$ of the particles, which can occur in the simulation

$$1 > (v_{e,i}^{max} \Delta t)/\Delta x > 0.1 \,. \tag{7.44}$$

No particle should be able to leap over one grid cell within one time step. On the other hand, it makes no sense to keep one particle too long in one cell, since the charge density in that cell remains then unchanged.

Given a system length l, a plasma density n_{pl}, the masses m_e and m_i, and temperatures T_e and T_i the parameters of the numerical simulation can be chosen as

$$\begin{aligned}
\Delta x &= 0.2\sqrt{\epsilon_o k_B T_e k_B T_i/[n_{pl} e^2 (k_B T_e + k_B T_i)]} \\
\Delta t &= 0.5\frac{\Delta x}{\sqrt{k_B T_e/m_e}} \\
N^{max} &= 200\frac{l}{\sqrt{\epsilon_o k_B T_e k_B T_i/[n_{pl} e^2 (k_B T_e + k_B T_i)]}}
\end{aligned} \tag{7.45}$$

since $\Delta t = 0.1/\omega_{ce} \simeq 0.1/\omega_{pl} = 0.1\lambda_D/v_e = 0.5\Delta x/v_e$. Note that under the conditions of today's fusion experiments, $\lambda_D \approx \rho_e$ and $\omega_{pl} \approx \omega_{ce}$.

To get an idea of the computational costs, the repetition number N^{cycle} of the main cycle (steps 3–6 in Sect. 7.6.1) can be estimated. Let the calculation region be 10 times larger than the ion gyro-radius, i.e., $l = 10\,\rho_{ci}$, and the simulation last 10 times the ion gyration time, i.e., $t^{max} = 10\,T_{ci}$. Then, the number of repetition is about

$$\begin{aligned}
N^{cycle} &= N_t \cdot 2 \cdot N^{max} = \frac{t^{max}}{\Delta t}\, 2\, N^{max} \\
N^{cycle} &= \left(\frac{10 T_{ci}\sqrt{k_B T_e/m_e}}{0.1\lambda_D}\right) 2 \left(\frac{200 \cdot 10\rho_{ci}}{\lambda_D}\right) \\
N^{cycle} &= \left(\frac{10 \cdot 2\pi\, m_i\sqrt{k_B T_e/m_e}}{0.1\lambda_D\, eB}\right) 2 \left(\frac{200 \cdot 10\rho_{ci}}{\lambda_D}\right) \\
N^{cycle} &\simeq 2\times 10^6 \left(\frac{m_i}{m_e}\right)\frac{\rho_{ce}\rho_{ci}}{\lambda_D^2} \propto \frac{m_i^{3/2}}{\sqrt{m_e}}\, n_{pl} \,.
\end{aligned} \tag{7.46}$$

The factor of 2 appears because two species (positively and negatively-charged particle) are involved. Essential are the mass ratio and the density. For a (pure academic) mass ratio of $m_i/m_e = 1$ and the set of parameters: $l = 0.02\,$m,

$T_e = T_i = 10\,\text{eV}$, $n_{pl} = 10^{14}\,1/\text{m}^3$, $m_i = m_e = 1\,\text{amu}$, $Q_i = e$, $Q_e = -e$, and $B = 0.2\,\text{T}$, we have $\Delta x = 3.3 \times 10^{-4}\,\text{m}$, $\Delta t = 5.3 \times 10^{-9}\,\text{s}$, $N^{\text{max}} = 200 \cdot l/\lambda_D \simeq 2400$, and $N^{\text{cycl}} = 4 \times 10^6$. On a common PC, this takes not longer than one minute having a CPU time of roughly $10^{-5}\,\text{s}$ for one cycle. For more realistic parameters, i.e., $m_i/m_e = 1836$ and $n_{pl} = 10^{20}\,1/\text{m}^3$, the effort increases by five orders of magnitude. Such simulations are very time consuming and often simplifying approximations have to be applied, even when using supercomputer systems. Note that an extension of the model to two or three dimensions is readily programmable, but the effort increases with the power of the dimension number.

8

Power Coupling

The electric sheath established via the contact of a plasma with a material surface acts as an energy transfer zone. While the electron and ion energy fluxes vary with distance to the surface, the sum of both contributions remains independent of the position above the surface, since there are no sources or sinks of energy in the sheath. The ions gain their energy from the electrons by being accelerated in the electric field, which is generated at the expense of the electron kinetic energy. The transfer of the electron kinetic energy (out of the tail of their Maxwellian velocity distribution) via the potential energy of the electric field in the sheath to the kinetic energy of the ions can be adequately described only by numerical simulations, as detailed in Sect. 7.6. The heat flux through the sheath (7.23) is determined by the particle distribution function, which is strongly distorted in the sheath, and the concept of temperature becomes, therefore, questionable. Nevertheless, the simplified analytical relations given in the following sections help to understand the main effects and dependencies, even though the transformation of the velocity distributions cannot be addressed there.

8.1 Heat Flux Densities

The heat flux density from the plasma to the wall $P_{\mathrm{w}} = P_{\mathrm{i}} + P_{\mathrm{e}}$ is given by the contribution of ions

$$\begin{aligned} P_{\mathrm{i}} = {} & \Gamma_{\mathrm{i}} R_{\mathrm{N}}^{\mathrm{ion}} \left[(1 - R_{\mathrm{E}}^{\mathrm{ion}})(2k_{\mathrm{B}}T_{\mathrm{i}} + |e\phi_{\mathrm{w}}|) + I_{\mathrm{ion}} - W - E_{\mathrm{ex}}^{\mathrm{atom}}\right] \\ & + \Gamma_{\mathrm{i}}(1 - R_{\mathrm{N}}^{\mathrm{ion}})\big[2k_{\mathrm{B}}T_{\mathrm{i}} + |e\phi_{\mathrm{w}}| + I_{\mathrm{ion}} \\ & - W + (E_{\mathrm{diss}} - E_{\mathrm{ex}}^{\mathrm{molec}})/2 - E_{\mathrm{therm}}\big] \end{aligned} \tag{8.1}$$

and of electrons

$$\begin{aligned} P_{\mathrm{e}} = {} & \Gamma_{\mathrm{e}} R_{\mathrm{N}}^{\mathrm{ele}} 2k_{\mathrm{B}}T_{\mathrm{e}}(1 - R_{\mathrm{E}}^{\mathrm{ele}}) + \Gamma_{\mathrm{e}}(1 - R_{\mathrm{N}}^{\mathrm{ele}})(2k_{\mathrm{B}}T_{\mathrm{e}} + W) \\ & - \Gamma_{\mathrm{e}}\delta_{\mathrm{ESEE}}(E_{\mathrm{ESEE}} + W) f_{\mathrm{esc}}^{\mathrm{SEE}} \\ & - \Gamma_{\mathrm{i}}\delta_{\mathrm{ISEE}}(E_{\mathrm{ISEE}} + W) f_{\mathrm{esc}}^{\mathrm{SEE}} - \Gamma_{\mathrm{em}} E_{\mathrm{em}} f_{\mathrm{esc}}^{\mathrm{em}} \end{aligned} \tag{8.2}$$

with the ion and electron flux density according to (7.25) and (7.8)

$$\Gamma_{\mathrm{i}} = n_{\mathrm{pl}} c_{\mathrm{s}} \sin\alpha \;, \quad \Gamma_{\mathrm{e}} = \frac{n_{\mathrm{pl}}}{4} \langle v_{\mathrm{e}} \rangle \exp\left\{ \frac{e(\phi_{\mathrm{w}} - \phi_s)}{k_{\mathrm{B}} T_{\mathrm{e}}} \right\} \sin\alpha \qquad (8.3)$$

at the sheath entrance, respectively. The assumption of a sine dependence of the fluxes remains valid down to grazing angles of magnetic field line incidence, i.e., for $\alpha \geq 1^{\circ}$ [312]. The density at the sheath edge is simply set equal to the plasma density here.

Those ions which are reflected with a probability of $R_{\mathrm{N}}^{\mathrm{ion}}$ deposit only the part $(1 - R_{\mathrm{E}}^{\mathrm{ion}})$ of their thermal energy flux density of $\Gamma_{\mathrm{i}} 2 k_{\mathrm{B}} T_{\mathrm{i}}$ (here a Maxwellian distribution of incoming ions has been assumed) and of the energy gain in the sheath $|e\phi_{\mathrm{w}}|$ at the target. Equation (8.1) is valid for singly charged ions. A higher charge results in minor changes of the corresponding terms in (8.1). The ions recombine at the surface and release the ionization energy I_{ion} minus the work function W, since the recombination process is accompanied by electron extraction from the bulk material. Further, energy (E_{cx}) can be lost due to excitation of the reflected particles. Especially at low plasma temperatures, the recombination term in (8.1) is important and cannot be neglected [313].

Those ions which are implanted deposit their whole energy and are predominantly released as thermal molecules with an energy of E_{therm}, which corresponds to the surface temperature. In addition, half of the dissociation energy E_{diss} and excitation energy of the emitted molecule must be added.

Electrons which are not reflected release the work function W (see (8.2)), because of their transfer from a free to a bounded state in the bulk material. Emission of secondary electrons can be caused by ion impact (ISEE) and electron impact (ESEE). The last term in (8.3) corresponds to thermionic emission, where E_{em} is the energy of emitted electrons. The coefficients $f_{\mathrm{esc}}^{\mathrm{SEE}}$ and $f_{\mathrm{esc}}^{\mathrm{em}}$ describe the escape probability of emitted electrons in the case of oblique magnetic field (see Sect. 7.5). As a rough estimation, the approximation $f_{\mathrm{esc}} \simeq \sin\alpha$ can be used.

Often, a so-called energy transmission coefficient $\gamma_{\mathrm{E}} = P_{\mathrm{w}}/(\Gamma_{\mathrm{i}} k_{\mathrm{B}} T_{\mathrm{i}})$ is introduced. Neglecting electron emission, γ_{E} becomes simply

$$\gamma_{\mathrm{E}} = 2 + \frac{|e\phi_{\mathrm{w}}|}{k_{\mathrm{B}} T_{\mathrm{i}}} + \frac{2 k_{\mathrm{B}} T_{\mathrm{e}}}{k_{\mathrm{B}} T_{\mathrm{i}}} + \frac{I_{ion} + W}{k_{\mathrm{B}} T_{\mathrm{i}}} \simeq 7.5 + \frac{I_{ion} + W}{k_{\mathrm{B}} T_{\mathrm{i}}} \qquad (8.4)$$

with $\Gamma_{\mathrm{e}} = \Gamma_{\mathrm{i}}$ and the further assumption of $T_{\mathrm{i}} = T_{\mathrm{e}}$; for hydrogen plasmas ϕ_{w} is approximately equal to $3.5 k_{\mathrm{B}} T_{\mathrm{e}}$ as shown in (7.22).

With electron emission, the energy transmission coefficient can reach values up to 25–30 [216,314,315] until space charge limitation sets in. In order to explain this sudden increase in heat load, without referring to bulky relations, a reduced set of equations (8.1, 8.2) with only secondary electron emission by electrons should be considered here. Then, $\gamma_{\mathrm{E}} = (P_{\mathrm{e}} + P_{\mathrm{i}})/(\Gamma_{\mathrm{i}} k_{\mathrm{B}} T_{\mathrm{i}})$ yields

$$\gamma_E \simeq \frac{\Gamma_e(2k_BT_e - \delta_{ESEE}E_{ESEE}) + \Gamma_i(2k_BT_i + |e\phi_w|)}{\Gamma_i k_B T_i}$$
$$= \frac{\Gamma_e}{\Gamma_i}\,[2T_e/T_i - \delta_{ESEE}E_{ESEE}/T_i] + 2 + |e\phi_w|/k_BT_i$$
$$= \frac{1}{1-\delta_{ESEE}}\,[2T_e/T_i - \delta_{ESEE}E_{ESEE}/T_i] + 2 + |e\phi_w|/k_BT_i \qquad (8.5)$$

where the ratio Γ_e/Γ_i is obtained from the flux balance

$$\Gamma_i = \Gamma_e - \Gamma_e\delta_{ESEE} = \Gamma_e(1-\delta_{ESEE}). \qquad (8.6)$$

With increasing emission, the energy transmission coefficient increases too—due to the term $1/(1-\delta_{ESEE})$ in (8.5). In other words, each emitted electron is directed away from surface and is, therefore, equivalent to an ion directed toward the surface with respect to the current balance. Hence, with higher electron emission, less ions are required of being attracted to the surface and the surface potential becomes smaller in absolute value, but of course remains still negative. With smaller potential difference over the sheath, more plasma electrons are streaming to the surface (see (8.3)) and increase the heat load. At higher electron emission ($\delta_{ESEE} > 0.8$–0.9), space charge limitation occurs (see Sect. 7.4) and the power transmission coefficient saturates at the high level of $\gamma_E = 25$–30.

8.2 Change of Surface Temperature

The wall surface is cooled by heat conduction into the bulk of the wall, by thermal radiation, by emission of thermal electrons (already included in (8.2)), and by evaporation of the wall material. The density of heat flux conducted away from the surface P_s depends on the surface temperature T_s

$$P_s = k_{eff}(T_s - T_{bulk}) + \varepsilon_g\sigma_{SB}T_s^4 + \Gamma_{subl}E_{subl} \qquad (8.7)$$

where k_{eff} is the effective coefficient of heat transmission in $W/(m^2K)$, determined by the geometry and the heat conductivity, T_{bulk} is the temperature of the cooled side of the target, ε_g is the emissivity or grayness coefficient, σ_{SB} is the Stefan–Boltzmann constant, E_{subl} is the sublimation energy, and Γ_{subl} is the flux density of the sublimated particle, which depends also strongly on the surface temperature T_s according to (6.127). The first term on the right-hand side of (8.7) describes the energy losses due to heat conduction, and the second term the cooling by radiation. This highly non-linear equation can be solved by iteration in order to obtain the surface temperature in accordance with the heat flux balance of $P_s = P_w$.

In general, the heat conduction equation

$$c_p\,\rho\,\frac{\partial T}{\partial t} = \mathrm{div}(k\,\mathrm{grad}T) + Q_E \qquad (8.8)$$

should be solved together with adequate boundary conditions, which are usually highly non-linear according to the last two terms on the right-hand side of (8.7), to obtain the temperature distribution in the material. Neglecting the temperature and spatial dependence of the heat conductivity k in W/(m K) and having no inner heat sources, i.e., $Q_\mathrm{E} = 0$, this equation reduces to $\partial T/\partial t = a\nabla^2 T$ with the thermal diffusivity $a = k/(c_\mathrm{p}\rho)$ in m^2/s, the material density ρ in kg/m^3, and the heat capacity c_p in J/(kg K). The numerical solution of (8.8) can be obtained using the methods of finite differences or finite elements. In the case of actively cooled target structures of complex geometry, which consist of different materials and cooling media, the application of widely accepted commercial software packages such as ANSYS or FEMLAB is recommended. In special cases, an analytical description is possible, which can serve as a testbed for numerical analysis. Some of them are given in the next sections.

8.2.1 Heat Conduction in a Half-Infinite Medium

In the case of a half-infinite space, the one-dimensional time-dependent solution of (8.8) is

$$T(x,t) = T_\mathrm{o} + \frac{2P\sqrt{a\,t}}{k\sqrt{\pi}} \exp\left(-\frac{x^2}{4at}\right) - \frac{P\,x}{k}\left[1 - \mathrm{erf}\left(\frac{x}{2\sqrt{a\,t}}\right)\right] \tag{8.9}$$

for a constant heat flux density onto the surface

$$P = -k\left.\frac{\partial T}{\partial x}\right|_{x=0} \tag{8.10}$$

and an initial, uniform temperature of the bulk material $T = T_\mathrm{bulk}$ at $t = 0$; $\mathrm{erf}(x) = (2/\sqrt{\pi})\int_0^x \exp(-t^2)\mathrm{d}t$ is the error function. According to (8.9), the rise of temperature at the surface is proportional to $\sqrt{t}$ and given by the often used relation

$$T(x=0,t) = T_\mathrm{s}(t) = T_\mathrm{bulk} + \frac{2}{\sqrt{\pi c_\mathrm{p}\rho k}}\,P\sqrt{t}\,. \tag{8.11}$$

For many materials, the material parameter $2/(\pi c_\mathrm{p}\rho k)^{1/2}$ is about 10^{-4} K m^2 /(W s$^{1/2}$). The relations (8.9) and (8.11) can be used to describe inertial cooling of a material layer, which is thicker than the d_crit

$$d_\mathrm{crit} = \sqrt{2\,a\,t_\mathrm{max}} = \sqrt{\frac{2k}{c_\mathrm{p}\,\rho\,t_\mathrm{max}}}\,, \tag{8.12}$$

the depth reached by the thermal wave after a certain time t_max. As long as $d > d_\mathrm{crit}$, the influence of the boundary condition at the $x = d$ can be neglected.

8.2.2 Point-like Heat Load

If at point $\boldsymbol{r}_\text{o} = (0,0,0)$ at the surface a certain amount of energy W_th is instantly released, then the corresponding temperature distribution is given by

$$T(r,t) = \frac{2\,W_\text{th}}{c_\text{p}\,\rho\,(4\pi\ t)^{3/2}\sqrt{a_x\,a_y\,a_z}}\ \exp\left[-\frac{x^2}{4a_x t} - \frac{y^2}{4a_y t} - \frac{z^2}{4a_z t}\right] \tag{8.13}$$

where a_x, a_y, and a_z are the thermal diffusivities for the different directions. The loaded energy W_th is related to a heat flux density P

$$W_\text{th}\,[\text{J}] = \int\limits_{\Delta t}\int\limits_{\Delta S} P\left[\frac{\text{W}}{\text{m}^2}\right]\ \text{d}S\,\text{d}t \simeq P\Delta t\,\Delta S = P\Delta t\,\Delta y\Delta z \tag{8.14}$$

acting during a short time interval Δt on a small surface area $\Delta S = \Delta y\Delta z$.

In the case of a local, but constant heat load, the temperature distribution is obtained by integrating (8.13) over the exposure time

$$T(r,t) = \frac{2\,P\Delta S}{c_\text{p}\,\rho\,(4\pi\,a)^{3/2}}\int_0^t \frac{1}{(t-t')^{3/2}}\ \exp\left[-\frac{r^2}{4a(t-t')}\right]\,\text{d}t' \tag{8.15}$$

for $a = a_x = a_y = a_z$. With the substitution $\tau = 1/(t-t')$, (8.15) is transformed into an integral of type

$$\int \exp(-c\,\tau)\,\text{d}\tau/\sqrt{\tau} = \sqrt{\pi/c}\ \text{erf}(\sqrt{c\,\tau}) \tag{8.16}$$

leading to

$$T(r,t) = \frac{2P\,\Delta S}{\rho\,c_\text{p}\,4\pi\,a\,r}\left[1 - \text{erf}\left(\frac{r}{\sqrt{4at}}\right)\right] \tag{8.17}$$

with $r = \sqrt{x^2+y^2+z^2}$. Considering different diffusivities a_x, a_y and a_z, we have

$$T(r_*,t) = \frac{2P\,\Delta S}{\rho\,c_\text{p}\,4\pi\,r_*\sqrt{a_x a_y a_z}}\left[1 - \text{erf}\left(\frac{r_*}{\sqrt{4t}}\right)\right] \tag{8.18}$$

with

$$r_* = \sqrt{\frac{x^2}{a_x} + \frac{(y-y_\text{o})^2}{a_y} + \frac{(z-z_\text{o})^2}{a_y}} \tag{8.19}$$

if the heat load is released at the surface element ΔS at the position $\boldsymbol{r}_\text{o} = (0, y_\text{o}, z_\text{o})$.

8.2.3 Heat Conduction and Diffusion

The equation of diffusion

$$\frac{\partial n}{\partial t} = \text{div}(D\,\text{grad}\,n) \tag{8.20}$$

has the same mathematical form as the equation of heat conduction (8.8). The particle density (or concentration) corresponds to the temperature, the diffusion coefficient D to the thermal diffusivity $a = k/(c_\mathrm{p}\,\rho)$. Releasing instantly at the point $r = 0$ a certain amount of energy W_th, the temperature distribution evolves in time as (see (8.13))

$$T(r,t) = \frac{W_\mathrm{th}}{c_\mathrm{p}\,\rho\,(4\pi\,a\,t)^{3/2}}\,\exp\left[-\frac{r^2}{4at}\right]. \tag{8.21}$$

Analogously, the concentration profile of N particles, which start at $r = 0$ to diffuse, is described by

$$n(r,t) = \frac{N}{(4\pi\,D\,t)^{3/2}}\,\exp\left[-\frac{r^2}{4Dt}\right] \tag{8.22}$$

with the equivalence

$$N \equiv \frac{W_\mathrm{th}}{c_\mathrm{p}\,\rho}\,[\mathrm{Km}^3]\,. \tag{8.23}$$

8.3 Power Removal

Limiter and divertor plates are especially subjected to high power flux densities (up to several tens of $\mathrm{MW/m^2}$) due to ion and electron bombardment and by radiation. For design purposes, typically 80% of the produced α power is assumed to be intercepted by the divertor (240–480 MW) leading to a load of 40–60 $\mathrm{MW/m^2}$ (see Sect. 9.1.5). Using additional means, such as radiation, this value should be reduced to 5–10 $\mathrm{MW/m^2}$ [5]. While below 1 $\mathrm{MW/m^2}$ passive (inertial) cooling is sufficient, active cooling is required for 1–10 $\mathrm{MW/m^2}$.

According to (8.11), the rise of temperature at the surface is proportional to $\sqrt{t}$. The power flux density can be estimated for typical plasma parameters as

$$P = \gamma_\mathrm{E}\,c_\mathrm{s}\,n_\mathrm{e} k_\mathrm{B} T_\mathrm{e} \sin\alpha \simeq 3.5\,\mathrm{MW/m^2} \tag{8.24}$$

with $n_\mathrm{e} = 2.5 \times 10^{20}\,1/\mathrm{m}^3$, $T_\mathrm{e} = T_\mathrm{i} = 10\,\mathrm{eV}$, $m_\mathrm{i} = m_\mathrm{D}$, $\alpha = 2^\circ$, and an energy sheath transmission coefficient of $\gamma_\mathrm{E} \simeq 8$. For normal incidence of the magnetic field lines, i.e., $\alpha = 90^\circ$, a value of 100 $\mathrm{MW/m^2}$ would result. It is obvious from (8.11) that with $P > 1\,\mathrm{MW/m^2}$ a surface temperature close to the melting temperature is reached after some tens of seconds. Hence, active cooling is quite essential regarding steady state operation.

The heat removal capacity is limited by the thermal conductivity of the material. Since the heat flux density P through a layer of thickness d is constant in steady state, the integration of $P = -k(T)\,\mathrm{d}T/\mathrm{d}x$ =const. yields

$$\int_0^d P dx = \int_{T_d}^{T_0} k(T)\,\mathrm{d}T \;\rightarrow\; P_\mathrm{max} = \frac{1}{d}\int_{T_d}^{T_0^\mathrm{max}} k(T)\,\mathrm{d}T = \frac{\Lambda(T_0^\mathrm{max}) - \Lambda(T_d)}{d} \tag{8.25}$$

using the integral thermal conductivity $\Lambda(T)$, which includes the temperature dependence of the thermal conductivity $k(T)$. T_d is the temperature at the cooling side of the target plates. The temperature at the plasma-facing surface T_0 is limited to the temperature at which melting (metals) or strong sublimation (in the case of graphite) starts. The relation in (8.25) determines the power flux density, which can be removed in steady state by one-dimensional thermal conduction through a material of thickness d for a given temperature difference. The choice of the thickness is always a compromise between a safety margin with respect to thinning by erosion and still tolerable cooling properties. Comparing different materials and alloys, it was shown in [215,316] that the maximum power flux density, which can be carried off by conduction is 5 to 20 MW/m^2.

Using carbon-based materials such as CFC (carbon fiber composites) with their superior heat conductivity, much higher values can be achieved. However, at high radiation fluences degradation of the thermal properties of these materials occurs. Recently, robust technical solutions have been developed to handle steady state power flux densities up to 10 MW/m^2. Higher values can be tolerated only transiently and have to be avoided by aiming for plasma scenarios with tolerable plasma parameters near the surfaces.

8.4 Thermal Stress

Another critical subject is thermal stress built up due to temperature gradients in the material in a certain geometry. The thermal stress σ_{ts} (as force per unit area), which develops if a structure is completely constrained (not allowed to move at all), is the product of the coefficient of linear expansion α_{T}, the temperature difference ΔT, and Young's modulus E_{Y} for the material, i.e., $\sigma_{\mathrm{ts}} = \alpha_{\mathrm{T}} \Delta T E_{\mathrm{Y}}$. During heating, the arising stresses are compressive in the near surface region and tensile in the bulk region. A thermal stress criterion can be derived by expressing the heat flux density P simply by $P = k|\Delta T|/d$ [317]

$$\gamma_\sigma = \frac{\sigma}{\sigma_{\mathrm{y}}} = \frac{\alpha_{\mathrm{T}}\, E_{\mathrm{Y}}\, P\, d}{(1-\nu_{\mathrm{P}})k\, \sigma_{\mathrm{y}}} < 1 \tag{8.26}$$

where d is the thickness of the material, k is the thermal conductivity, σ_{y} is the yield strength (the stress at which plastic deformation starts), and ν_{P} is the Poisson ratio (the ratio of the contraction strain normal to the applied load divided by the extension strain in the direction of the load).

If the thermal stress limit (8.26) is exceeded ($\gamma_\sigma > 1$), then a part of the surface will be plastically deformed by thermal cycling. In addition, this deformation leads to residual tensile stresses in the surface region during the cooling phase after plasma exposure.

Almost all technical designs of target plates are based on the use of different materials, where additional stresses are induced owing to different thermal expansion properties. These stresses have to be kept well below the yield strength under all plasma conditions, i.e., also in the case of transient effects such as ELMs and disruptions. Extensive thermal testing of components subjected to repeated thermal loading is required for reliable lifetime predictions.

In the case of thin films, for example, having a tungsten layer on graphite, the initial stress distribution built up during the deposition process is also of importance. There are thermal and intrinsic stresses. Thermal stress occurs because the films are usually deposited above room temperature. Upon cooling from the deposition temperature to room temperature, the difference in the thermal expansion coefficients of the substrate and the film cause thermal stress. Intrinsic stress results from the microstructure created in the films as atoms are deposited on the substrate. Tensile stress results from microvoids in the thin films, and compressive stress results when heavy ions or energetic particles strike the film during deposition, leading to a more tightly packing of atoms, hence, to incomplete structural ordering. Obviously, the same processes of high-energy implantation come into play during the exposure in fusion experiments.

9

Impurity Problems in Fusion Experiments

In a burning fusion plasma, the concentration of impurities, i.e., other particles than D and T, must be kept sufficiently low for two reasons. Firstly, impurities dilute the reacting hydrogen plasma thereby reducing the available fusion power (see in Sect. 3.2). An impurity level of some percent would decrease the fusion power by factors. Secondly, highly-charged impurity ions can increase the radiation losses by emission of bremsstrahlung, recombination and line radiation up to a limit where ignition is impossible (in Sect. 9.1).

The impurity problem can be divided into the generation of impurities at the wall elements, the first ionization of the emitted impurity neutrals, and the transport of the impurity ions in the SOL and in the confined plasma region [318]. All three aspects are shortly considered below. Erosion phenomena which are specific to fusion experiments are described in Sect. 9.2. A criterion for the critical impurity concentration in the main plasma is derived in Sect. 9.4.

9.1 Impurity Radiation

If a fusion plasma radiates as a black body, a huge energy outflux of $P = \sigma_{\mathrm{SB}} T^4 = 5.7 \cdot 10^{-8} T^4 \simeq 10^{25}\,\mathrm{W/m^2}$ would be expected for $T = 10\,\mathrm{keV}$. Since the plasma is almost transparent for the emitted photons, the energy losses due to the different radiation processes have to be calculated separately and the simple Stefan–Boltzmann law cannot be applied.

The radiation losses due to collisions of plasma electrons with impurities of different charge state q

$$P_{\mathrm{v}} = \sum_q \left[P_{\mathrm{recomb}} + P_{\mathrm{brems}} + P_{\mathrm{cycl}} + P_{\mathrm{q}}\right] \tag{9.1}$$

has contributions from recombination, bremsstrahlung, cyclotron radiation, and line radiation.

The last term in (9.1) is the most important channel of power loss, especially in the plasma edge of fusion experiments. The radiation losses due to recombination (radiative, dielectronic, and three-body recombination) become important in cold plasmas with electron temperatures of only a few electron volts.

9.1.1 Line Radiation

The line radiation of hydrogenic neutrals is small compared to bremsstrahlung in the plasma core. The reason is that the core neutral density is quite low, even if some central fueling by neutral beams or pellet injection is provided. Line radiation of highly-charged impurities is, unfortunately, very effective and is $\propto q^n$ ($n = 4$–6), where q is the charge state. The ionization equilibrium for the various ions of a certain impurity element have to be included into the calculations of the radiation constants $L(q, T_e)$ that allow the following presentation of the radiation losses

$$P_q = n_e n_q L_q(T_e) \quad [\mathrm{W/m^3}] \tag{9.2}$$

where n_q is the density of the impurity in charge state q. Typically values of about $L_q \simeq 10^{-33}$–$10^{-32}\,\mathrm{Wm^3}$ are found in fusion experiments in the boundary plasma due to oxygen and carbon impurities [319, 320]. The radiation function for tungsten is about $5 \times 10^{-31}\,\mathrm{Wm^3}$ in the temperature range of $10 \leq T_e \leq 1000\,\mathrm{eV}$. For argon and krypton, we have about $L = 2 \times 10^{-31}\,\mathrm{Wm^3}$ at the maximum around $T_e = 15\,\mathrm{eV}$. Beryllium radiates with $L = 2 \times 10^{-32}\,\mathrm{Wm^3}$ at its maximum at $T_e = 2\,\mathrm{eV}$.

For many impurities, the radiation functions have been calculated and/or approximated (see [321, 322]). Tokar proposed the following approximation [323]:

$$L_q = \frac{C_1\sqrt{T_e}}{1 + C_2\sqrt{T_e} + C_3 T_e} \exp(-E/T_e) \quad [\mathrm{eV/(m^3 s)}] \tag{9.3}$$

where the electron temperature should be given in electron volts. The fitting coefficients C_1, C_2, C_3, and E are listed in Table 9.1 for carbon and oxygen ions. However, relation (9.3) is suitable only for a limited number of impurity ions since the different noble gas configurations are not considered.

9.1.2 Bremsstrahlung

Bremsstrahlung occurs during collisions of electrons with ions in a hot thermonuclear plasma, and is one of the mechanisms by which the plasma loses energy. It can also be used for plasma diagnostics. The atoms of light elements are present in fully ionized form as bare nuclei. Electron bremsstrahlung on these positive ions (on pure Coulomb centers) has been studied in great detail before and presents no theoretical problem. In calculating bremsstrahlung on ions that contain electrons, we encounter a many-body problem, which so

Table 9.1. Coefficients to be used for calculation of radiation constants according to (9.3) (from [323])

Species	E (eV)	C_1 (eV$^{1/2}$m^3/s)	C_2 (1/eV$^{1/2}$)	C_3 (1/eV)
C^+	7.76	5.6×10^{-13}	0.213	0.0143
C^{2+}	7.59	7.27×10^{-13}	0.265	0.023
C^{3+}	5.48	7.54×10^{-13}	0.483	0.0565
C^{4+}	257	2.23×10^{-14}	0	0.0034
C^{5+}	250	4.63×10^{-15}	0	0.00058
O^+	6.23	1.04×10^{-13}	0	0.025
O^{2+}	7.94	1.59×10^{-13}	0	0.0282
O^{3+}	7.61	1.41×10^{-13}	0	0.0095
O^{4+}	9.7	2.7×10^{-13}	0	0.0143
O^{5+}	4.13	1.05×10^{-13}	0.2	0.0134
O^{6+}	400	6.17×10^{-15}	0	0.000362
O^{7+}	420	2.38×10^{-15}	0	0.000321

far has no solution. Both the incident and bound electrons participate in the emission.

The bremsstrahlung losses can be much easier estimated than the energy losses due to line radiation. Contrary to the latter, this loss channel remains in the case of fully stripped ions. It is well-known that one electron radiates with a power

$$P = \frac{e^2(\dot{v})^2}{6\pi\epsilon_o c^3}\ [\mathrm{W}] \tag{9.4}$$

due to acceleration $\dot{v}$. Having the attracting force on that electron in the field of a positive ion with charge $q_i e$, $F_r = (q_i e^2)/(4\pi\epsilon_o r^2)$, where r is the distance between the electron and the ion, yields

$$P = \frac{2}{3}\frac{q_i^2 e^6}{(4\pi\epsilon_o)^3 m_e^2 r^4 c^3} \tag{9.5}$$

if one takes simply $F_r = m_e\dot{v}$, thus neglecting the real trajectory of the electron. With the electron density n_e and the impurity density n_z, the emitted power per volume element is $\mathrm{d}P_v = P\, n_e\, n_i\, \mathrm{d}V$. Introducing the collision parameter ρ_p and using the estimation $l = vt \simeq v\,(\rho_p/v) = \rho_p$, where t is the reaction time, the volume element can be expressed as $\mathrm{d}V = 2\pi\,\rho_p\,\mathrm{d}\rho_p\, l$. Integration of (9.5) gives

$$P_v = \frac{4\pi}{3}\frac{q_i^2 e^6 n_e n_i}{(4\pi\epsilon_o)^3 m_e^2 c^3}\int_{\rho_{p,\min}}^{\infty}\frac{1}{\rho_p^2}\,\mathrm{d}\rho_p = \frac{4\pi}{3}\frac{q_i^2 e^6 n_e n_i}{(4\pi\epsilon_o)^3 m_e^2 c^3 \rho_{p,\min}}\left[\frac{\mathrm{W}}{\mathrm{m}^3}\right]. \tag{9.6}$$

Taking the deBroglie wavelength as the minimum collision parameter

$$\rho_{p,\min} = \frac{\hbar}{m_e v} = \frac{\hbar}{m_e\sqrt{8k_B T_e/(\pi m_e)}} = \frac{\hbar}{\sqrt{(8/\pi)k_B T_e m_e}} \tag{9.7}$$

and taking the mean velocity of the Maxwellian distribution, we obtain finally

$$P_{\rm v} = \frac{4\sqrt{8}\pi q_{\rm i}^2 e^6 n_{\rm e} n_{\rm i}}{3\sqrt{\pi}(4\pi\epsilon_{\rm o})^3 m_{\rm e}^{3/2} c^3 \hbar} \sqrt{k_{\rm B}T_{\rm e}} = \frac{4\sqrt{8\pi}}{3} \frac{q_{\rm i}^2 e^6 n_{\rm e} n_{\rm i}}{(4\pi\epsilon_{\rm o})^3 m_{\rm e}^{3/2} c^3 \hbar} \sqrt{k_{\rm B}T_{\rm e}}\,. \quad (9.8)$$

The exact solution based on quantum mechanics

$$P_{\rm brems} = c_{\rm br} n_{\rm e}^2 Z_{\rm eff} \sqrt{k_{\rm B}T} \left[\frac{\rm W}{\rm m^3}\right] \text{ with}$$

$$c_{\rm br} = \frac{16\sqrt{2\pi}\gamma_{\rm G}}{3\sqrt{3}(4\pi\epsilon_{\rm o})^3 m_{\rm e}^{3/2} c^3 \hbar} = 3.84 \times 10^{-29} \gamma_{\rm G} \frac{{\rm Wm}^2 s}{\sqrt{\rm kg}} \quad (9.9)$$

differs only by the prefactor $16\sqrt{2\pi}/(3\sqrt{3}) = 7.72\gamma_{\rm G}$ instead of $4\sqrt{8\pi}/3 = 6.68$ from the estimation (9.8). Thus, the bremsstrahlung losses are proportional to $n_{\rm e}^2 Z_{\rm eff}\sqrt{T_{\rm e}}$.

9.1.3 Cyclotron Radiation

Similar to the estimation of the bremsstrahlung losses, we use (9.4) and take for the acceleration $\dot{v} = eBv_\perp/m_{\rm e} = \omega_{\rm c} v_\perp$

$$P = \frac{e^2}{6\pi\epsilon_{\rm o}c^3} \omega_{\rm c}^2 v_\perp^2 \text{ [W]}\,. \quad (9.10)$$

This relation transforms to

$$P = \frac{e^2}{6\pi\epsilon_{\rm o}c^3} \frac{\omega_{\rm c}^2 v_\perp^2}{1-(v/c)^2} \quad (9.11)$$

for the relativistic case, where v is the absolute value of the electron velocity and $v_\perp$ its component perpendicular to the magnetic field B. The emitted power per volume element is $P_{\rm cycl} = n_{\rm e}P$ [W/m^3], and therefore proportional to $T_{\rm e}B^2$ (see (9.10)).

Cyclotron radiation due to gyration in the magnetic field with the gyrofrequency $\omega_{\rm c}$ is important only well above 25 keV and can usually be neglected [22]. In addition, taking proper account of self-absorption of cyclotron radiation (including the effect of wall reflection), it can be concluded that cyclotron radiation is a minor loss channel for ignited DT plasmas (as shown by Trubnikov in 1958), although for DD and DHe^3 fusion reactors it is more important, but does not preclude ignition [23].

9.1.4 Radiation Phenomena

On a number of tokamaks, intense edge radiation zones called MARFE (multifaceted radiation from the edge) are observed. MARFEs are radiation condensation instabilities that appear in tokamaks as the density limit is approached.

They form a poloidally localized and toroidally symmetric ring usually located near an outer flux surface on the high field side of the torus, i.e., near the inner wall [324].

The radiation functions $L_q(T_e)$ (see (9.2)) as a function of the electron temperature usually have several maxima. In regions with a negative derivative $\mathrm{d}L_q(T_e)/\mathrm{d}T_e < 0$, a local temperature decrease would lead to an increase of radiation losses and to an increase of the heat conduction to that region along the magnetic field lines. If the radiation losses cannot be compensated by heat conduction, the local volume becomes thermally unstable, and a major disruption can result. The density threshold for the appearance of a MARFE has been found for a wide range of tokamak devices, determined by the simple relationship [325]

$$n_e[1/\mathrm{m}^3] = \frac{10^{20}\, C\, I_p[\mathrm{MA}]}{\pi a_t^2[\mathrm{m}]} \tag{9.12}$$

and scales linearly with the plasma current I_p. There is no explanation for the processes leading to such a dependence. In some cases, the MARFE is a precursor to a density limit disruption. In (9.12), a_t is the minor radius and C a constant to be taken in the interval C=0.4–0.7, typically 0.55.

MARFEs can be avoided in sufficiently pure hydrogenic plasma, but small fractions of low-Z impurities are sufficient for their formation. In principle, a tokamak reactor could take advantage of this tendency by using MARFEs localized near the X-point to radiate a significant fraction of the power leaving the plasma [50]. However, MARFE control is limited due to its intrinsically unstable behavior. In addition, repeated switching of the MARFE between the X-point and the divertor region is observed. Radiation from a MARFE may lead to significant peak loads on the wall in its vicinity.

9.1.5 Benefits of Radiation

The fusion power produced in the plasma center is transported to the target via a rather thin scrape-off layer with a radial decay length of about 1 cm. The resulting heat flux toward the divertor plates is reduced by a factor of 10 by the tilt of the magnetic field lines, by a factor of 2 when using two divertors and a factor of 2 to 4 by expansion of the flux in the divertor. The heat load onto the divertor plates of about 30–100 MW/m^2 in the case of a 1.5 GW fusion power reactor is still too large and cannot be handled by conventional materials and cooling technologies [5]. Thus, before reaching the plate a large fraction of the incoming power should be distributed over a large area, e.g., by means of radiation.

Power exhaust by radiation at the plasma boundary could abate the problems of the plasma–wall interaction in fusion devices, since damage of wall elements due to high heat loads can be avoided by distributing the power uniformly over the first wall [326]. The spreading of the heat flux crossing the separatrix over a larger area is a new intention, in contrast to the simple reduction of near-target temperatures in order to avoid erosion [53].

However, if the energy has to be provided along field lines by heat conduction, it is difficult to convert it into other transport channels, e.g., radiation, at low temperatures. The properties of parallel heat conduction imply that in order to pass a given heat flux, a much steeper T_e gradient has to exist at low temperatures. Hence, the size of the region corresponding to a particular temperature interval will decrease with temperature [327]. This implies of course that by increasing the impurity concentration in the edge and raising the edge density by additional strong gas puffing, which unfortunately also implies higher core densities, more power can be radiated.

While advanced confinement modes such as the H-mode requires a certain power flux through the separatrix, it is important to achieve efficient radiation in the scrape-off layer. In addition to the injection of noble gases, an intrinsic impurity such as carbon (if used as a divertor material) can help to establish a self-regulating loop. With higher plasma temperature, the erosion level increases and so does the radiation by the released carbon. This radiation cools the plasma, hence, the erosion is reduced, and less material is emitted, which in turn reduces the radiation level, and the plasma again becomes hot. The following set of equations describes this situation [328, 329]:

$$\frac{\mathrm{d}n_{\mathrm{imp}}}{\mathrm{d}t} = \frac{\Gamma_{\mathrm{ero}}}{l} - \frac{n_{\mathrm{imp}}}{\tau} , \tag{9.13}$$

$$\frac{3}{2}\frac{\mathrm{d}(n_{\mathrm{e}}k_{\mathrm{B}}T_{\mathrm{e}})}{\mathrm{d}t} = \frac{P_{\mathrm{in}} - P_{\mathrm{surf}}}{l} - n_{\mathrm{e}}\, n_{\mathrm{imp}}\, L_{\mathrm{imp}}(T_{\mathrm{e}}) , \tag{9.14}$$

$$n_{\mathrm{e}} = n_{\mathrm{i}} + q_{\mathrm{imp}}\, n_{\mathrm{imp}} = n_{\mathrm{e}}^{\mathrm{o}} + q_{\mathrm{imp}}\, n_{\mathrm{imp}} . \tag{9.15}$$

The change of impurity density n_{imp} is given by the erosion flux density Γ_{ero} from the surface and the losses due to transport out of the considered volume, which are approximately described by the second term in the right-hand side of (9.13) by introducing an average confinement time τ. The change of energy is determined by the power input P_{in} into the region, the power losses P_{surf} to the surface, and the radiation losses represented by the third term in (9.14) where $L_{\mathrm{imp}}(T)$ is the radiation function. The electron density n_{e} can be determined by the charge neutrality condition (9.15). Far away from the radiating zone, the impurity density is assumed here to be zero and, therefore, the plasma ion density n_{i} is simply equal to the electron density $n_{\mathrm{e}}^{\mathrm{o}}$; l is the length of the radiating region along the magnetic field lines. Under steady state conditions, (9.13) yields

$$n_{\mathrm{imp}} = \Gamma_{\mathrm{ero}}\frac{\tau}{l} = n_{\mathrm{e}}c_{\mathrm{s}}Y\frac{\tau}{l} \approx n_{\mathrm{e}}Y \tag{9.16}$$

assuming that the impurity ions are accelerated up to the ion sound speed c_{s} due to friction with the plasma ions, thus taking roughly $l/\tau \simeq c_{\mathrm{s}}$. The erosion flux $\Gamma_{\mathrm{ero}} = n_{\mathrm{e}}c_{\mathrm{s}}Y$ is the product of the ion flux to the surface $\Gamma_{\mathrm{ion}} = n_{\mathrm{e}}c_{\mathrm{s}}$ and the erosion yield Y. Substituting (9.15) into (9.16) gives

$$n_{\mathrm{imp}} = \frac{n_{\mathrm{e}}^{\mathrm{o}}}{1/Y - q_{\mathrm{imp}}} . \tag{9.17}$$

With $P_{\text{surf}} = \gamma_{\text{E}} c_{\text{s}} n_{\text{e}} k_{\text{B}} T_{\text{e}}$ where $\gamma_{\text{E}} \simeq 7$–$10$ is the energy transmission factor (Sect. 8.1), (9.14) defines the electron temperature

$$k_{\text{B}} T_{\text{e}} = \frac{P_{\text{in}} - l\,(n_{\text{e}}^{\text{o}} + q_{\text{imp}} n_{\text{imp}}) n_{\text{imp}} L_{\text{imp}}}{(n_{\text{e}}^{\text{o}} + q_{\text{imp}} n_{\text{imp}}) c_{\text{s}} \gamma_{\text{E}}} \,. \tag{9.18}$$

Aiming for high radiation losses, the ratio

$$\begin{aligned} \gamma_{\text{rad}} &= \frac{l(n_{\text{e}}^{\text{o}} + q_{\text{imp}} n_{\text{imp}}) n_{\text{imp}} L_{\text{imp}}}{P_{\text{in}}} \\ &= \frac{l[n_{\text{e}}^{\text{o}} + q_{\text{imp}} n_{\text{e}}^{\text{o}}/(1/Y - q_{\text{imp}})] n_{\text{e}}^{\text{o}} L_{\text{imp}}/(1/Y - q_{\text{imp}})}{P_{\text{in}}} \\ &\simeq \frac{l(n_{\text{e}}^{\text{o}})^2 L_{\text{imp}} Y}{P_{\text{in}}} \end{aligned} \tag{9.19}$$

should be as large as possible. In the derivation of (9.19), the erosion yield Y is usually small, i.e., $Y \ll 1$. For typical values of $Y = 0.02$, $n_{\text{e}}^{\text{o}} = 10^{20}\,1/\text{m}^3$, $L_{\text{imp}} = 10^{-32}\,\text{Wm}^3$, and $l = 10\,\text{m}$, the "radiated power flux density" $l(n_{\text{e}}^{\text{o}})^2 L_{\text{imp}} Y$ is about $20\,\text{MW/m}^2$. Self-sputtering with sputtering yields close to unity can contribute to the erosion flux leading to increased impurity concentration in the plasma as well as thermal sublimation, which depends on the surface temperature. With respect to the desired radiation in the near-surface region, the erosion of the surface turns out to be a favorable effect. Indeed, additional injection of, for example, noble gases would be required in the case of zero or small erosion of the target plates.

There is a rather subtle balance between the benefits of a high radiation level at the edge and the disadvantages due to radiation losses from the center and fuel dilution, since impurities at the edge might increase the impurity flux into the core region. By an appropriate choice of certain low-Z materials (e.g., C, Ne, or Ar), the corresponding radiation zones can be localized in plasma regions where their radiation functions are at maximum [50]. The radial profiles of low-Z impurities are rather flat or even hollow, in contrast to the peaked profiles of the electrons and deuterons, thus generating a high impurity level at the edge and a low impurity level in the center [25]. Low-Z materials radiate strongly at low plasma temperatures characteristic for the edge region, whereas high-Z materials mainly contribute to the radiation in the plasma center where the plasma temperature is much higher.

9.2 Erosion Phenomena in Fusion Experiments

In addition to the elementary erosion processes (see Sect. 6.5), phenomena observed in present fusion experiments such as plasma disruptions, ELMs, hot spots, dust production, and erosion due to runaway electrons, alpha particle, and charge-exchange neutrals can generate large influxes of impurities into the plasma.

9.2.1 Plasma Disruption

Rapid plasma termination events called disruptions are usually the result of reaching one of the operational limits in tokamaks. Stellarators have no toroidal plasma current, therefore, disruptions do not occur. Here, the plasma extinguishes rather smoothly when the radiation losses are not compensated for. However, prevention of neoclassical tearing modes, which degrade the plasma confinement, is a major challenge. Tearing modes are magnetic islands formed by the topological rearrangement of magnetic field lines through reconnection, while ideal modes can seed neoclassical tearing modes through forced reconnection [330]. Neoclassical tearing modes can be stabilized by driving current inside the islands.

In tokamaks, the thermal quench, i.e., the loss of thermal energy, takes about 1–10 ms, and is followed by a current quench due to the increasing resistivity at lower temperature. The thermal energy divided by the area wetted by the plasma and relating it to its time duration yields the energy flux density. Disruptions can cause significant damage such as deformation of in-vessel structures, short circuits in external supplies due to induced eddy currents, as well as melting and vaporization of wall materials [331].

A vapor cloud above the surface will form as result of the sudden energy deposition due to direct impact of plasma particles from the disrupted plasma. After a short time, almost all plasma particles are completely stopped in that vapor cloud and their kinetic energy is transformed into radiation. The heat flux arriving at the material surface at this stage is determined by the transport of radiation. The vapor shielding effect can significantly reduce the erosion of the targets and, therefore, prolong their lifetime [332].

The modeling of the dynamics and evolution of the vapor shield includes the consideration of the plasma–material and plasma–vapor interactions as well as the simulation of the radiation transport. Three moving boundaries are involved: the vapor front, the receding (liquid) target surface, and the solid/liquid interface [334]. The complex interlinked processes have been studied with two-dimensional radiation–magnetohydrodynamic models such as the HEIGHTS code coupled with the solution of the time-dependent heat conduction equation (A*THERMAL-S code) [335–337] and the FOREV-2 code [338] to assess damage caused by disruption and ELMs. Validation against disruption simulation experiments performed at plasma gun facilities such as TRINITI Troitsk gave more confidence that the modeling covers important aspects of vapor shield properties, and that the basic effects are adequately described. In particular, the stability of the cold and dense region of the vapor shield has been confirmed [338], i.e., turbulence might be neglected.

The strong magnetic field leads to a compression of the vapor cloud closer to the surface, thereby enhancing the shielding action but increasing the layer thickness of the melted material. Such effects as well as the excitation of instabilities (Rayleigh–Taylor and Kelvin–Helmholtz instabilities, $\boldsymbol{E} \times \boldsymbol{B}$ motion effects) in the melt layer are the topics of further investigations. The

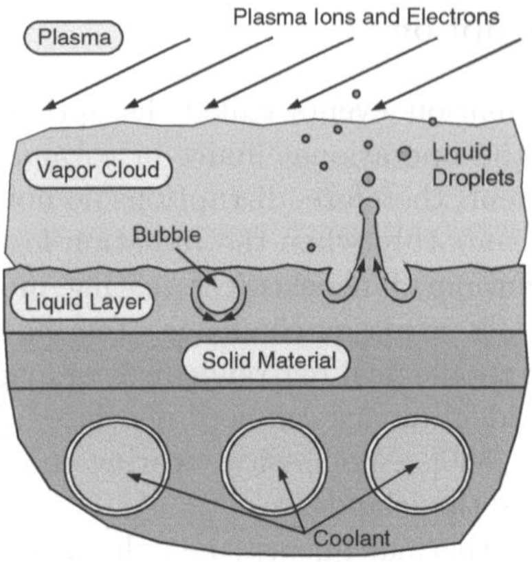

Fig. 9.1. Different interaction zones during large heat loads onto the material, for example, during plasma disruptions or ELMs (see Fig. 1 of [333])

achievement of erosion of melted material is impossible in today's fusion devices and experimental investigations are restricted to plasma gun experiments.

Brittle destruction has been observed to be critical for carbon-based materials [339] and is enhanced under cyclic heat load. Volumetric heating in the bulk produces inner cracks, while surface heating results in crack propagation into the depth.

The erosion of metals is mainly attributed to melt layer loss, where the driving forces are gravitation and the Lorentz force, the latter being able to trigger a pronounced motion of the melted layer. A large fraction of eroded mass (up to 20%) is splashed away by droplets due to formation and boiling of vapor bubbles inside the liquid layer.

Melt flow and droplet splashing during disruptions form surfaces with considerable roughness and drastically change thermophysical properties [340]. Surface irregularities are responsible for hot spot erosion during further plasma operation.

Disruptions with 10–100 MJ/m^2 in 1–10 ms will cause melting and ablation of any material [341]. Therefore, disruption mitigation is required and different methods such as strong gas injection or using so-called killer pellets containing high-Z elements as well as liquid hydrogen and helium jets have been tested in order to increase the radiation losses in the whole volume.

9.2.2 Edge Localized Modes (ELMs)

ELMs are highly non-linear magnetohydrodynamic events and are characterized by a periodic expulsion of particles and thermal energy (with up to 3–10% of the core thermal energy) from the inner region at the separatrix into the edge plasma and finally—parallel along the magnetic field lines—onto the divertor and/or wall surfaces [342–344]. The energy pulse caused by the ELM is usually larger than the plasma energy content in the divertor. Typical values of the ELM energies deposited at the divertor plates are 0.01–0.05 MJ/m^2 for ASDEX-Upgrade, 0.1–0.5 MJ/m^2 for JET and as predicted for ITER 1–5 MJ/m^2 [345]. As a consequence of the high peak load during typically 0.1–1 ms, processes such as melting, evaporation, ejection of clusters and droplets, release of adsorbed or codeposited hydrogen isotopes, and electron emission are initiated due to increased surface temperatures. Long-term effects such as the degradation of the thermophysical properties due to the cyclic heat load are also a matter of concern. The ELM energy losses are determined by the local plasma (pedestal) parameters at the separatrix, which drop at each ELM burst. ELMs contribute strongly to the global power and particle balance of the plasma. However, ELMs could have a significantly favorable application in a fusion reactor, since they could be used to exhaust the He ash out of the confined region, sustaining at the same time a high fusion power production in the core. Investigations on how to control the cycle, duration, and energy content of ELMS are ongoing [346]. Recently, some progress has been made by using pellet (frozen deuterium of about 1 mm in diameter) injection into the pedestal region characterized by strong gradients of the plasma parameters [347].

9.2.3 Runaway Electrons

Since the collision frequency ν in a fully ionized plasma decreases strongly with increasing velocity ($\nu \propto 1/v_{\mathrm{e}}^3$), some of the electrons of the high-velocity tail in the Maxwellian distribution are practically unaffected by friction and, therefore, may be accelerated in the electric field of the tokamak loop voltage. Equating the collisional friction force $F_{\|} = m v_{\mathrm{e}} \nu \simeq -n_{\mathrm{e}} e^4 \ln \Lambda / (2\pi\epsilon_{\mathrm{o}}^2 m_{\mathrm{e}} v_{\mathrm{e}}^2)$ [35] along the magnetic field lines to the force of the externally-induced electric field $F_{\mathrm{E}} = -eE$, a critical electric field E_{c}

$$E_{\mathrm{c}} = \frac{n_{\mathrm{e}} e^3 \ln \Lambda}{2\pi\epsilon_{\mathrm{o}}^2 m_{\mathrm{e}} v_{\mathrm{e}}^2} = \frac{e \ln \Lambda}{4\pi\lambda_{\mathrm{D}}^2} \frac{k_{\mathrm{B}} T_{\mathrm{e}}}{m_{\mathrm{e}} v_{\mathrm{e}}^2/2} \tag{9.20}$$

can be deduced at which runaway starts. Here, λ_{D} is the Debye length, n_{e} is the electron density, T_{e} is the electron temperature, and $\ln \Lambda$ is the Coulomb logarithm. It is obvious that even for small values of the electric field electrons from the tail of the Maxwellian distribution with velocities much larger than the thermal velocity, i.e., $v_{\mathrm{e}} \gg \sqrt{8 k_{\mathrm{B}} T_{\mathrm{e}}/(\pi m_{\mathrm{e}})}$, could overcome the retarding collision force. These electrons can reach energies up to several MeV (20–300 MeV [348]) after having performed several million toroidal revolutions and cause severe damage by impinging on wall elements by the localized high heat load deposition, leading to melting and/or vaporization and the occurrence of thermal stress cracks. The maximum energy is limited either by synchrotron radiation or by the time available for acceleration. It has been estimated that the deposited energy flux densities could reach values up to 80 MJ/m^2 over areas of a few square centimeters [317, 349].

A commonly accepted approximation for the runaway generation rate $\gamma_{\rm RE}$ including the effect of multiple ionic species with an effective charge state q

$$\gamma_{\rm RE} = C \left(\frac{E_{\rm c}}{E}\right)^{3(q+1)/16} n_{\rm e}\nu \exp\left[-\frac{E_{\rm c}}{4E} - \sqrt{(q+1)\frac{E_{\rm c}}{E}}\right] \quad (9.21)$$

has been proposed based on the solution of the Fokker–Planck equation by perturbation methods [350]. C is a weak function of q in the range between 0.13 and 0.43 for q values between 1 and 10 [348].

Unfortunately, beams of runaway electrons can be generated also following the onset of plasma disruptions reaching values of up to 50% of the initial plasma current [351]. The drop in plasma temperature after the start of disruption causes an increase in resistance that leads to a loop voltage rise up to several kilovolts from its normal value of about 1 V. The plasma current drops and the collapsing poloidal magnetic field inductively sustains the high loop voltage for a certain time [348].

In contrast to common surface effects during plasma disruptions, generated runaway electrons have a large penetration capability resulting in a significant heating of cooling structures and tubes underneath the actual target plates. Local melting and damage induced by rising temperature gradients may be the consequences. Critical areas at the wall components must be protected from runaway electrons, for example, by carbon tiles.

The energy losses of electrons penetrating matter are predominantly due to ionization and bremsstrahlung [352]. The strong magnetic field in fusion devices, which is aligned nearly parallel to the surface, returns reflected electrons back to the surface, therefore, leading to an increase of energy deposition by a factor of about 3. On the other hand, the penetration depth of the electrons is significantly reduced by the action of the magnetic field, since the electrons are forced to gyrate inside the material [353,354].

9.2.4 Erosion by Energetic Alpha Particles

The alpha particles generated in fusion reactions should transfer their energy (3.5 MeV) to the plasma particles for self-sustaining operation of the burning fusion plasma. Firstly, the electrons are heated, then in turn the plasma ions through electron–ion collisions. The slowed down alpha particles (helium ash) have to be removed from the core region to prevent fuel dilution and radiation losses. On the other hand, energetic alpha particles may leave the plasma due to poor confinement before they could fully transfer their energy. Since the alpha particles supply several hundred MW in a fusion reactor, the loss of even a small fraction of them leads to large heat loads and damage localized at the point of impact onto wall components.

The discreteness of toroidal field coils, resulting in the occurrence of toroidal field ripple, can lead to the loss of energetic helium nuclei before they are thermalized. In addition, anomalous transport due to collective instabilities can also contribute to the loss of alpha particles. Fast particles may be efficiently trapped in banana orbits, which depend sensitively on the magnetic field configuration. By crossing regions with ripple wells, collisionless drifts occur resulting in different channels for loss.

Striking the surface of wall components, the alpha particles with their rather small mean free path (in comparison with neutrons) heat a rather thin surface layer. The localized heat load leads to melting and evaporation in these hot spots. More experimental and modeling work is required to estimate the effect of hot spot generation by lost alpha particles on the lifetime of plasma-facing components.

9.2.5 Hot Spots or Carbon "Blooming"

In high-power fusion experiments often a rapid increase of impurity concentration is observed, usually leading to a plasma termination [275, 355–357]. This effect is related to the occurrence of so-called hot spots, i.e., small localized regions with high surface temperatures each extending over an area up to $1\,\text{cm}^2$. The emission of material atoms is mainly due to thermal sublimation.

Local emission of electrons could lead to a drop of potential in the electric sheath (see Sect. 7), which is followed by an enhancement of hot electron flow from the plasma to the wall. This additional heat load results in higher surface temperature and, subsequently, higher thermionic emission of electrons [216, 358]. This self-enhancing loop is stopped when the electric field near the surface becomes zero and a further increase of electron current from the surface is no longer possible (see Chap. 7.4). A small dip in the potential distribution (on the order of the energy of the emitted electrons) appears near the surface and prevents the penetration of additional thermal electrons into the plasma. The wall surface is cooled by heat conduction into the bulk, by thermal radiation, by emission of thermal electrons, and by evaporation of the wall material. It has been found that the steady state condition of equal power input by plasma ions and electrons and power losses as mentioned can be fulfilled for different surface temperatures. Thus, areas with different temperatures at the same surface are predicted [216,359]. However, relatively high surface temperatures ($\geq 3200\,\text{K}$ in the case of carbon) are required for sufficient thermionic emission currents, while the inset of carbon blooming has been observed already at temperatures of about 2800 K. Furthermore, with usual grazing incidence of the magnetic field lines the emitted electrons have a good chance to return immediately to the surface during the gyration, therefore, reducing their influence on the potential distribution.

To explain the hot spot phenomenon, an additional heating mechanism has been proposed in [360, 361]. If the ionization of emitted material atoms occurs inside or just outside the electric sheath, the ionized target ions are accelerated back to the surface gathering a considerable amount of energy on their way through the sheath. This energy is deposited in the direct vicinity and heats the surface. Of course, this mechanism can only play a role if the ionization length is on the order of the sheath thickness. It is noteworthy that in the case of grazing incidence of the magnetic field lines the sheath thickness is about the plasma ion gyro-radius, therefore, much larger than the Debye length (see Sect. 7.5).

9.2.6 Flake and Dust Production

The formation of dust and flakes has been observed in many fusion experiments [362–365]. Their composition represents the materials used in the device. Formation mechanisms appear to be fatiguing and thermal overloading of wall components, arcing, flaking of redeposited layers, which are mechanically weak and have bad thermal conduction, and the loosening and flaking of wall conditioning films after long exposure of the vessel walls to air [366].

These small particles (1–30 µm) can grow under certain conditions analogous to reactive plasmas [367]. In the plasma, the dust becomes charged and can be transported inside the vacuum vessel [368, 369]. The problems are its high chemical reactivity with steam and air, possible toxicity, and radiological hazard [5].

9.2.7 Erosion by Charge-Exchange Neutrals

Low-energy neutrals (usually molecules) emitted from wall elements are subjected with a certain probability to charge-exchange reactions with energetic ions in the edge plasma. For low edge temperatures, the first step is molecular break-up into two Franck–Condon atoms with a few electron volts of kinetic energy, while, at higher edge temperatures, direct ionization to form a molecular ion is the more likely first step, rapidly followed by break-up. Recent measurements have shown that recycled molecules may be in highly-excited vibrational or rotational states, which facilitate the dissociation process, releasing atoms with energies as low as 0.3 eV [370]

$$e + H_2 \rightarrow H_2^+ + 2e \; . \tag{9.22}$$

The rate coefficients of ionization and charge exchange of deuterium ions are approximated by the formulae [371]

$$\langle \sigma v \rangle_{\mathrm{iz}} = 0.73 \times 10^{14} \sqrt{T_{\mathrm{e}}[\mathrm{eV}]}\, \frac{\exp(-13.6/T_{\mathrm{e}}[\mathrm{eV}])}{1 + 0.01 T_{\mathrm{e}}[\mathrm{eV}]} \quad [\mathrm{m}^3/\mathrm{s}] \tag{9.23}$$

and

$$\langle \sigma v \rangle_{\mathrm{cx}} = 10^{14}\, T_{\mathrm{i}}^{0.3} \quad [\mathrm{m}^3/\mathrm{s}] \; . \tag{9.24}$$

Because the charge-exchange rate coefficient is about 2 to 3 times that of the ionization rate coefficient in plasmas with $T_{\mathrm{i}} \simeq T_{\mathrm{e}}$, a neutral hydrogen atom that enters the plasma will probably result in a charge-exchange hydrogen atom

$$H^+ + H^o \rightarrow H^o + H^+ \,. \tag{9.25}$$

The formed energetic neutrals are not confined by the magnetic and electric fields and cause physical sputtering at the places of impact. Experimental results show clearly the importance of charge-exchange sputtering [372]. Further studies on ASDEX-Upgrade [196, 373] identified charge-exchange sputtering at the wall as the main source of impurities in the central plasma. It is important to keep the recycling neutrals away from the main plasma, e.g., operating with complete recycling in the divertor region.

9.2.8 Erosion by Arcing

Arcing is commonly observed in fusion experiments and occurs predominantly during the current rise and shutdown phases when the plasma is unstable. If the potential drop in the electric sheath at the plasma-facing surface exceeds the arc voltage threshold of about 10–30 V, a localized discharge, an arc, can be initiated. These rather short events provide, nevertheless, a high current density resulting in a rapid local heating and evaporation of material. Not only are atoms emitted, but also small clusters with sizes up to a micrometer are ejected by the high-pressure vapor produced in the arc discharge [374–376].

An arc burns usually between the cathode spot on the surface and the plasma acting as an anode, and is called unipolar, since only one solid electrode is involved. The metal atoms for the arc plasma are provided by the cathode spot. After ignition, the arc moves randomly over the surface. Having a magnetic field, retrograde motion is observed. The arc moves opposite to the Lorentz force $\boldsymbol{j} \times \boldsymbol{B}$ with $\boldsymbol{j}$ being the current density in the arc. The retrograde motion seems to be controlled by plasma jets as the consequence of instabilities in the plasma of the arc. New spots are formed exactly in the jet direction. Measurements with a time resolution of 50 ns revealed that the motion of an arc spot should be rather understood as a jump-like generation, and extinction of spots that are characterized by strong inner dynamics resulting in fragmentation and merging. The effect is superimposed on the random motion of the spots, caused by the dynamics of inner fragments [377].

In various fusion experiments, the consequences of arc erosion were observed on graphite and metal surfaces after plasma exposure: tracks of about 10 µm in depth, 10–100 µm in width and 5–10 mm in length, resulting in 10^{17}–10^{18} atoms of eroded material per arc event [375]. In order to avoid the initiation of arcs, the plasma has to be stable and the potential difference across the sheath, hence the electron temperature near the plasma-facing components has to be low.

9.2.9 Non-Linear Erosion due to Impurities

Several processes such as preferential sputtering, segregation, and recoil mixing leading to material modification have already been considered in Sect. 6.3. In the environment of fusion plasmas, some additional effects arise due to plasma impurities, which are always present at a certain level. Erosion investigations of material exposed to such a "dirty" plasma have been performed mostly with the assumption that the measured erosion is the result of the summarized sputtering due to the various plasma species. It was shown in experiments (JET [378], TEXTOR [379], ASDEX-Upgrade [173], PISCES [380,381]) that even at very low impurity concentrations of about a few percents in the plasma, this assumption about the linear superposition of the sputtering yields cannot be maintained.

During the bombardment of the target surface, the impurity ions are implanted in the near-surface region. They alter the surface composition and, subsequently, the sputtering yields during the exposure. Usually, the erosion of the target is reduced due to the presence of plasma impurities such as carbon ions. Despite the fact that the sputtering of the target material is more effective by the heavier carbon ions than by the plasma ions (for example by deuterium), the protection by the growing deposition layers prevails (Fig. 9.2).

Sputtering, as the collisional removal of surface atoms by energetic impinging particles, depends strongly on the composition profile in the target. The key parameters are the impurity concentration in the plasma, which defines the amount of material being deposited, and the plasma temperature, which determines the erosion rate. An analytical model has been developed, which can predict the erosion behavior of such systems [172,176,382].

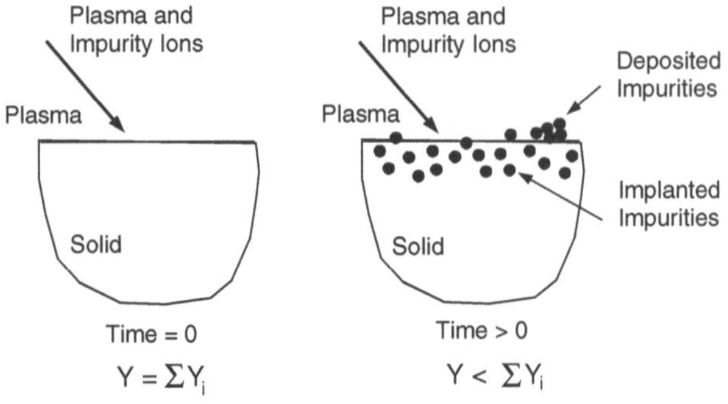

Fig. 9.2. Protection effect due to implantation and deposition of impurities such as carbon or beryllium. The resulting sputtering yield of the solid material is, shortly after the exposure starts, a function of the transient surface composition. The total sputtering yield is usually much smaller than the yield obtained by simply superpositioning the individual contributions

Assuming, a target of the material "M" is exposed to a plasma with only two species, "i" and "I", where the index "I" indicates the plasma impurity (for example carbon) and the index "i" the plasma ions (for example deuterium ions). The amount of material deposited or eroded can be expressed by the thickness d

$$d = \frac{n^{\mathrm{I}}}{n_{\mathrm{o}}^{\mathrm{I}}} + \frac{n^{\mathrm{M}}}{n_{\mathrm{o}}^{\mathrm{M}}} \tag{9.26}$$

where $n_{\mathrm{o}}^{\mathrm{I}}$ and $n_{\mathrm{o}}^{\mathrm{M}}$ denote the atomic densities for the materials "I" and "M", respectively. The change of eroded amount of the target material n^{M}, given as a surface concentration (in particles/m^2),

$$\frac{\mathrm{d}n^{\mathrm{M}}}{\mathrm{d}t} = -f_{\mathrm{i}}\Gamma_{\mathrm{e}}\hat{Y}_{\mathrm{i}\to\mathrm{M}} - f_{\mathrm{I}}\Gamma_{\mathrm{e}}\hat{Y}_{\mathrm{I}\to\mathrm{M}} \tag{9.27}$$

is determined by the erosion due to plasma ions and impurity ions. The concentration of the two plasma species with respect to the electron density must satisfy the quasineutrality condition in the plasma

$$f_{\mathrm{i}} + q_{\mathrm{I}} f_{\mathrm{I}} = 1 \tag{9.28}$$

where q_{I} is the charge state of the impurity ions. A similar equation can be written for the amount of deposited material

$$\frac{\mathrm{d}n^{\mathrm{I}}}{\mathrm{d}t} = f_{\mathrm{I}}\Gamma_{\mathrm{e}}(1-\hat{R}) - f_{\mathrm{i}}\Gamma_{\mathrm{e}}\hat{Y}_{\mathrm{i}\to\mathrm{I}} - f_{\mathrm{I}}\Gamma_{\mathrm{e}}\hat{Y}_{\mathrm{I}\to\mathrm{I}}\,. \tag{9.29}$$

The first term on the right-hand side of (9.29) describes the deposition of the incoming impurities, which stick to the surface with a probability $(1-\hat{R})$, where $\hat{R}$ is the reflection coefficient of the impurity ions with respect to the target.

The sputtering yields $\hat{Y}$ as well as the reflection coefficient $\hat{R}$ depend on the surface composition and can be expressed in a first approximation by the formulae (illustrated in Fig. 9.3)

$$\begin{aligned}
\hat{Y}_{\mathrm{i}\to\mathrm{M}}(t) &= Y_{\mathrm{i}\to\mathrm{M}}\left(1 - \frac{n^{\mathrm{I}}(t)}{n_{\mathrm{o}}^{\mathrm{I}}\Delta_{\mathrm{M}}}\right)\\
\hat{Y}_{\mathrm{I}\to\mathrm{M}}(t) &= Y_{\mathrm{I}\to\mathrm{M}}\left(1 - \frac{n^{\mathrm{I}}(t)}{n_{\mathrm{o}}^{\mathrm{I}}\Delta_{\mathrm{M}}}\right)\\
\hat{Y}_{\mathrm{i}\to\mathrm{I}}(t) &= Y_{\mathrm{i}\to\mathrm{I}}\,\frac{n^{\mathrm{I}}(t)}{n_{\mathrm{o}}^{\mathrm{I}}\Delta_{\mathrm{I}}}\\
\hat{Y}_{\mathrm{I}\to\mathrm{I}}(t) &= Y_{\mathrm{I}\to\mathrm{I}}\,\frac{n^{\mathrm{I}}(t)}{n_{\mathrm{o}}^{\mathrm{I}}\Delta_{\mathrm{I}}}\\
\hat{R}(t) &= R_{\mathrm{I}\to\mathrm{M}} - (R_{\mathrm{I}\to\mathrm{M}} - R_{\mathrm{I}\to\mathrm{I}})\,\frac{n^{\mathrm{I}}(t)}{n_{\mathrm{o}}^{\mathrm{I}}\Delta_{\mathrm{R}}}
\end{aligned} \tag{9.30}$$

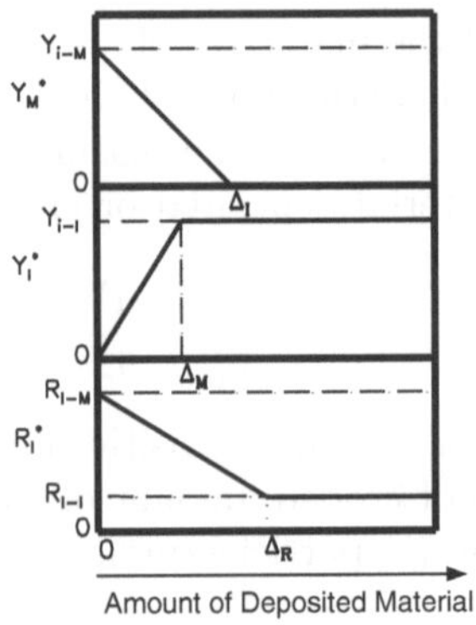

Fig. 9.3. Approximated sputtering yields and particle reflection coefficient as a function of the deposited depth of material "I"

assuming initially a linear increase of the impurity concentration at the surface. For higher amounts of deposited material, i.e., $n^{\mathrm{I}}(t) \geq n_{\mathrm{o}}^{\mathrm{I}} \Delta$, the parameters are assumed to saturate as one would expect, since both processes (sputtering and reflection) are restricted to the near-surface region of half-infinite materials. The various characteristic thicknesses $\Delta = n(t)/n_{\mathrm{o}}$, which specify the transition from linear dependence to saturation are usually ordered as $\Delta_{\mathrm{I}} \leq \Delta_{\mathrm{M}} \leq \Delta_{\mathrm{R}}$. Due to the discontinuous approximation (9.30), the set of equations (9.27,9.29) is obtained as

$$
\begin{aligned}
n^{\mathrm{M}}(t) = {} & \left(\frac{C_3\, C_4}{C_1} - C_2\right) t - \frac{C_3\, C_4}{(C_1)^2}\left[1 - \exp(-C_1\, t)\right] \\
& + \frac{C_4 n^{\mathrm{I}}(t=0)}{C_1}\left[1 - \exp(-C_1\, t)\right] + n^{\mathrm{M}}(t=0)
\end{aligned}
\tag{9.31}
$$

$$
n^{\mathrm{I}}(t) = \frac{C_3}{C_1}\left[1 - \exp(-C_1\, t)\right] + n^{\mathrm{I}}(t=0)\exp(-C_1\, t) \tag{9.32}
$$

for $n^{\mathrm{I}}(t) \geq n_{\mathrm{o}}^{\mathrm{I}} \Delta_{\mathrm{I}}$. The coefficients used in (9.31) are given by

$$
\begin{aligned}
C_1 &= \frac{f_{\mathrm{i}} \Gamma_{\mathrm{e}} Y_{\mathrm{i} \to \mathrm{I}}}{n_{\mathrm{o}}^{\mathrm{I}} \Delta_{\mathrm{I}}} + \frac{f_{\mathrm{I}} \Gamma_{\mathrm{e}} Y_{\mathrm{I} \to \mathrm{I}}}{n_{\mathrm{o}}^{\mathrm{I}} \Delta_{\mathrm{I}}} - \frac{f_{\mathrm{I}} \Gamma_{\mathrm{e}} (R_{\mathrm{I} \to \mathrm{M}} - R_{\mathrm{I} \to \mathrm{I}})}{n_{\mathrm{o}}^{\mathrm{I}} \Delta_{\mathrm{R}}} \\
C_2 &= f_{\mathrm{i}} \Gamma_{\mathrm{e}} Y_{\mathrm{i} \to \mathrm{M}} + f_{\mathrm{I}} \Gamma_{\mathrm{e}} Y_{\mathrm{I} \to \mathrm{M}} \\
C_3 &= f_{\mathrm{I}} \Gamma_{\mathrm{e}} (1 - R_{\mathrm{I} \to \mathrm{M}}) \\
C_4 &= C_2 / (n_{\mathrm{o}}^{\mathrm{I}} \Delta_{\mathrm{M}})\,.
\end{aligned}
\tag{9.33}
$$

If $n_{\mathrm{o}}^{\mathrm{I}} \Delta_{\mathrm{I}} < n^{\mathrm{I}}(t) < n_{\mathrm{o}}^{\mathrm{I}} \Delta_{\mathrm{R}}$, then the following expressions for the coefficients C_1 and C_2 have to be used

$$
\begin{aligned}
C_1 &= -\frac{f_{\mathrm{I}} \Gamma_{\mathrm{e}} (R_{\mathrm{I} \to \mathrm{M}} - R_{\mathrm{I} \to \mathrm{I}})}{n_{\mathrm{o}}^{\mathrm{I}} \Delta_{\mathrm{R}}} \\
C_3 &= f_{\mathrm{I}} \Gamma_{\mathrm{e}} (1 - R_{\mathrm{I} \to \mathrm{M}}) - f_{\mathrm{i}} \Gamma_{\mathrm{e}} Y_{\mathrm{i} \to \mathrm{I}} - f_{\mathrm{I}} \Gamma_{\mathrm{e}} Y_{\mathrm{I} \to \mathrm{I}}\,.
\end{aligned}
\tag{9.34}
$$

The initial conditions have to be changed consistently to ensure that the solutions in the different regions of validity match each other. If $n^{\mathrm{I}}(t) \geq n_{\mathrm{o}}^{\mathrm{I}} \Delta_{\mathrm{M}}$, then $n^{\mathrm{M}}(t)$ =const. and

$$\Delta n^{\mathrm{I}} = \left[f_{\mathrm{I}} \Gamma_{\mathrm{e}} (1 - R_{\mathrm{I}\to\mathrm{I}}) - f_{\mathrm{i}} \Gamma_{\mathrm{e}} Y_{\mathrm{i}\to\mathrm{I}} - f_{\mathrm{I}} \Gamma_{\mathrm{e}} Y_{\mathrm{I}\to\mathrm{I}} \right] \Delta t \,. \tag{9.35}$$

In the case of "I"="M", i.e., the plasma impurities and target material are equal, the complicated non-linear dependence is reduced to a linear solution

$$d(t) = \frac{(C_3 - C_2)t}{n_{\mathrm{o}}^{\mathrm{I}}} \tag{9.36}$$

where the plasma concentration f_{I} defines whether erosion or deposition prevails. With

$$f_{\mathrm{I}} < f_{\mathrm{I}}^* = \frac{Y_{\mathrm{i}\to\mathrm{I}}}{1 - R_{\mathrm{I}\to\mathrm{I}} + q_{\mathrm{I}} Y_{\mathrm{i}\to\mathrm{I}} - Y_{\mathrm{I}\to\mathrm{I}}} \tag{9.37}$$

the target material will be eroded, whereas for $f_{\mathrm{I}} > f_{\mathrm{I}}^*$ deposition will prevail.

Having two different materials, the erosion/deposition behavior can change during the bombardment, i.e., a transition from an erosion phase to a deposition phase or vice versa is possible. At the beginning of the bombardment, material will be deposited if

$$f_{\mathrm{I}} < f_{\mathrm{I}}^* = \frac{Y_{\mathrm{i}\to\mathrm{M}}}{(1 - R_{\mathrm{I}\to\mathrm{M}}) n_{\mathrm{o}}^{\mathrm{M}} / n_{\mathrm{o}}^{\mathrm{I}} + q_{\mathrm{I}} Y_{\mathrm{i}\to\mathrm{M}} - Y_{\mathrm{I}\to\mathrm{M}}} \,. \tag{9.38}$$

In the opposite case $f_{\mathrm{I}} > f_{\mathrm{I}}^*$, the target will be eroded at time $t \geq 0$.

Setting the derivation of $d(t)$ to zero, a critical time t^*

$$t^* = -\frac{1}{C_1} \ln \left[\frac{C_1 C_2 / C_3 - C_4}{C_1 n_{\mathrm{o}}^{\mathrm{M}} / n_{\mathrm{o}}^{\mathrm{I}} - C_4} \right] \tag{9.39}$$

can be calculated, at which a maximum or minimum in the time dependence, i.e., a transition between erosion and deposition phases, occurs, while $n^{\mathrm{I}}(t) \leq \Delta_{\mathrm{I}} n_{\mathrm{o}}^{\mathrm{I}}$. As seen from (9.39), the condition

$$0 < \left[\frac{C_1 C_2 / C_3 - C_4}{C_1 n_{\mathrm{o}}^{\mathrm{M}} / n_{\mathrm{o}}^{\mathrm{I}} - C_4} \right] < 1 \tag{9.40}$$

must be satisfied for the existence of a minimum or maximum in the time dependence.

The impurity concentration sensitively influences the time behavior. Low concentrations lead finally to a stationary erosion phase. For higher concentrations, deposition prevails over erosion. The interplay of deposition and erosion also depends of course on the plasma temperature. Four different cases of deposition/erosion behaviors during the bombardment can be distinguished. Using the conditions (9.40) and (9.38), the regions of validity can be defined and are shown schematically in Fig. 9.4 for the case of tungsten.

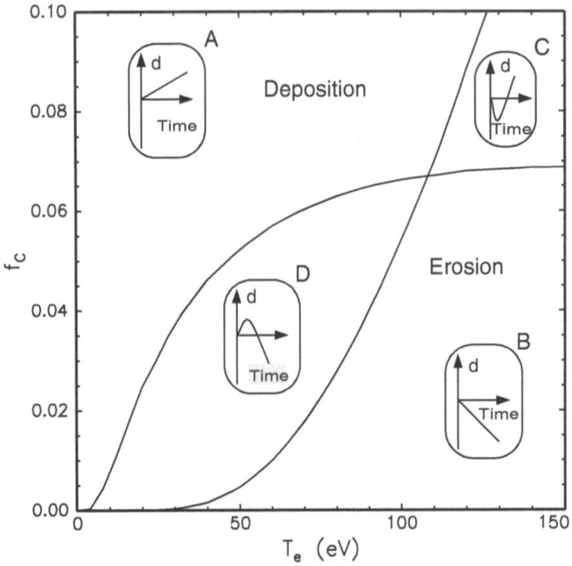

Fig. 9.4. The critical value f_I^* according to (9.40) and (9.38) as a function of the plasma temperature T_e for a tungsten target and triply-charged carbon ions as impurities in a deuterium plasma. The four possible time developments (A–D) are schematically indicated in the figure in their regions of validity

Transitions from an erosion phase to a deposition phase, and vice versa, have been observed during the bombardment in both analytical and computer calculations. As expected, for low plasma temperature, i.e., low erosion, and high impurity concentrations, i.e., high deposition, more and more impurity material will be deposited during the bombardment (case A, Fig. 9.4). On the other hand, for high plasma temperature and low impurity concentrations the target will be eroded just from the beginning of the bombardment (case B).

The cases C and D represent more complex situations. In case C, the process starts with an erosion phase. The plasma temperature and thus the energy of the incident plasma ions is high enough to sputter tungsten at the beginning. During the bombardment more and more impurity ions (here carbon) are deposited, which cannot be fully removed. Consequently, the sputtering of the W atoms from the surface decreases (see also Fig. 9.3) and with higher fluences deposition prevails. The deposited carbon atoms act as a protection for the tungsten target and reduces the resulting erosion significantly up to zero, and even deposition occurs later on. This protection effect has been confirmed in experiments. In addition, with a higher surface concentration of carbon the sticking of the carbon ions becomes more efficient, since the reflection coefficient for initially pure tungsten is larger than that for the evolving tungsten/carbon mixture.

In case D, the plasma temperature is too low to remove the incoming carbon, and deposition occurs. The amount of the deposited carbon increases and, subsequently, the erosion of these atoms increases as well, since the sputtering yield is proportional to the surface concentration according to (9.30). A situation is established, where the amount of deposited atoms can be eroded during each time interval. The carbon concentration at the surface now stays constant. At this time, the target, i.e., the carbon and the tungsten atoms, are eroded due to impact of the plasma and impurity ions. The tungsten erosion is almost fully caused by the heavier carbon ions.

The main point is that even with constant plasma parameters the erosion of the exposed target changes during the bombardment due to a change in the surface composition. The time that is necessary to achieve steady state behavior is very long in some cases, dependent on the plasma parameters. Under the condition of today's fusion experiments, they are on the order of the discharge time.

This model is restricted to collisional processes; diffusion, and segregation effects are not included. Chemical erosion could be considered in this model by adding to the physical sputtering yield the chemical sputtering yield. The W/C system reveals many interesting features and has been investigated in detail in experiments [383] and by computer simulations [175].

9.3 Impurity Transport

To have a large source of impurities is only one side of the problem. Whether the impurity concentration in the plasma core of fusion experiments is affected by this source, and to which degree, is determined by transport processes.

Even with respect to the lifetime of components, a large impurity source is not always critical. A favorable transport of impurities back to the place of origin would result in a small net effect. Using the toroidal symmetry, a redeposition to the same radial position would already be sufficient (see Sect. 12.4.2).

Almost all impurity atoms leave the surface as neutrals despite the large variety of different erosion processes. Collisions with other neutrals can often be neglected due to the low neutral density in the edge plasma. Therefore, the main loss process for neutrals, which are penetrating the plasma, is ionization by electron impact. Processes, such as three-body and dissociative recombination, photoionization and radiative recombination, charge transfer, autoionization, and dielectronic recombination can also play a significant role under certain plasma conditions (Sect. 9.3.2).

Atoms eroded as neutrals from the plasma-facing surfaces are released with a certain angular and energy distribution. They move along straight trajectories in the plasma until they become ionized. Only in the case where the neutrals are ionized near the source location, can the influx rate be inferred simply from the measured line emission intensity integrated over the entire ionization/excitation region. In other situations, where the optics do not collect all the photons or not all the atoms are ionized before leaving the observed space, the knowledge about the spatial distribution of the neutrals is necessary. The spatial position of their ionization depends on the initial energy, the plasma parameters, and on the statistical nature of the ionization process itself. Having a source, i.e., the flux of eroded atoms and a loss term, i.e., the ionization, a steady state density distribution of the neutral impurities in front of the surface is established. Analytic expressions of the spatial distribution of neutrals have been derived in [384] for the case of a narrow beam injection of impurities. The spatial distribution of neutrals for different angular and energy functions of emission can be derived analytically for homogeneous sources as well as for local sources and are given in [385] (see Sect. 9.3.1).

Once ionized, the transport of impurities is inhibited perpendicular to the magnetic field lines by the gyro-motion. A complete theoretical understanding of the transport processes is not yet available for impurities or any other particle and energy flows in a fusion plasma, and no reliable predictions can be made [73]. Only by assuming a rather large diffusion coefficient ($D_\perp = 0.1$–$1\,\mathrm{m^2/s}$) and an inward directed drift velocity (pinch velocity), can the form of the measured profiles be described. The estimated radial transport based on collisions is several orders of magnitude too low as observed in the experiments. The theory of neoclassical transport, caused by Coulomb collision in toroidal magnetic field geometry, is well developed, but not able to explain many features of the observed transport phenomena. In most cases, the transport is dominated by "anomalous" transport, e.g., transport driven by plasma turbulence. While a plasma is able to excite a variety of different instabilities, i.e., perturbations of the electric and magnetic fields that can lead to drift motion across the magnetic field, one of the liable candidates is electrostatic fluctuation resulting in an unstable drift motion of particles.

The parallel transport is fast but irrelevant in the confined region. It becomes, however, important in the SOL. The force on an impurity ion along the magnetic field is given by [59, 73]

$$F_{\mathrm{total}} = -\frac{1}{n}\frac{\partial p}{\partial s} + m\,\frac{(v_{\mathrm{pl}} - v)}{\tau_{\mathrm{s}}} + q\,e\,E + \alpha_{\mathrm{e}}\,\frac{\partial T_{\mathrm{e}}}{\partial s} + \beta_{\mathrm{i}}\,\frac{\partial T_{\mathrm{i}}}{\partial s} \tag{9.41}$$

owing to pressure gradient, friction with the streaming plasma ions moving with v_{pl}, electric field, and temperature gradients, respectively.

The complexity of impurity transport in the edge plasma can only be addressed by multi-fluid two-dimensional SOL codes, which have to be benchmarked against existing experiments [53, 327]. In these codes, assumptions with regard to the anomalous terms are supposed, i.e., diffusion, drift, and viscosity are made. The geometry of the experiment, the magnetic field configuration, and the dynamics of neutrals (using neutral transport codes) can be very realistically taken into account in the simulations. By taking measured plasma parameters as a background, an efficient analysis of impurity transport, especially near the material surfaces, is possible by using Monte Carlo methods (Sect. 9.3.9).

9.3.1 Spatial Distributions of Neutrals

For many processes, the angular distribution of emission obey a cosine law—also called Lambert's law. It states that the number of particles emitted into a solid angle interval $(\theta + \mathrm{d}\theta, \varphi + \mathrm{d}\varphi)$ (Fig. 9.5) is proportional to

$$\cos\theta \, \sin\theta \, \mathrm{d}\theta \, \mathrm{d}\varphi \, . \tag{9.42}$$

The derivation is relatively short and usually not given in textbooks, but it shows some aspects which might be of interest in various applications. Having a certain volume consisting of a large number of homogeneously distributed point sources of isotropic emission would also result in an isotropic emission for the whole volume element—just by superposition (if the distance between the volume element and the detector is sufficiently large). To obtain the cosine law, an additional condition is required. Introducing a certain free path length l_p of the emitted particles (or radiation) in the material, only those particles are now counted under a certain emission angle θ, which are generated no deeper than $z \leq l_\mathrm{p} \cdot \cos\theta$ (Fig. 9.6). Thus, the effective volume, which is proportional to the number of emitted particles, is reduced to

$$\mathrm{d}V' = l_\mathrm{p} \cos\theta \, \mathrm{d}S \tag{9.43}$$

in comparison to the volume $\mathrm{d}V = l_\mathrm{p} \, \mathrm{d}S$ (Fig. 9.6). A cosine angular distribution results, if the following conditions are fulfilled:

1. All processes of generation and scattering are characterized by isotropy.
2. The free path in the material l_p is small with respect to the surface element ΔS, i.e., $l_\mathrm{p} \ll \sqrt{\Delta S}$.
3. The surface should be smaller than the detector surface so that the position of emission is not important by counting the particles under a certain direction of emission.

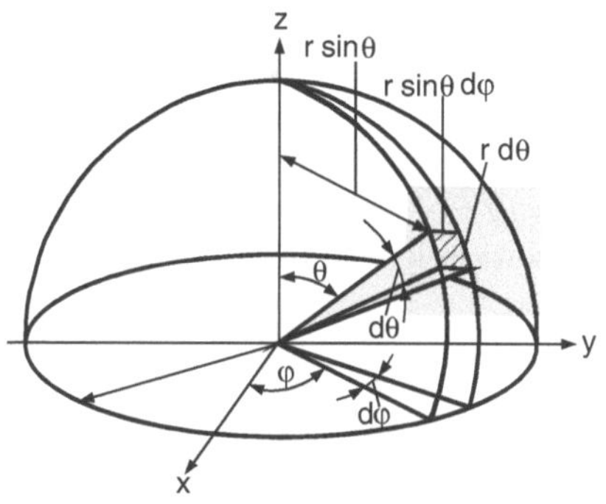

Fig. 9.5. Geometry of emission and notation of angles ($\mathrm{d}S = r^2 \sin\theta \mathrm{d}\theta \mathrm{d}\varphi$)

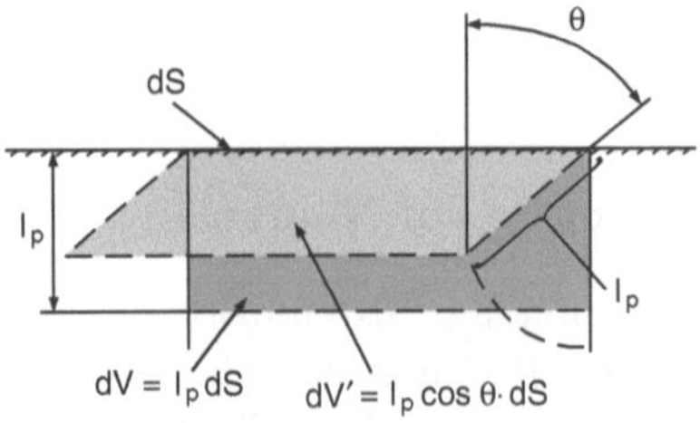

Fig. 9.6. The cosine law of emission

In the case of large emitting surfaces, an integration over all small surface elements, each emitting with a cosine, has to be performed. While the last two conditions are non-critical, often the first assumption is not met. For example, backscattering at high energies is determined by only several strong collisions in the material, and the angular distribution of emission shows a strong anisotropy.

The number of particles sputtered from a surface element $\mathrm{d}S$ during a time interval $\mathrm{d}t$ is $\Gamma\,\mathrm{d}S\,\mathrm{d}t$, where Γ is the flux of the sputtered atoms. Taking a cosine distribution as the angular distribution of the emission process, one obtains for the number of particles emitted into the solid angle $\mathrm{d}\Omega$

$$\Gamma\,\mathrm{d}S\,\mathrm{d}t\,\frac{\cos\theta}{\pi}\,\mathrm{d}\Omega \tag{9.44}$$

with θ being the angle with respect to the surface normal. During the same time interval $\mathrm{d}t$, the moving particles form the volume $\mathrm{d}\Omega\,r^2\,\mathrm{d}r = \mathrm{d}\Omega\,r^2\,v\mathrm{d}t$, where v is the velocity of the sputtered particles. Following, the neutral density $n(r,\theta)$ can be calculated as

$$n(r,\theta) = \frac{\Gamma\,\mathrm{d}S\,\mathrm{d}t\,\cos\theta\,\mathrm{d}\Omega}{\pi\,r^2\,v\,\mathrm{d}t\,\mathrm{d}\Omega} = \frac{\Gamma\,\mathrm{d}S\,\cos\theta}{\pi\,r^2\,v} \tag{9.45}$$

where r is the distance to the point of emission. Let us assume that all particles are ionized after traveling a certain distance λ_{iz} in the plasma. By integration over the surface S, with a dimension larger than the ionization length ($\sqrt{S} \gg \lambda_{\mathrm{iz}}$), the density of the neutral impurities $n(z)$ for $z \le \lambda_{\mathrm{iz}}$ becomes

$$n(z) = \int\limits_{\varphi}\int\limits_{r'} \frac{\Gamma\,\cos\theta}{\pi\,r^2\,v}\,r'\,\mathrm{d}\varphi\,\mathrm{d}r' = \frac{\Gamma z}{v\pi}\int\limits_0^{2\pi}\mathrm{d}\varphi\int\limits_z^{\lambda_{\mathrm{iz}}}\frac{\mathrm{d}r}{r^2} = \frac{2\Gamma}{v}\left(1-\frac{z}{\lambda_{\mathrm{iz}}}\right) \tag{9.46}$$

with $z = r\cos\theta$, $r' = \sqrt{r^2-z^2}$, $\mathrm{d}r' = r\mathrm{d}r/\sqrt{r^2-z^2}$ and thus $r'\mathrm{d}r' = r\mathrm{d}r$. Using (9.46), the ionization length distribution can be taken into account by the integration

$$\begin{aligned} n(z) &= \int\limits_z^{\infty} \frac{2\Gamma}{v}\left(1-\frac{z}{\lambda'_{\mathrm{iz}}}\right)\frac{\exp\left(-\lambda'_{\mathrm{iz}}/\lambda_{\mathrm{iz}}\right)}{\lambda_{\mathrm{iz}}}\,\mathrm{d}\lambda'_{\mathrm{iz}} \\ &= \frac{2\Gamma}{v}\left[\exp\left(-\frac{z}{\lambda_{\mathrm{iz}}}\right) - \frac{z}{\lambda_{\mathrm{iz}}}E_1\left(\frac{z}{\lambda_{\mathrm{iz}}}\right)\right] \end{aligned} \tag{9.47}$$

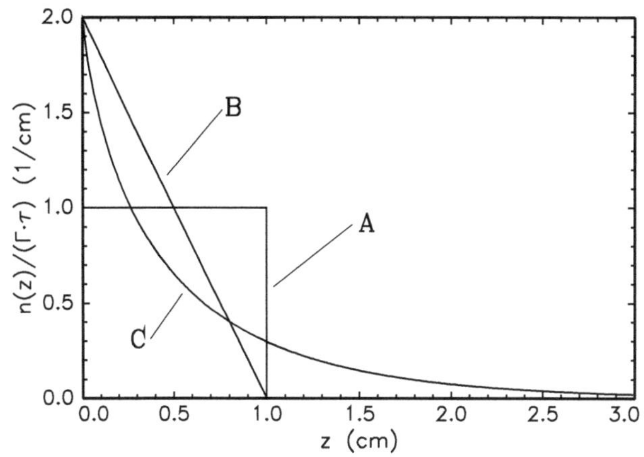

Fig. 9.7. Normalized density distributions of the neutrals emitted from a infinite extended surface vs the distance from the surface, z (A: for $\theta = 0$ and $\lambda_{\rm iz} = 1$ cm; B: (9.46); C: (9.47))

where $E_1(x) = \int_x^\infty \exp(-t)/t\,{\rm d}t$. As a good approximation, the following expression can be used [386]: $E_1(x) = [\ln(1 + 1/x) - 0.4/(1 + x)^2]\exp(-x)$. Figure 9.7 shows the normalized density distributions for the average value of the ionization length, $\lambda_{\rm iz} = 1$ cm. Furthermore, the energy of the neutrals E influences, via the corresponding ionization length, the neutral density distribution. Two main energy distributions for the emitted particles can be distinguished, the Thompson distribution ${\rm d}N^{\rm TH}/{\rm d}E$ (characterizing physical sputtering [170])

$$\frac{{\rm d}N^{\rm TH}}{{\rm d}E} = \frac{2\,E_{\rm s}\,E}{(E + E_{\rm s})^3} \tag{9.48}$$

with surface temperature $T_{\rm s}$ and the Maxwellian distribution ${\rm d}N^{\rm MW}/{\rm d}E$

$$\frac{{\rm d}N^{\rm MW}}{{\rm d}E} = \frac{2}{k_{\rm B}T_{\rm s}}\left(\frac{E}{\pi k_{\rm B}T_{\rm s}}\right)^{1/2}\exp\left(-\frac{E}{k_{\rm B}T_{\rm s}}\right) \tag{9.49}$$

caused by thermal evaporation or thermal desorption. $E_{\rm s}$ is the surface binding energy. From (9.48), the following expressions can be easily obtained using $\lambda_{\rm iz} = \tau_{\rm iz}\sqrt{2\,E/m}$

$$\frac{{\rm d}N^{\rm TH}}{d\lambda_{\rm iz}} = \frac{8\,E_{\rm s}}{m\,\tau_{\rm iz}}\,\frac{(\lambda_{\rm iz}/\tau_{\rm iz})^3}{\left[(\lambda_{\rm iz}/\tau_{\rm iz})^2 + 2\,E_{\rm s}/m\right]^3} \tag{9.50}$$

$$\frac{{\rm d}N^{\rm MW}}{{\rm d}\lambda_{\rm iz}} = \frac{4}{\lambda_{\rm iz}\sqrt{\pi}}\left(\frac{\varepsilon}{k_{\rm B}T_{\rm s}}\right)^{3/2}\exp\left(-\frac{\varepsilon}{k_{\rm B}T_{\rm s}}\right) \tag{9.51}$$

with $\varepsilon = m\,\lambda_{\rm iz}^2/(2\,\tau_{\rm iz}^2)$; $\tau_{\rm iz}$ is the average ionization time and m the mass of the sputtered atoms. Assuming that all particles are emitted normally from the surface, the effect of the different energy distributions on $n(z)$ can be calculated by

$$n(z) = \int_z^\infty \left(\frac{\Gamma}{v}\right) \left(\frac{\mathrm{d}N}{\mathrm{d}\lambda_{\mathrm{iz}}}\right) \mathrm{d}\lambda_{\mathrm{iz}} \,. \tag{9.52}$$

One obtains for the Thompson distribution

$$n(z) = \Gamma \left(\frac{1}{2\sqrt{2\,E_\mathrm{s}/m}} \left[\frac{\pi}{2} - \arctan\left(\frac{\gamma_{\mathrm{iz}}}{\sqrt{2\,E_\mathrm{s}/m}} \right) \right] + \frac{2\gamma_{\mathrm{iz}} E_\mathrm{s}/m - \gamma_{\mathrm{iz}}^3}{2\,(\gamma_{\mathrm{iz}}^2 + 2\,E_\mathrm{s}/m)^2} \right) \tag{9.53}$$

and in the case of a Maxwellian distribution

$$n(z) = \Gamma \sqrt{\frac{2\,m}{\pi\, k_\mathrm{B} T_\mathrm{s}}} \exp\left(-\frac{m}{2k_\mathrm{B} T_\mathrm{s}}\, \gamma_{\mathrm{iz}}^2 \right) \tag{9.54}$$

with $\gamma_{\mathrm{iz}} = z/\tau_{\mathrm{iz}}$. In order to describe the neutral density distribution in the case of physical sputtering, (9.47) and (9.50) should be combined by the integration

$$n^{\mathrm{TH}}(z) = \frac{16\,\Gamma E_\mathrm{s}/m}{\tau_{\mathrm{iz}}} \int_0^\infty \left[\exp\left(-\frac{z}{\lambda_{\mathrm{iz}}}\right) - \frac{z}{\lambda_{\mathrm{iz}}} E_1\left(\frac{z}{\lambda_{\mathrm{iz}}}\right) \right] \times \frac{(\lambda_{\mathrm{iz}}/\tau_{\mathrm{iz}})^2}{\left[(\lambda_{\mathrm{iz}}/\tau_{\mathrm{iz}})^2 + V\right]^3}\, \mathrm{d}\lambda_{\mathrm{iz}} \,. \tag{9.55}$$

The analytical distributions for the different cases are presented in the normalized form $n(z)/\int_0^\infty n(z)\,\mathrm{d}z = n(z)/(\Gamma\,\tau_{\mathrm{iz}})$ in Fig. 9.8.

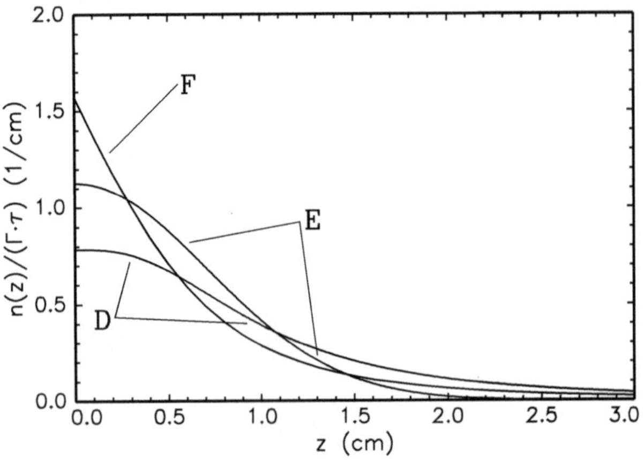

Fig. 9.8. As in Fig. 9.7 (D: (9.53); E: (9.54); F: (9.55))

A similar integration as in (9.55) for the Maxwellian distribution (9.51) can be carried out. The values at the surface $z = 0$ can be useful

$$n^{\mathrm{TH}}(0) = \Gamma \sqrt{\frac{m\,\pi^2}{8\,E_{\mathrm{s}}}} \quad \text{and} \quad n^{\mathrm{MW}}(0) = \Gamma \sqrt{\frac{8\,m}{k_{\mathrm{B}} T_{\mathrm{s}} \pi}}\,. \tag{9.56}$$

It should be noted that the ionization length and the decay length of the neutral density distribution differ significantly in contrast to the often made assumption that they are equal.

The spatial distribution of the sputtered neutral atoms can be obtained experimentally by measuring the line-of-sight photon intensities of the neutral line emission [387].

In order to relate the photon intensity to a local density of the neutral atoms, the branching ratio for the certain observed spectral line and the excitation rate coefficient must be known. However, for materials of interest for example plasma-facing components in fusion devices (such as Be, C, W), these values are known only with large uncertainties. In well-defined plasma experiments, these coefficients can be defined by means of the derived analytical expressions given below. If the excitation rate coefficients are known with a satisfactory accuracy, the flux of eroded particles can be determined in these experiments. Such an experiment with Li has been performed in the linear plasma generator PSI-1 [385].

One has to calculate the spatial distribution of the neutral atoms using (9.45) and taking the energy distribution of the emitted neutrals into account. The integration over a circular marker surface with the radius r_{F} gives

$$n(z, r_{\mathrm{P}}) = \frac{\Gamma}{\pi} \int_0^{2\pi} \int_0^{r_{\mathrm{F}}} \frac{z\,r'}{r^3}\,\mathrm{d}r'\,\mathrm{d}\varphi \int_r^{\infty} \frac{\tau_{\mathrm{iz}}}{\lambda_{\mathrm{iz}}} \frac{\mathrm{d}N}{\mathrm{d}\lambda_{\mathrm{iz}}}\,\mathrm{d}\lambda_{\mathrm{iz}} \tag{9.57}$$

with $r^2 = r'^2 + r_{\mathrm{P}}^2 - 2r' r_{\mathrm{P}} \cos\varphi \sqrt{1-(z/r_{\mathrm{P}})^2}$ and r_{P} being the distance from the center of the marker spot to the certain spatial point where the density should be determined. For the inner integral in (9.57), the analytical expressions (9.53,9.54) can be used. In order to calculate the photon emissivity for the transition from the level k to level i, which is actually measured in the experiment, we have to know the density of the excited neutral atoms of level n_{k}. For the one-dimensional case (only motion in direction of z is considered and the neutrals, which are in the ground state, leave the surface with the same velocity v), it can be shown that

$$n_{\mathrm{k}}(z) = \frac{\Gamma\,n_{\mathrm{e}}\,\langle\sigma v\rangle_{\mathrm{ex}}}{v\left(\sum_{i\leq k} A_{k\to i} - v/\lambda_{\mathrm{iz}}\right)} \left[\exp\left(-\frac{z}{\lambda_{\mathrm{iz}}}\right) - \exp\left(-\frac{\sum_{i\leq k} A_{k\to i}}{v}\,z\right)\right] \tag{9.58}$$

with the maximum at $z_{max} = \ln(\lambda_{\mathrm{iz}} \sum_{i\leq k} A_{k\to i}/v)/(A/v - 1/\lambda_{\mathrm{iz}})$; λ_{iz} is the average ionization length, $\langle\sigma v\rangle_{\mathrm{ex}}$ is the rate coefficient for the excitation of level k from the ground state, and $\sum_{i \le k} A_{k \to i}$ is the spontaneous transition probability from level k into all lower states i. The main parameter is $\gamma_\zeta = \lambda_{\mathrm{iz}} \sum_{i \le k} A_{k \to i} / v$, usually much greater ($\gamma_\zeta \simeq 15000$), hence, the so-called corona population equilibrium approximation can be used. We obtain finally for the photon production in a certain line-of-sight volume at the distance from the target z

$$P(z) = \frac{\pi\, d_{\mathrm{s}}^2}{4} \int\limits_{-d_{\mathrm{p}}/2}^{d_{\mathrm{p}}/2} \langle \sigma v \rangle_{\mathrm{ex}}\, \gamma_{\mathrm{B}}\, n_{\mathrm{e}}\, n(z, r_P)\, \mathrm{d}x \quad [\text{photons/s}] \tag{9.59}$$

where γ_{B} denotes the branching ratio $\gamma_{\mathrm{B}} = A_{k \to i} / \sum_{i \le k} A_{k \to i}$, d_{s} is the diameter of the line-of-sight, d_{p} is the diameter of the plasma column, and $r_{\mathrm{P}} = \sqrt{z^2 + x^2}$.

9.3.2 Atomic Processes in Impure Plasmas

Properties of multiply-charged ions can be studied in so-called isoelectronic sequences. These are sequences of ions with equal numbers of electrons. For example, H, He^+, and Li^{2+} form a sequence of hydrogen-like ions, and He, Li^+, and Be^{2+} a sequence of helium-like ions. The cross-sections of multiply-charged ions can be calculated using perturbation theory. Important reactions in impure plasmas are [3]:

1. Impact ionization and three-body recombination

$$X + e \Leftrightarrow X^+ + e + e \tag{9.60}$$

2. Associative ionization and dissociative recombination

$$X + Y \Leftrightarrow XY^+ + e \tag{9.61}$$

3. Photoionization and radiative recombination

$$X + h\nu \Leftrightarrow X^+ + e \tag{9.62}$$

4. Charge exchange

$$X^+ + Y \Leftrightarrow X + Y^+ \tag{9.63}$$

5. Autoionization and dielectronic recombination

$$X^{**} \Leftrightarrow X^+ + e \tag{9.64}$$

The forward reaction of (9.62) requires a photon energy of $h\nu$ (ν is the photon frequency), which is larger than the ionization energy of the atom or molecule represented by X. In (9.64), X^{**} represents an atom or molecule in which two electrons are in excited states and the total potential energy

Table 9.2. Rate coefficients of ionization for different ions $\langle\sigma v\rangle_{\mathrm{iz}}$ given in m^3/s [391–393]

Particle	$T_e = 5\,\mathrm{eV}$	$T_e = 50\,\mathrm{eV}$	$T_e = 100\,\mathrm{eV}$	$T_e = 1000\,\mathrm{eV}$
$C \to C^+$	3.7×10^{-15}	8.9×10^{-14}	1.1×10^{-13}	1.1×10^{-13}
$C^+ \to C^{2+}$	1.3×10^{-16}	2.0×10^{-14}	2.8×10^{-14}	2.9×10^{-14}
$C^{2+} \to C^{3+}$	2.9×10^{-19}	3.5×10^{-15}	6.2×10^{-15}	8.4×10^{-15}
$Al \to Al^+$	9.1×10^{-14}	2.9×10^{-13}	2.9×10^{-13}	1.9×10^{-13}
$Si \to Si^+$	4.6×10^{-14}	2.4×10^{-13}	2.5×10^{-13}	1.7×10^{-13}
$Si^+ \to Si^{2+}$	2.3×10^{-15}	5.3×10^{-14}	6.3×10^{-14}	5.3×10^{-14}
$Si^{2+} \to Si^{3+}$	2.0×10^{-17}	1.4×10^{-14}	2.1×10^{-14}	2.2×10^{-14}
$V \to V^+$	1.1×10^{-13}	8.4×10^{-13}	7.8×10^{-13}	5.4×10^{-13}
$Mo \to Mo^+$	4.3×10^{-14}	2.8×10^{-13}	3.2×10^{-13}	2.5×10^{-13}
$W \to W^+$	4.5×10^{-14}	2.5×10^{-13}	2.6×10^{-13}	1.7×10^{-13}
$W^+ \to W^{2+}$	4.0×10^{-15}	1.2×10^{-13}	1.5×10^{-13}	1.3×10^{-13}
$W^{2+} \to W^{3+}$	5.0×10^{-16}	9.1×10^{-14}	1.2×10^{-13}	1.2×10^{-13}
$W^{3+} \to W^{4+}$	2.9×10^{-17}	5.2×10^{-14}	8.2×10^{-14}	1.1×10^{-13}

exceeds the ionization energy. The process is possible in plasmas as well as in metal vapor.

A convenient expression for the ionization rate coefficient $\langle\sigma v\rangle_{\mathrm{iz}}$ is given by [388]

$$\langle\sigma v\rangle_{\mathrm{iz}} = C\,\exp(-I_{\mathrm{ion}})\sqrt{\frac{k_{\mathrm{B}}T_{\mathrm{e}}}{I_{\mathrm{ion}}}} \tag{9.65}$$

where the coefficient C, which can be used to fit (9.65) to experimental data, is on the order of $10^{-13}\,\mathrm{m}^3/\mathrm{s}$. A large compilation of ionization rate coefficients measured in experiments for almost all atoms of the periodic system together with approximation formulae are summarized in [389–394]. As an example, some of them are listed in Table 9.2.

The equilibrium distribution of plasma ions with their different charge states is given by balancing the processes of impact ionization and radiative recombination for each ionization state. This state is called coronal equilibrium.

Dissociative recombination is an important electron loss mechanism, because of its relatively large cross-section. Dielectronic recombination involves transitions by two electrons. In this recombination process, a free electron is captured into an excited state and the excess energy is used for excitation of a different bound electron to a higher state. Both electrons then decay to the ground state by emitting photons. Burgess [395, 396] performed quantitative calculations and showed that the efficiency of dielectronic recombination in high-temperature plasmas is considerably larger than that of radiative recombination. In high-density plasmas, however, the effect of three-body recombination dominates.

Charge exchange is the transfer of an electron from an atom to an ion when they collide. The cross-section of charge exchange can be large, and the process becomes important especially in the plasma edge of fusion experiments. After dissociation of hydrogen molecules emitted from the wall surfaces with thermal energies, two atoms with equal and opposite momenta and each with energy of about 3 eV are produced. With such energy, they are able to penetrate deeper into the plasma than the slow molecules. In charge-exchange reactions, a high energetic neutral is suddenly generated with a high ability to penetrate into the plasma. With such high energy, on average equal to the plasma ion temperature, these neutrals contribute significantly to the erosion of plasma-facing components (Sect. 9.2.7).

9.3.3 Prompt Redeposition

A favorable effect with respect to net erosion is the effect of prompt redeposition. In the case of a small ionization length and comparable gyro-radius, an ionized particle has a rather high probability to be redeposited during its first gyro-motion that is very close to its point of emission [397] (Fig. 9.9).

The underlying mechanism is simple and the analytical derivation, given first by Fussmann [398], is straightforward. Neglecting the electric field in the thin sheath above the surface, i.e., assuming that the ionization occurs well above the sheath region, the equations of motion are

$$m\frac{\mathrm{d}v_x}{\mathrm{d}t} = 0\,, \quad m\frac{\mathrm{d}v_y}{\mathrm{d}t} = qv_zB\,, \quad m\frac{\mathrm{d}v_z}{\mathrm{d}t} = -qv_yB\,. \tag{9.66}$$

The magnetic field $\boldsymbol{B} = (B_x, 0, 0) = (B, 0, 0)$ is here directed along the x-axis to simulate the almost grazing incidence of the magnetic field lines onto the divertor targets. Integrating the second equation in (9.66) over time yields

$$v_y = v_{y\mathrm{o}} + \omega(z - z_\mathrm{o}) \tag{9.67}$$

where $\omega = qB/m$. Using this relation in the energy conservation equation

$$\frac{m}{2}(v_x^2 + v_y^2 + v_z^2) = \frac{m}{2}(v_{x\mathrm{o}}^2 + v_{y\mathrm{o}}^2 + v_{z\mathrm{o}}^2) \tag{9.68}$$

gives

$$v_z^2 = v_{y\mathrm{o}}^2 + v_{z\mathrm{o}}^2 - [v_{y\mathrm{o}} + \omega(z - z_\mathrm{o})]^2 \tag{9.69}$$

since $v_x = v_{x\mathrm{o}}$ as seen from the first equation in (9.66).

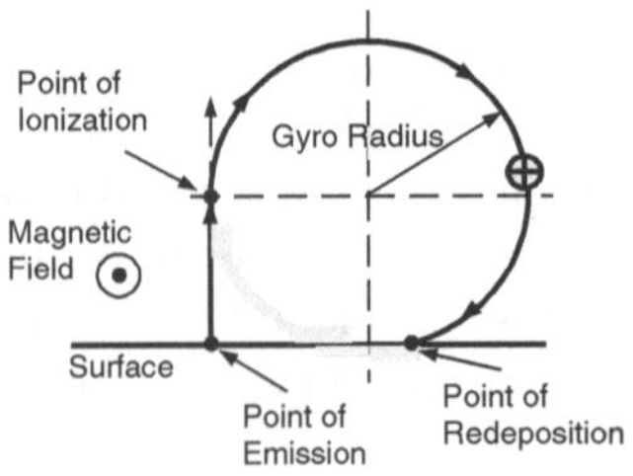

Fig. 9.9. Schematic depiction of prompt redeposition during the first cycle of gyration around the magnetic field line. The magnetic field is directed toward the viewer as indicated

Further, the initial velocities in (9.69) can be replaced by the following presentation in spherical coordinates

$$v_{xo} = v \sin\theta \cos\varphi \ , \quad v_{yo} = v \sin\theta \sin\varphi \ , \quad v_{zo} = v \cos\theta \tag{9.70}$$

yielding

$$v_z^2 = v^2 \cos^2\theta + v^2 \sin^2\theta \sin^2\varphi - [v \sin\theta \sin\varphi + \omega(z - \lambda_{iz}\cos\theta)]^2 \tag{9.71}$$

where $z_o = \lambda_{iz} \cos\theta$ is the distance from the plate at which the ionization happens, and $\lambda_{iz} = v/(\langle\sigma v\rangle n_e)$ is the ionization length.

The condition $v_z = 0$ and $z = 0$ applied to (9.71) defines, after some rearrangement, the corresponding angle of emission θ^*

$$\theta^* = \arctan\left(\frac{\gamma_p^2 - 1}{2\gamma_p \sin\varphi}\right) \tag{9.72}$$

for a given azimuthal angle of emission φ and a given ratio $\gamma_p = \lambda_{iz}/(v/\omega) = \lambda_{iz}/\rho$ (ρ is the gyro-radius) that leads to the grazing of the surface during the gyro-motion. Assuming now a $f_\theta = \cos\theta/\pi$ distribution for emitted particles, which is realistic for most of the erosion processes, the angle θ^* defines a cut in angular space, thus, determining whether a particle strikes the surface during its motion or not. The probability of prompt redeposition is, therefore, given by

$$\begin{aligned} P_{pr}(\gamma_p < 1) &= \int_0^{\pi}\int_0^{\pi/2} f_\theta \sin\theta \mathrm{d}\theta \mathrm{d}\varphi + \int_{\pi}^{2\pi}\int_0^{\theta^*} f_\theta \sin\theta \mathrm{d}\theta \mathrm{d}\varphi \\ &= \frac{1}{2} + \frac{1}{2\pi}\int_{\pi}^{2\pi} \sin^2\theta^* \mathrm{d}\varphi = \frac{1}{1+\gamma_p^2} \end{aligned} \tag{9.73}$$

and

$$P_{pr}(\gamma_p > 1) = \int_0^{\pi}\int_{\theta^*}^{\pi/2} f_\theta \sin\theta d\theta d\varphi = \frac{1}{2\pi}\int_0^{\pi} \left(1 - \sin^2\theta^*\right) d\varphi = \frac{1}{1+\gamma_p^2} \tag{9.74}$$

for the two cases $\gamma_p < 1$ and $\gamma_p > 1$, respectively. In these calculations, the relations $\sin\theta^* = 1/\sqrt{1+\cot^2\theta^*}$ in (9.73) and $(1 - \sin^2\theta^*) = 1/(1+\tan^2\theta^*)$ in (9.74) have been used together with (9.72).

If the ionization length is equal to the gyro-radius, i.e., $\gamma_p = 1$, the probability of prompt redeposition is $P_{pr} = 0.5$, which can be more readily obtained by geometrical arguments. Despite their complex appearance, both integrals (9.73) and (9.74) have the same simple result

$$P_{pr} = \frac{1}{1+\gamma_p^2} \ . \tag{9.75}$$

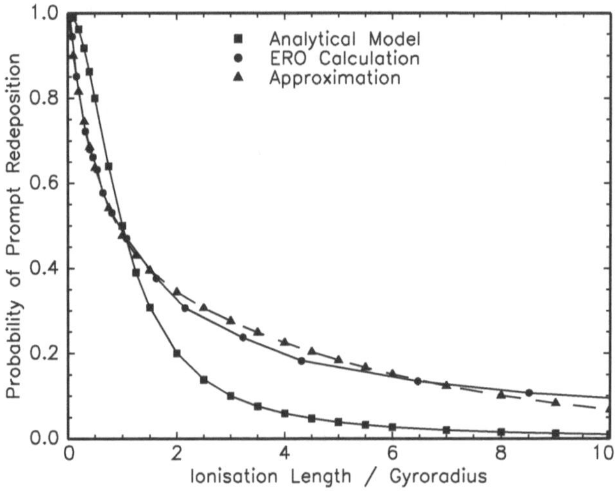

Fig. 9.10. Probability of prompt redeposition. The calculated values using the ERO code [397] are compared with the analytical relation (9.75) and the approximation (9.76)

Unfortunately, effects such as the multi-step ionization during the first gyration (important especially for high-Z elements) and the supporting action of the sheath electric field can only be analyzed by numerical simulations. Nearly all particles that are ionized within the electric sheath will be strongly attracted back to the plate. While this effect enhances the redeposition probability, dilution of the electrons in the sheath region tends to reduce it due to a lower ionization probability. For $\gamma_p \ll 1$, there is a high probability of multiple ionization. The particles are ionized a second or even third time before completing their first orbit. In this case, the gyro-radii become smaller and thus the probability of prompt redeposition decreases (Fig. 9.10). Based on calculations with the Monte Carlo code ERO [397] where all these effects are included, the following approximation

$$P_{\mathrm{pr}} = \frac{1}{2}\left[\exp\left(-2\gamma_{\mathrm{p}}\right) + \exp\left(-\frac{\gamma_{\mathrm{p}}}{5}\right)\right] \tag{9.76}$$

has been suggested, and can be applied, if the surface of emission is sufficiently large, i.e., $F \geq \max(\rho, \lambda_{\mathrm{iz}})^2$.

One has to note that prompt redeposition does not occur at target plates with normal incidence of the magnetic field lines. Much more restrictive is the fact that for large parameters $\gamma_{\mathrm{p}} = \lambda_{\mathrm{iz}}/\rho$ the corresponding probability of prompt redeposition is very small. For example, particles emitted from the wall usually travel a long distance up to the separatrix until they become ionized. In this case, the large gyro-radius of high-Z elements does not help.

The effect of prompt redeposition has been confirmed in experiments [399, 400].

9.3.4 SOL Screening Efficiency

Even though the impurity generation cannot be fully avoided, the aim is to keep the impurities far away from the central plasma region. Whether this will be possible depends to a large degree on where the impurities are emitted and where they become ionized. As discussed above, there is a good chance for impurities generated in the divertor region to fulfill their recycling cycle there. However, impurities emitted from wall elements or limiter plates have a much larger probability to reach the plasma core. If they are ionized inside the SOL, most of them are immediately swept away by the plasma flow streaming along the magnetic field lines toward a target, before these impurity ions could reach the confined region by cross-field transport. In the case of ionization beyond the LCFS, this favorable effect disappears, and their radial motion is determined mainly by the anomalous high cross-field transport.

It has been found by code calculations [401] that for an open divertor geometry and for plasma conditions typical of low-density operations, a wall source is 3 to 5 times more effective in contaminating the core plasma than a divertor target source of equal strength.

There is no widely accepted definition of the concept of fueling efficiency, SOL screening efficiency or retention (see [2,318,372]). The recycling behavior of the different impurity species makes the definition very difficult, but the most critical factor governing retention is the location at which the impurity neutrals are first ionized. The fraction of neutral atoms, launched at a radius r_w at the wall with velocity v that reach the radius r_L without being ionized is given by

$$f = \exp\left(-\int_r^{r_\mathrm{w}} \frac{n_\mathrm{e}(r)\,\langle\sigma v\rangle}{v}\right)\,\mathrm{d}r \tag{9.77}$$

where $\langle\sigma v\rangle$ is the ionization rate coefficient. Assuming an exponential decay of the plasma density in the SOL, characterized by the decay length λ_SOL, one obtains [319]

$$f = \exp\left\{-\frac{\lambda_\mathrm{SOL}}{\lambda_\mathrm{iz}}\left[1-\exp\left(-\frac{\Delta}{\lambda_\mathrm{SOL}}\right)\right]\right\} \tag{9.78}$$

with the average ionization length λ_iz and $\Delta = r_\mathrm{w} - r$.

Once ionized, the transport of now ionized particles is much more challenging to follow. After ionization, their velocity parallel to the magnetic field lines v_imp increases owing to friction with the streaming plasma flow with $v_\mathrm{imp} \leq c_\mathrm{s}$. In the time τ_SOL, the impurity ions flowing along magnetic field lines reach the target plates. During the same time, the length they travel across the magnetic field lines is $\lambda_\mathrm{SOL} = \sqrt{D_\perp\,\tau_\mathrm{SOL}} = \sqrt{D_\perp\,L_\mathrm{c}/v_\mathrm{imp}}$, which thereby defines the decay length of the SOL.

Several models based on one-dimensional analytical estimations (see [2, 319, 402–404]) have been developed in the past to calculate the probability of reaching the plasma core. The most comprehensive (analytical) analysis is given by Fussmann [318] based on the exact solution of the one-dimensional continuity equation in cylindrical geometry including diffusion, parallel losses, and realistic sources of impurities for the SOL region and the inner region of the closed magnetic surfaces.

Due to the complex, usually three-dimensional, geometry in fusion experiments given by the wall geometry, the spatial distributions of the plasma parameters, and sources, numerical code calculations are necessary [401, 405–408] and helpful to obtain more realistic values of the screening efficiency of the SOL. Nevertheless, the quantitative relation between the screening efficiency and the central impurity concentration remains unclear [409].

9.3.5 Accumulation of High-Z Impurities

It is very difficult to maintain purity in the case of highly-charged ions due to their strong tendency to diffuse into the plasma core region [410].

Under improved confinement conditions, when the anomalous diffusion is strongly reduced, the proton–impurity and impurity–impurity collisions may become important. Together with a pronounced central peaking of the electron density profile, impurity accumulation is possible due to neoclassical transport effects [73].

Studying the impurity transport in the plasma core, Tokar [326, 411] estimates a critical plasma density on the axis at which the radiation instability sets in. A spontaneous increase of the impurity density on the tokamak axis leads to an increase of radiation and flattening of the temperature profile. As a result, part of the neoclassical velocity, proportional to $\partial T/\partial r$ and directed toward the edge, decreases. A net flow of impurities towards the axis arises and the impurity concentration further increases. For lower plasma densities, the critical spontaneous perturbations of the impurity concentration and the temperature are suppressed by changes in the anomalous fluxes of particles and energy. For higher densities, the effect of radiation and neoclassical transport prevails. Accumulation could be mitigated by the appropriate shaping of the plasma parameter profiles, in particular, a peaked density profile should be avoided.

9.3.6 Transport Barriers

Different types of improved confinement such as the "H-mode" found in the ASDEX experiments with divertor configuration [412, 413] have been observed in today's fusion experiments. They are all characterized by steep gradients of the plasma parameters, just at the inside of the plasma boundary determined by the separatrix. A strong negative radial electric field arises in that narrow zone extending over a scale of a few centimeters causing sheared plasma rotation in the poloidal and toroidal direction. Poloidal rotation in present tokamaks is observed to be significant only very close to the plasma edge [414].

The importance of sheared flow for suppression of edge turbulence and for improved confinement was emphasized by several authors (see in [83]). Hinton [415] pointed out that mainly the ion pressure profile induces the gradient of the electric field. However, it has become clear from experiments that regimes with improved energy confinement tend to confine impurities as well as lead to a potential conflict between energy confinement and impurity control [53].

Better energy confinement is good for fusion, but means less diffusion in the edge, hence, less outflux of impurities. Thus, the central concentration rises. As mentioned before, ELMs, which usually accompany better confinement modes, could provide an additional mean of pushing out the impurities by spontaneous cyclic destruction of the transport barrier at the edge with only partial sacrifice of the energy confinement improvement. The time-averaged effect of ELMs can be viewed as an outwardly directed drift term in the energy and particle balance equation.

9.3.7 Sawteeth as Plasma Cleaner

As a result of MHD plasma activity, "sawtooth" oscillations of the plasma parameters are observed in a region approximately within the $q = 1$ surface. Under high current operation in tokamaks, the affected region can be large (about half the plasma radius). Since the time between two sawteeth is smaller than the confinement time, the ignition/burning condition, and also the He power deposition, are affected. According to the Kadomtsev model, this plasma core effect is the result of magnetic field reconnection caused by the $m = n = 1$ resistive kink instability. In the resistive MHD model, the magnetic energy is converted into kinetic energy via the resistivity. However, the observed energy conversion from magnetic to kinetic forms occurs in the experiments on a much smaller time scale characterized by the Alvén time, which is given by a characteristic length divided by the Alvén velocity $v_{\mathrm{A}} = B/(\mu_{\mathrm{o}} m_{\mathrm{i}} n_{\mathrm{pl}})^{1/2}$. Several features are still open for discussion, e.g., the survival of so-called snakes (i.e., helical filaments of high density plasma localized at the $q = 1$ surface) during the crash is not explained yet [416]. Attempts to stabilize the oscillations by means of ICRH ion cyclotron resonance heating, unfortunately, led to giant or monster sawteeth [417].

These sawtooth oscillations might be useful to clean the central plasma periodically, thus helping to remove impurities and He out of the burning region [418]. This would especially be desirable in the case of high-Z elements, which tend to accumulate in the plasma center. This is not possible in stellarators, since there such oscillations do not occur.

A similar *cleaning service* could provide the edge localized modes (ELMs), since particles and energy are transported during these short events from the region close to the separatrix into the SOL and further to the target plates.

9.3.8 Deposition of Impurities

Once emitted from plasma-facing components into the plasma, the impurity atoms become ionized and start their motion in the plasma governed by electric and magnetic fields, and collisions. Sooner or later, they come back to a material surface due to transport processes. According to their charge state, they are accelerated in the electric sheath established in front of the surface and strike the surface with significant energy, causing there sputtering and material modification. Non-volatile impurity ions such as carbon or beryllium have a relatively low probability to be reflected from the surface. Most of them are implanted in the near-surface region. This deposition process of impurities can reduce the net erosion of wall elements significantly.

The sticking coefficient s, the complement of the particle reflection coefficient, i.e., $s = 1 - R_{\mathrm{N}}$, is usually well-known for simple atomic impurities. In the case of molecular ions such as the family of the hydrocarbons $\mathrm{C}_x\mathrm{H}_y$, the situation is much more uncertain. There are no direct measurements of their sticking coefficients in the important energy region of several electron volts. The measurements using plasma discharges in fusion experiments suffer from a large number of more or less uncertain parameters and processes, beginning with the spatial profiles of plasma parameters such as $T_{\mathrm{e}}, T_{\mathrm{i}}, n_{\mathrm{e}}$ and ending with the difficulty to provide controlled sources of hydrocarbons, not forgetting the transport of the hydrocarbons, their composition and energy distribution in the plasma. The measurements in laboratory experiments are restricted to thermal energies again due to the lack of reliable, well-quantified sources of radicals with higher energies.

Recently, molecular dynamics simulations indicated a significant energy dependence of the sticking coefficient for hydrocarbon radicals. Having a very small sticking probability at thermal energies, the radicals stick quite well at impact energies of several electron volts. This seems also to be true for CH_3. If the impact energy of the hydrocarbons exceeds the internal binding energy, the radical $\mathrm{C}_x\mathrm{H}_y$ is usually broken upon impact with the surface. The carbon atoms will stick at these low energies, whereas the free hydrogen atoms cause chemical erosion. Hence, the sticking process of hydrocarbons with higher energies is rather a process of transformation and release of different hydrocarbons as the result of chemical sputtering.

A simple and smart method to obtain sticking coefficients is the application of so-called cavities [184, 419, 420]. Once the particle enters through a small hole or a narrow slit into the cavity (Fig. 9.11), it is scattered several times between the upper and lower inner surfaces, until it sticks at one of the two surfaces—in the ideal case. The sticking at the side walls and a possible loss of the particles back through the aperture is usually neglected. The hole or slit is designed as small as possible as a compromise between the undesired loss of particles out of the cavity and the achievable sensitivity to measure the amount of deposited particles after a certain exposure time. With an almost closed aperture, the problem of particle loss does not arise, since only a few

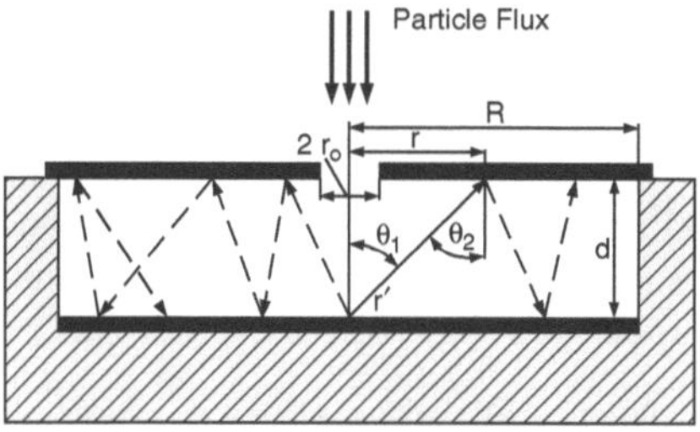

Fig. 9.11. Geometry of a cavity experiment

particles will be able to enter the cavity. To reduce the influence of the side walls, the ratio of the cavity height to its length is chosen to be very small.

But even these effects can readily be considered using computer simulation. Nevertheless, some simple analytical derivation provides the main results. Out of N_p particles, which entered the cavity, the following number of particles

$$\begin{aligned} N_\mathrm{bottom} &= N_1 + N_3 + N_5 + \ldots \\ &= sN_\mathrm{p} + s(1-s)^2 N_\mathrm{p} + s(1-s)^4 N_\mathrm{p} + \ldots \\ &= sN_\mathrm{p} + N_\mathrm{p} \sum_{i=1}^{\infty} s(1-s)^{2i} \end{aligned} \tag{9.79}$$

will be deposited onto the lower inner side. For the number of particles deposited at the upper inner side, we have correspondingly

$$\begin{aligned} N_\mathrm{top} &= N_2 + N_4 + N_6 + \ldots \\ &= s(1-s)N_\mathrm{p} + s(1-s)^3 N_\mathrm{p} + s(1-s)^5 N_\mathrm{p} + \ldots \\ &= N_\mathrm{p} \sum_{i=1}^{\infty} s(1-s)^{2i-1} \,. \end{aligned} \tag{9.80}$$

The subscript is used here to indicate the number of interaction events with one of the surfaces. Since

$$\sum_{i=1}^{\infty} (1-s)^{2i} = \frac{(1-s)^2}{s(2-s)} \,, \quad \sum_{i=1}^{\infty} (1-s)^{2i-1} = \frac{1-s}{s(2-s)} \,, \tag{9.81}$$

and

$$\sum_{i=1}^{\infty} (1-s)^{2i} + \sum_{i=1}^{\infty} (1-s)^{2i-1} = \sum_{i=1}^{\infty} (1-s)^{i} = \frac{1-s}{s} \tag{9.82}$$

the relations (9.79) and (9.80) can be rewritten as

$$N_\mathrm{bottom} = sN_\mathrm{p} + \frac{N_\mathrm{p}\,(1-s)^2}{(2-s)} \tag{9.83}$$

$$N_\mathrm{top} = \frac{N_\mathrm{p}(1-s)}{(2-s)} \,. \tag{9.84}$$

As expected, the global balance states

$$N_{\mathrm{p}} = N_{\mathrm{bottom}} + N_{\mathrm{top}} \; . \tag{9.85}$$

Measuring the total amount of deposited particles on the upper surface N_{top} and on the lower surface N_{bottom}, the sticking coefficient s can be readily obtained from the ratio

$$\frac{N_{\mathrm{bottom}}}{N_{\mathrm{top}}} = \frac{s(2-s)}{1-s} + (1-s) = \frac{1}{1-s} \tag{9.86}$$

yielding

$$s = 1 - \frac{N_{\mathrm{top}}}{N_{\mathrm{bottom}}} \; . \tag{9.87}$$

If the deposition is equally distributed, i.e., $N_{\mathrm{top}} \simeq N_{\mathrm{bottom}}$, then sticking occurs with a probability of about zero. The reflection coefficient is given by the ratio $N_{\mathrm{top}}/N_{\mathrm{bottom}}$. In all cases, $N_{\mathrm{top}} \leq N_{\mathrm{bottom}}$.

For normal incidence into the cavity, the deposition distribution on the upper inner side after the reflection from the lower inner side from point $r = 0$ is given by

$$G(r) = s(1-s)N_{\mathrm{p}} \frac{\cos^2\theta}{\pi\,(d^2+r^2)} = \frac{s(1-s)N_{\mathrm{p}}\,d^2}{\pi\,(d^2+r^2)^2} \quad \left[\frac{\mathrm{particles}}{\mathrm{m}^2}\right] \tag{9.88}$$

in the hole geometry. Relation (9.88) is derived by assuming a cosine angular distribution of emission from the point $r = 0$ into the upper half-space. Integration of (9.88) yields the expected result

$$\begin{aligned} \frac{s(1-s)N_{\mathrm{p}}d^2}{\pi} \int\limits_0^\infty \frac{2\pi r\,\mathrm{d}r}{(d^2+r^2)^2} &= s(1-s)N_{\mathrm{p}}\,2d^2 \left[-\frac{1}{2(d^2+r^2)}\right]_0^\infty \\ &= s(1-s)N_{\mathrm{p}} \; . \end{aligned} \tag{9.89}$$

In the case of a slit geometry, we have for the number of particles coming through the aperture

$$N_{\mathrm{p}}^{\mathrm{l}} = \Gamma\, l\, \Delta y\, \Delta t \tag{9.90}$$

where l is the slit length, Δy is the width of the slit, Δt is the exposure time, and Γ is the particle flux density. If $l \to \infty$, we can simplify to $G(x,y) = G(x=0,y) = G(y)$ and, thus,

$$G(y) = \frac{s(1-s)\Gamma\,\Delta y\,\Delta t\,d^2}{2(d^2+y^2)^{3/2}} \; . \tag{9.91}$$

The relations (9.88) and (9.91) describe the deposition profile on the top just after two interactions: firstly, at the bottom of the cavity, and, secondly, at the top. From more extended analytical calculation the ratio of the maxima of both deposition distributions (after an infinite number of interactions) is found to be

$$\left.\frac{G_{\text{top}}(r=0)}{G_{\text{bottom}}(r=0)}\right|_{\text{hole}} = \frac{3}{1-s} \tag{9.92}$$

and

$$\left.\frac{G_{\text{top}}(y=0)}{G_{\text{bottom}}(y=0)}\right|_{\text{slit}} = \frac{16}{3\pi(1-s)} \tag{9.93}$$

for the two geometries. The contribution of the first deposition onto the lower inner side of the cavity just after incidence is excluded from G_{bottom}. In practice, the relation (9.87) is more helpful than (9.92,9.93), since the maximum of the deposition profile at the top is located at the point $r = 0$ or $y = 0$, i.e., at the location of the hole or slit, and can hardly be measured.

A problem arises in experiments due to the uncertainty in determining the sometimes very broad angular distribution of the incoming particles, which strongly influence the shape of the deposition profile. In addition, more than one particle species usually are deposited in the cavity, for example, CH_3, C_2H_3, and C_2H_5. It is not easy to extract from one carbon deposition profile the different sticking coefficients of the involved species [419].

9.3.9 Modeling of Erosion and Redeposition

A large number of experiments dedicated to erosion and deposition studies, mostly using small surface probes, have been realized in the last twenty to thirty years. It became clear that a thorough analysis of the results obtained is not feasible without extended code simulation. By applying only analytical models, it is not possible to assess the specific role of the various processes and effects being involved. However excellent the experimental techniques are, the manifold of processes such as transport, deposition, dynamical change of surface composition, are strongly interlinked and can hardly be *dissected* and studied separately. In addition, complex geometries in present fusion experiments in connection with more or less uncertain plasma parameters (and their spatial distributions) make the analysis difficult.

On the other hand, the role of computer simulation should not be overstressed. There are still uncertain parameters in describing the processes in detail, for example, the sticking coefficients, transport coefficients, or reaction rate coefficients. For example, in one case good agreement has been achieved by analyzing deposition profiles in two different devices by using absolutely different models for sticking. In the first analysis, nearly zero-sticking has been assumed, in the other one, complete sticking of all species. How can contradictory models yield consistent results? One answer may be that the net effects, and only those that are observable in most of the cases, are usually smaller than the gross effects by several orders of magnitude. Whether net deposition or net erosion occurs depends therefore rather sensitively on the parameters.

Already a slight change in plasma impurity concentration can determine the experimental results (Sect. 9.2.9). In particular, this parameter is not measured in the erosion/redeposition experiments to the required high degree of accuracy.

It is absolutely necessary to use simulation tools with care when comparing their results with experimental values, since often more or less uncertain parameters are used as fit parameters in the simulation.

There have been several codes developed and are widely used in erosion and deposition studies, such as the REDEP code [421], the LIM code [422], the DIVIMP code [423], the WBC code [424, 425], the ERO code [173, 210, 378], the ERO-TEXTOR code [426, 427], and the EDDY code [428, 429]. The main steps in performing an erosion/redeposition simulation are described below:

1. Definition of the simulation region including the plasma region of interest and the affected plasma-facing components; specification of the surface temperature. It is very unusual and not economical to track the particles over the whole device. Rather, a much smaller region, for example, the near-surface region above the divertor plates or limiter elements, is considered. A numerical grid has to be defined for all surfaces in order to obtain spatial distributions.
2. Definition of the background plasma. Though a direct coupling to a plasma code such as B2-Eirene [430] is in principle possible, the corresponding computational costs make it difficult to handle such a code package. Given profiles of various plasma parameters, among them the ion temperature, the electron temperature, and the plasma density are used instead and taken either from measurements or plasma code simulation. In addition, the strength and direction of the magnetic and electric fields should also be given for every point in the simulation region. Furthermore, the impurity concentration in the plasma as well as the influx of impurities into the region must be specified. For some applications, the neutral density and the flux of charge-exchange neutrals, which contribute to surface erosion, must be given as well. The plasma flow pattern is required to calculate the friction forces exerted by the plasma ions, which are streaming toward the surface, on the impurity ions.
   A major uncertainty is introduced into the analysis due to the lack of measured values of the electric field distribution in the plasma. Exposed surface probes themselves alter the electric field distribution in their neighborhood.
   Near the surface, the potential profile (along the surface normal) can be approximated by [424]
   $$\phi(z) = \phi_1 \exp(-z/\lambda_D) + \phi_2 \exp(-z/\rho_{pl}) \tag{9.94}$$
   with $\phi_w = \phi_1 + \phi_2$ (Sect. 7.3, (7.22)). The electric field is then

$$E(z) = \frac{\phi_1}{\lambda_D} \exp(-z/\lambda_D) + \frac{\phi_2}{\rho_{pl}} \exp(-z/\rho_{pl}) \tag{9.95}$$

and the plasma density in the sheath is given by

$$n(z) = n_s \exp\left(\frac{e\phi(z)}{k_B T_e}\right) \tag{9.96}$$

where n_s is the plasma density at the sheath entrance and ρ_{pl} is the gyro-radius of the plasma fuel ions. The dual structure of (9.94) expresses the fact that the overall potential difference in the sheath is almost independent of the incidence angle of the magnetic field lines with respect to the surface α. However, for each part, i.e., for the electric and magnetic sheath, the potential difference varies with that angle. The values of ϕ_1 and ϕ_2 have to be specified according to α. The electric field in the sheath can be calculated—with some effort—using a PIC simulation (Sect. 7.6.1).

3. Calculation of the flux density of plasma ions (including impurity ions) onto a small surface element by using the given plasma parameters and the results of the calculation itself (from step 7); definition of the emission processes to be considered, for example physical sputtering and chemical erosion; calculation of the corresponding yields and specification of the emitted species. If thermal sublimation is significant, the heat conduction equation has to be solved with appropriate boundary conditions (Sect. 8.2). In the case of chemical sputtering, different hydrocarbons are emitted. Their probability of emission depends on the surface temperature, the energy of the incident hydrogen ion, and the surface bond structure. The present surface composition (step 8) should be taken into account by calculating the erosion yields.
4. Start of particles from the surface element; definition of the initial energy E_o and direction $\Omega_o = (\theta_o, \varphi_o)$.
   According to the calculated number of emitted particles from the surface element during a defined time interval Δt_1, a certain number of representative particles is chosen. The number of these particles should be sufficiently large to describe the corresponding angular and energy distribution of emission. Random numbers are applied to generate these distributions. For example, a cosine distribution results, if the angles are calculated by the relations

$$\theta_o = \arcsin\left(\sqrt{\mathrm{RND}_1}\right)$$
$$\varphi_o = 2\pi \mathrm{RND}_2$$

   using two random numbers RND_1 and RND_2 out of the interval $[0, 1]$.
5. Transport of neutral particles in the plasma. After emission, the particles are predominantly neutral and move on rather straight trajectories. Often, neutral–neutral collisions are neglected. It is decided after

a time step Δt_2, whether a certain reaction occurs or not. If a random number RND is smaller than the probability P for a reaction, i.e.,

$$\mathrm{RND} \leq P = 1 - \exp(-\Delta t_2/\tau) \tag{9.97}$$

then a reaction, for example ionization, occurs. The average time between two collisions/reactions is

$$\tau = \frac{1}{n_\mathrm{e} \sum_i \langle \sigma v \rangle_i} \tag{9.98}$$

where n_e is the electron density at the present position of the particle and $\langle \sigma v \rangle_i$ the reaction rate coefficient for a certain reaction i. If the condition (9.97) is fulfilled, then a certain reaction k has to be chosen in accordance to the probability $\langle \sigma v \rangle_k / \sum_i \langle \sigma v \rangle_i$.

6. Transport of impurity ions. After ionization, an impurity ion of mass M and with charge Q moves under the action of the Lorentz force according to the equation of motion

$$M \frac{\mathrm{d}\boldsymbol{v}}{\mathrm{d}t} = \boldsymbol{F} = Q\left(\boldsymbol{E} + \boldsymbol{v} \times \boldsymbol{B}\right) + M \left.\frac{\partial \boldsymbol{v}}{\partial t}\right|_{\text{collisions}} \tag{9.99}$$

where $\boldsymbol{v}$ is the velocity of the particle. This equation has to be solved numerically using time step Δt_3. The numerical procedure is described in detail in Sect. A.2. Especially near the surface, the assumption of guiding center motion fails and the exact solution of (9.99) is required. As an example, trajectories of C, Si, and W atoms emitted into different directions are shown in Fig. 9.12.
With the last term in (9.99), collision effects such as friction, parallel diffusion, and cross-field diffusion have to be taken into account. Possible reactions such as ionization, recombination, and charge exchange are treated in the same way as above.
7. Deposition of impurity ions. Each particle is followed until it leaves the simulation region or impinges on a certain surface element. The deposition is counted, when an additional random number is smaller than the sticking probability, i.e., if RND $\leq s$. Otherwise, the particle leaves the surface again and is followed further on. Better statistics can be achieved by using particle splitting techniques. Then, only a *part* of the particle is deposited, while the other part with the weight $(1 - s)$ is reflected.
8. Change of surface composition. The material composition of the considered surface element is changed in accordance with the deposited amount of particles. Here, some additional assumptions about material mixing and implantation depths are required. In principle, codes such TRIM or TRIDYN can be used to simulate the modifications in the first surface layer (Sect. 6.10.2).

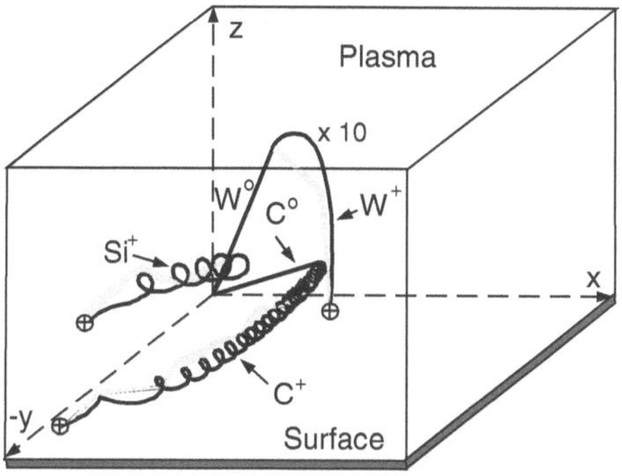

Fig. 9.12. Calculated trajectories for three different elements (C, Si, W enlarged by a factor of 10) [173]. For the high-Z element W, the gyro-radius is usually larger than the ionization length and prompt redeposition occurs with a high probability during the first gyro-motion (Sect. 9.3.3). The C and Si ions perform many gyro-periods in the magnetic field until they reach the surface. The acceleration of these ions due to friction with the plasma flow that is streaming toward the plate is visible. The $\boldsymbol{E} \times \boldsymbol{B}$-force acting on the impurity ion near the plate, i.e., in the electric sheath, shift the ions in the $(-x)$-direction (the magnetic field is directed toward the viewer)

9. Rerun of steps 3–8. An integration over the exposure time t_{exp} with time steps Δt_1 is performed as well as an (inner) integration over all surfaces.
10. Output and diagnostics of the results. The main result is the change of surface composition in time. In the simulation the net effects can be clearly distinguished from the gross ones. Often, the density distribution of impurities in the plasma is of interest.

The time step Δt_1 is on the order of 0.1–1 s, the time step Δt_2 is about 10^{-7}–10^{-6} s, and the time step Δt_3 varies in the interval $[10^{-11}, 10^{-8}]$, dependent on the strength of the electric field at the current position of the particle. A full simulation according to the described scheme can be very time-consuming. To reduce the effort, the simulation programs are adapted and optimized for the certain application under investigation.

9.4 Critical Impurity Concentration

The restrictions imposed by the unavoidable presence of helium in a burning fusion plasma are further enhanced due to impurities [431, 432]. They contribute to fuel dilution and power losses by radiation. For estimation purposes, a rough simplification of the radiation functions L_{i} will be used (valid around $k_{\mathrm{B}}T = 10\,\mathrm{keV}$)

$$P_{\mathrm{rad}} = n_{\mathrm{e}} n_{\mathrm{imp}} L_i \approx n_{\mathrm{e}} n_{\mathrm{imp}} 10^{-38}\, Z_{\mathrm{imp}}^4 \quad [Wm^3] \qquad (9.100)$$

with the atomic number Z_{imp}—not to be confused with the charge state q_{imp}. Only one impurity species is considered, of course, besides helium. For low atomic numbers, bremsstrahlung clearly dominates the radiation losses. A cutoff function in (9.100) for fully stripped impurities (low-Z materials) is not necessary. Relation (9.100) is based on an approach proposed by Vernickel and Bohdansky (see [432]). Of course, detailed collisional-radiative calculations including all relevant atomic physics processes are required [322,431] to obtain reliable data for each impurity species under consideration. In addition to this complexity, the charge state distribution for a given temperature—as an input for such calculations—is usually affected by transport processes in the plasma and can only roughly be described by the coronal equilibrium assumption. In this model, ionization and recombination are simply balanced

$$\frac{n_{\mathrm{imp}}^{q+1}}{n_{\mathrm{imp}}^{q}} = \frac{S_q(T_{\mathrm{e}})}{\alpha_{q+1}(T_{\mathrm{e}})} \tag{9.101}$$

where $S_q(T_{\mathrm{e}})$ and $\alpha_q(T_{\mathrm{e}})$ are the ionization and recombination rate coefficients for a certain charge state q, respectively.

For given $\gamma = \tau_{\mathrm{He}}^*/\tau_{\mathrm{E}}$ and temperature, (3.28) has one solution for f_{imp} in the available range, which is determined by the condition (3.31). Results are shown in Fig. 9.13 for $k_{\mathrm{B}}T = 12\,\mathrm{keV}$, $f_{He} = 0.1$ and average charge states as a function of T [322].

Obviously, a value of $\gamma = 0$, i.e., each produced helium particle is immediately removed from the reaction zone, would allow for the maximum impurity concentration tolerable in the plasma. In this case, (3.28) converts

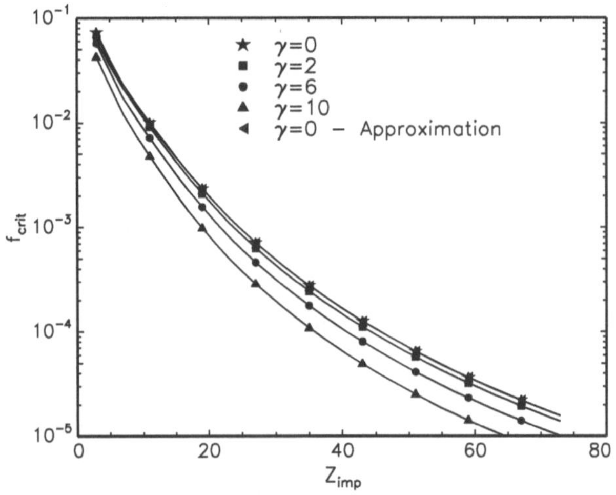

Fig. 9.13. Critical impurity concentration as a function of atomic number Z_{imp} for $k_{\mathrm{B}}T = 12\,\mathrm{keV}$ and different γ as indicated. The approximation (9.102) for $\gamma = 0$ is also shown

to a quadratic equation in respect to f_{imp} and has only one physically relevant solution

$$f_{\mathrm{crit}} = \frac{\langle \sigma v \rangle_{\mathrm{DT}} E_{\alpha}(1 - 2f_{\mathrm{He}})^2 - 4c_{\mathrm{br}}\sqrt{k_{\mathrm{B}}T}(1 + 2f_{\mathrm{He}})}{2\langle \sigma v \rangle_{\mathrm{DT}} E_{\alpha}(1 - 2f_{\mathrm{He}})q_{\mathrm{imp}} + 4c_{\mathrm{br}}\sqrt{k_{\mathrm{B}}T}(q_{\mathrm{imp}}^2 - q_{\mathrm{imp}}) + 4L_{\mathrm{imp}}} \tag{9.102}$$

which is also shown in Fig. 9.13. In the derivation of (9.102), the approximation $\sqrt{1 - \varepsilon} \approx 1 - \varepsilon/2$ with $\varepsilon \ll 1$ has been applied. It is worth noting that the impurity criterion

$$f_{\mathrm{imp}} < f_{\mathrm{crit}} \tag{9.103}$$

yields the most optimistic estimate. Nevertheless, it raises very challenging demands, especially, for the use of high-Z materials such as tungsten, since their permissible concentration in the plasma is very low. With increasing impurity concentration, the operational window for a fusion reactor becomes smaller and finally closes [432], resulting in plasma disruption or quench.

Part III

Operation Limits and Criteria

10

The Problem of Plasma Density Control

Regardless of the concept of magnetic confinement, the transport across the magnetic field lines cannot and should not be reduced to zero. Even if it is possible, the helium ash of the fusion reactions must be removed from the plasma core.

In practice, the transport losses are much larger than desired due to the anomalously high diffusion of energy and particles across the magnetic field. Particles from the main plasma come to the plasma edge and impinge on wall elements. At the material surface, a fraction of the particles is reflected as fast particles (in the backscattering process) and the others are implanted in the material. Some of the latter may be subsequently re-emitted as molecules, when the surface is saturated, while the remaining part is retained for a longer time in the material. Such behavior was observed in isotope changeover experiments in JET [433]. This retention is characterized by the time being of the same order as the global plasma particle confinement time [434,435]. However, within a few minutes after a discharge most of the gas input is released by outgassing from the walls. Thermal desorption and/or ion-induced release can play a significant role during the discharge.

In materials with high solubility, trapping occurs at interstitial positions in the lattice. Higher temperatures lead to diffusion. Very high amounts of trapped particles can be reached until the whole material is saturated. For materials with negative heat of solution, i.e., low solubility for the gas, trapping takes place predominantly at defects and vacancies, dislocations, or grain boundaries. Once saturation is reached in the implantation region, one particle is released for each incident ion. Diffusion into the bulk is not observed. Thermal detrapping and release occur, if the surface temperature is high enough, since then the particles are able to overcome the surface potential.

In general, the first wall represents simultaneously a large sink and source of fuel gas, since the amount of implanted and adsorbed hydrogen isotopes in the upmost surface layers can finally exceed the whole particle content in the plasma by a few orders of magnitude. Under quasisteady state wall conditions, recycling usually dominates plasma fueling by external particle sources, e.g., gas puffing or neutral beam injection. The efficiency with which the plasma is directly fueled by external sources is, therefore, only of minor importance. Starting the discharge with well "out-baked" and clean surfaces, heavy wall pumping is observed but remains a more or less transient effect and cannot be sustained in longer discharges.

Almost all particles, backscattered atoms or re-emitted molecules leave the surface as neutrals and enter the plasma where they become ionized, starting a new cycle. In one second of plasma operation, this cycle is repeated many times. Without recycling, the plasma density would rapidly drop. The ratio of global fluxes off the wall and onto it is defined as the recycling coefficient R_{cyc}.

In order to maintain a constant plasma density, a fueling flux is required to compensate the losses due to burn-up, deposition into wall components, and pumping. The global balance equation for the plasma fuel ions is

$$\begin{aligned}\frac{\mathrm{d}N}{\mathrm{d}t} &= Q - \frac{N}{\tau_{\mathrm{p}}} + \frac{N}{\tau_{\mathrm{p}}}\left(1 - f_{\mathrm{wall}} - f_{\mathrm{pump}}\right) - \frac{N^2}{V_{\mathrm{plasma}}}\langle\sigma v\rangle_{\mathrm{DT}} \\ &= Q - \frac{N}{\tau_{\mathrm{p}}}\left(f_{\mathrm{wall}} + f_{\mathrm{pump}}\right) - \frac{N^2}{V_{\mathrm{plasma}}}\langle\sigma v\rangle_{\mathrm{DT}} \qquad (10.1)\end{aligned}$$

where N is the number of fuel particles, Q is the external gas inflow rate in atoms/s, $\langle\sigma v\rangle_{fusion}$ is the fusion rate coefficient, τ_{p} is the particle confinement time, V_{plasma} is the plasma volume, $f_{\mathrm{wall}}(N/\tau_{\mathrm{p}})$ is the fraction of the flux out of the plasma N/τ_{p}, which is implanted or codeposited, i.e., lost in the wall, and $f_{\mathrm{pump}}(N/\tau_{\mathrm{p}})$ is the part of the flux N/τ_{p} pumped out. The term $(1 - f_{\mathrm{wall}})(N/\tau_{\mathrm{p}}) = R_{\mathrm{cyc}} \cdot N/\tau_{\mathrm{p}}$ is called the recycling flux. Combining the term $-N/\tau_{\mathrm{p}}$ from the right-hand side of (10.1) with the recycling term, one obtains $-(N/\tau_{\mathrm{p}})+R_{\mathrm{cyc}}\cdot N/\tau_{\mathrm{p}} = -N(1-R_{\mathrm{cyc}})/\tau_{\mathrm{p}} = -N/\tau_{\mathrm{p}}^*$ where the effective confinement time τ_{p}^* has been used (compare with (3.17)) in Sect. 3.2.

From (10.1) it is clear that efficient control of the plasma density is only possible by establishing a strong sink using a pump system and having a source of fuel gas. The fuel lost owing to burn-up, transport, and pumping has to be replenished by gas puffing or injection of pellets of frozen hydrogen. Simple gas puffing at the plasma edge, however, leads to generation of charge-exchange neutrals and thereby to increased sputtering. On the other hand, deep pellet fueling is not desirable, since density fluctuations in the core of the main plasma can generate substantial power excursions. Holding the main plasma density constant would lead to a reduction of the recycling flux in the plasma edge in the case of pellet fueling (acting as external source). Thus, the density there would also decrease in contrast to the desired situation of high local recycling.

Zero-dimensional models in terms of particle balance equations have been quite successful in explaining the general behavior of the fuel cycle in fusion experiments. In these models different reservoirs are considered for example the main plasma, the plasma edge, the divertor, and the wall [51, 436]. Using different confinement times τ (or residence times) for the respective particle reservoirs, the transport of fuel ions and neutrals and the losses due to external pumping are approximately described by replacing the term $\mathrm{div}\Gamma = \mathrm{div}(n\boldsymbol{v})$ simply by n/τ. The confinement times serve as a link between a volume-integrated source intensity and the global content of a quantity. Confinement times also depend not only on diffusion coefficients or drift velocities, but also on the spatial distribution of the sources.

The three main parameters are [434]: (1) the plasma particle confinement time determined by particle transport, (2) the neutral particle screening of the plasma edge governed by multiple reflection and screening, e.g., through ionization, at the plasma–material interface, and (3) the recycling coefficient describing the recycling at the material surfaces facing the plasma including the reflection of plasma particles, the desorption of particles, and diffusion processes in the material.

Besides the plasma density control, the control of plasma composition according to the burning requirements in the main plasma is an additional problem facing controlled thermonuclear fusion, which is not adequately addressed in fusion experiments today.

10.1 Long-Term Operation

In a pulsed reactor, even with 1000-second discharges as planned in ITER, many of the smaller components in the divertor, at the first wall, and blanket will have thermal time constants significantly shorter than the time between discharges. Hence, significant temperature excursions occur. Thus, thermal cycling and fatigue considerations for the pulsed reactor, based on the tokamak concept, are significantly more severe than for a steady state reactor, possibly based on the stellarator concept.

With respect to steady state operation, different time constants can be distinguished. MHD stability phenomena have a time scale of milliseconds. Transport processes are characterized by energy and particle confinement times on the order of seconds. It can also take several seconds to achieve an equilibrium between the current and pressure profile.

Using the tokamak concept, a large fraction of the plasma current ($\approx 70\%$) should be driven by the bootstrap current. This current is due to collisional diffusion of particles fluttering between trapped–untrapped trajectories and is expected to be strong when a sharp edge pressure pedestal is formed like in the H-mode [437]. Therefore, control of the current and edge pressure profile is necessary for high β operation with a high bootstrap fraction. Neutral beam injection can be used as a non-inductive current drive. Shallow pellet injection could serve as a control mean, which allows a suppression of the edge pedestal.

Stellarators inherently offer the possibility of steady state operation, thus depending mainly on the heating capability. Here, in striking contrast to tokamaks, additional effort is required to minimize the bootstrap current.

The impurity confinement times are on the same order as the particle confinement times of fuel ions, i.e., on the order of seconds. However, in the case of impurity accumulation, especially critical for high-Z elements, the core impurity density does not show saturation. As a consequence, the burn condition is finally violated and the plasma collapses due to fuel dilution and plasma cooling by impurity radiation. It is not yet clear whether helium tends to have an inward pinch. In this case, efficient He pumping, which is crucial for a reactor, would be almost impossible.

It takes minutes of operation to achieve equilibrium of power deposition onto wall elements and cooling. Mainly the surface temperature determines the amount of erosion and the gas release. During 10-second discharges in modern fusion experiments this balance cannot be reached and the density control is based on an ever-pumping wall. Almost after each discharge, wall conditioning procedures are applied to release absorbed gases. Often, strong gas puffing is necessary during the discharge to achieve transient plasma effects. It will therefore be a challenge to ensure density control in long-term operation with saturated material surfaces.

The fueling scheme in a power reactor has some constraints [438]. Simple gas puffing into the main plasma would cause strong sputtering erosion of the first wall by energetic charge-exchange neutrals. Shallow pellet injection of frozen hydrogen/deuterium is a possibility to avoid this by providing fueling just inside the separatrix. The speed, size, and frequency of the pellet injection has to be optimized in order to be consistent with burn-up compensation, density control, and the forced flow requirement, along with edge bootstrap current suppression—perhaps too many demands to be met by a single mean.

A large DT forced flow to the divertor plate is required to suppress the impurity backflow from the divertor region and to enhance the He ash exhaust. Since temperature gradients along the field lines are necessary to get a cold and dense divertor plasma, the impurity backflow due to thermal forces must be suppressed by the fuel-impurity friction force.

Having solved these problems, the key issues concerning long-term operation are the lifetime of wall components and tritium retention.

Without the benefits of redeposition, frequent replacement of the target plates would be necessary. As shown in Sect. 12.4.2, the net erosion can be minimized by an adequate arrangement of the target plates with respect to the magnetic field geometry. This allows for the use of the impurity drift in the electric sheath, which brings the impurity ions back to their (radial) position of emission. However, under the conditions of a cold divertor plasma in modern plasma scenarios, the gross erosion there is reduced and the main source of impurities appears to be the wall where the redeposition probability is much smaller. Most of the material eroded at the wall is transported into the divertor, where it is deposited on the neutralizer plate. Not only C, but also Be and W can be codeposited together with hydrogen isotopes, among them tritium [149, 439].

The maximum tritium inventory in ITER should not exceed the specified amount of several hundred grams. Besides the implantation of tritium, redeposited material can form thick layers with a high tritium concentration. Such layers tend to flake off from the bulk material.

10.2 Wall Conditioning

During a discharge of several seconds, each particle is subjected to many cycles of emission from the wall, dissociation, ionization in the plasma, recombination at the wall, and again emission. The reservoir in the surface layer with a thickness of about 0.1 μm (taking graphite with its atomic density n_C and a ratio of hydrogen to carbon of $c = 0.4$ [440]) amounts to

$$\begin{aligned} N_\mathrm{surface} &= c\, n_\mathrm{C}\, S_\mathrm{divertor}\, d \\ &\approx 0.4 \cdot 11.3 \times 10^{28}\mathrm{m}^{-3} \cdot 100\mathrm{m}^2 \cdot 10^{-7}\mathrm{m} = 4.5 \times 10^{23}\,. \end{aligned} \quad (10.2)$$

This takes only the divertor plates with an area of $100\, m^2$ into account. This then exceeds the inventory in the plasma

$$N_\mathrm{plasma} = n_\mathrm{plasma} V_\mathrm{plasma} \approx 10^{20}\mathrm{m}^{-3} \cdot 800\mathrm{m}^3 = 8 \times 10^{22} \quad (10.3)$$

by a factor of 5. A small change of the surface temperature can, therefore, be sufficient to initiate a strong gas puff, which easily affects the plasma behavior by disturbing the sensitive balance near the density limit (see Sect. 11). The wall can release more fuel as provided by external fueling (gas puffing and pellet injections); this could lead to an uncontrolled rise in density. DT active pumping in a fusion reactor with long operation times is therefore indispensable.

As mentioned above, up to now the density control in fusion experiments with rather short discharge times relies almost entirely on an effective pumping wall. High-temperature bake and cleaning discharges, e.g., glow discharge in He, are used to deplete hydrogen trapped in the walls and to pump off prior to normal plasma operation. In addition, various techniques to getter impurities such as oxygen, i.e., to bind them chemically at the surface, are applied. Thus, the oxygen impurity concentration in the plasma can be kept below 1%, and chemical erosion by oxygen is reduced. For this, the wall elements are covered by low-Z coatings such as Be (by evaporation), C (carbonization), B (boronization), or Si (siliconization), as well as metals such as Ti and Li, which adsorb oxygen and hydrogen isotopes very effectively.

During carbonization, thin hydrogenated amorphous carbon films (aC:H) are deposited on all plasma-facing surfaces, protecting especially metal parts in order to obtain plasmas that are essentially free of metal impurities [441].

Oxygen in the plasma results mainly from water vapor desorbing from surfaces in the vessel in remote areas. Glow discharges in He can deplete the surface layer from implanted oxygen by the release of CO, but this effect is only transient because of the persisting H_2O desorption [441]. Boron and silicon reduce the oxygen plasma contamination by the formation of strongly-bound oxides (getter effect). In addition, a significant reduction of chemical erosion is observed due to small amounts of boron and silicon in the upper graphite surface [442].

Thin films such as boron films on top of the material surfaces are attractive, for example, as a protecting layer against impact of energetic charge-exchange neutrals [443]. The chemical erosion of graphite would be reduced and most of the hydrogen isotope atoms are re-emitted at temperatures of 400 degrees centigrade [443].

Many advances in plasma performance have been achieved only with the help of improved wall conditioning, since the confinement depends sensitively on edge recycling. Here are some of the results.

Using boronization and siliconization in tokamak TEXTOR [444] allowed for the increase of density to 1.7 times the Greenwald density limit (see in [445]). The reduction of carbon release directly after this wall conditioning led to a suppression of MARFEs, which resulted in a higher density limit, and demonstrates the importance of local recycling and impurity release on the high field side for the development of the MARFE.

Electron cyclotron resonance (ECR) discharge cleaning with hydrogen and helium glow discharge cleaning and Ti gettering were successfully applied in the torsatron CHS [446].

The effect of wall conditioning on plasma performance has been investigated in the stellarator W7-AS by the routine use of He glow discharges, carbonization, and boronization. Due to the very localized regions of plasma contact, these techniques had only short-term effects [447].

Several methods have been used in tokamak TFTR to introduce lithium into the plasma vessel: lithium pellet injection, vacuum evaporation of lithium from an inserted oven, lithium crucible heating by scrape-off plasmas, lithium borohydride discharges, and laser-controlled lithium aerosol injection [448]. It was seen that with a deposition of just a few milligrams of Li on the limiter leads to a considerable improvement of the energy confinement time. Lithium conditioning in TFTR produced an increase of the fusion triple product by an order of magnitude [449].

Si and Li coating techniques by means of a plasma-assisted deposition are used in HL-1M. A frequent deposition of fresh silicon or lithium films is found to be necessary, in particular after each exposure to air. The results are strong pumping, low carbon and oxygen concentration in the main plasma, as well as very low hydrogen recycling [450]. Combined Li-Si coatings in HL-1M are found to be superior to Li coating and Si coating alone [451]. Serious problems associated with repetitive deposits, i.e., multiple and thick silicon or lithium layers are reported. These deposits tend to peal off, contributing to dust and flake production.

A new technique for wall conditioning using the high toroidal field has been developed at the superconducting tokamak HT-7 by launching ion cyclotron resonant waves (ICRF) into plasmas [452]. ICRF-produced helium discharges are shown to be applicable also in the presence of a confining stellarator magnetic field [453]. In superconducting fusion devices, conventional glow discharge conditioning cannot be used in the presence of continuous high magnetic fields. With an ICRF antenna, break-down is easily achieved at very low power over a wide range of gas pressures and the discharges can be sustained and heated.

The adaptability of the developed wall conditioning methods in a fusion reactor, i.e., under quasisteady state conditions, seems quite restricted. All these methods offer rather transient effects, which sooner or later disappear: either the deposited thin films are lost and could not be maintained by redeposition, or the material surfaces simply show saturation effects.

11

Plasma Operation Limits

Other than the requirements of particle and power balances in a fusion reactor, as discussed thus far, there are a number of conditions and constraints to be regarded for stable plasma operation.

First of all, the electrons in a fusion plasma should be well-confined by the magnetic field. This is achieved if their gyro-radius $\rho = v_{\perp} m_e/(Be) \simeq \sqrt{k_B T_e m_e}/(Be)$ is smaller or equal to the Debye length $\lambda_D = \sqrt{\epsilon_o k_B T_e/(n_e e^2)}$ —for when there are no parallel losses [62]. In this case, the magnetic field is superior to electric forces caused by natural plasma oscillations. Indeed, by comparing the magnetic force ev_eB acting on one electron with the electric force eE, where the intermittently arising electric fields can be estimated by $E \approx k_B T_e/(e\lambda_D)$, one obtains

$$ev_e B = eE = \frac{k_B T_e}{\lambda_D} \quad \rightarrow \quad 1 = \frac{m_e v_e}{eB\lambda_D} = \frac{\rho_e}{\lambda_D} \tag{11.1}$$

with $k_B T_e \simeq m_e v_e^2$.

Introducing the magnetization parameter ξ [63,454], the following criterion results:

$$\left(\frac{\rho_e}{\lambda_D}\right)^2 = \left(\frac{\omega_p}{\omega_{ce}}\right)^2 \simeq \frac{n_e m_e}{\epsilon_o B^2} = \xi_e \leq 1 \tag{11.2}$$

with $\omega_{ce} = eB/m_e$ and $\omega_p = \sqrt{n_e e^2/(m_e \epsilon_o)}$. Since $m_i \gg m_e$ this criterion cannot be fulfilled for the plasma ions, thus $\xi_i = (\rho_i/\lambda_D)^2 \gg 1$ when $\xi_e \approx 1$. It is sufficient to magnetically confine only one component of the plasma (the electrons), whereas the other component (the plasma ions) is held (close to the electrons) by immediately arising electric fields in the case of any disturbance of plasma neutrality.

Rearranging (11.2) the magnetic field, B can be presented as a function of the plasma density, $B^{crit} = \sqrt{n_e m_e/(\xi_e \epsilon_o)}$ for $\xi_e = \xi_e^{crit} = 1$. This value has been calculated in a wide range of plasma densities. As shown in Fig. 11.1, the criterion (11.2) is marginally met in operating plasma and fusion experiments, i.e., the used magnetic field is close to the critical field, $B \geq B^{crit}$. In linear

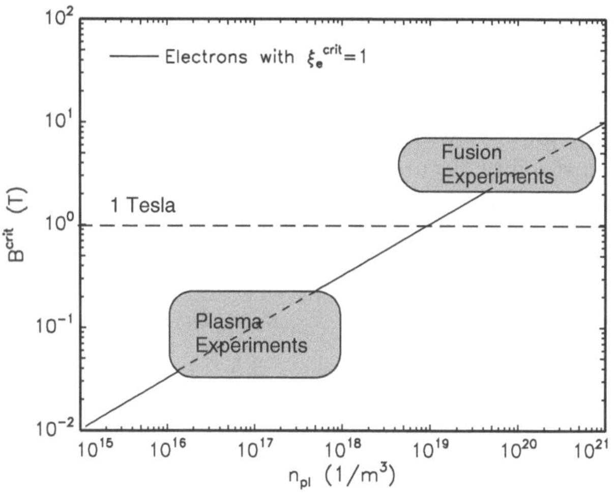

Fig. 11.1. B^{crit} vs plasma density. The parameter regions (B,n_{e}) of plasma and fusion experiments are indicated

plasma devices, where the magnetic field is directed along the axis, the plasma confinement is clearly dominated by the end losses at the plates (parallel to the magnetic field lines), i.e., the criterion (11.2) does not apply.

For a fixed value of the magnetic field, (11.2) gives a general density limit

$$n_{\mathrm{e}} \leq n_{\mathrm{c}} = B^2 \epsilon_{\mathrm{o}}/m_{\mathrm{e}} \tag{11.3}$$

for toroidal devices, i.e., valid for both tokamak and stellarator systems, under stable plasma conditions. This condition explains the threshold phenomenon observed on Alcator by Granetz [455], where disruptions occurred by reaching a certain density limit $n_{\mathrm{c}} \propto B^2$. Not only the dependence ($\propto B^2$) but also the values as given by (11.3) are in agreement with the experimental data in the region of B=1–10 T. Granetz found a best-fit expressed by [455]

$$n_{\mathrm{c}}^{\mathrm{Granetz}} = 8 \times 10^{18}\, B^2 \left[\frac{1}{\mathrm{m}^3}\right] \tag{11.4}$$

with the magnetic field B given in tesla. The prefactor 8×10^{18} in (11.4) is close, but smaller than the prefactor $\epsilon_{\mathrm{o}}/m_{\mathrm{e}} = 9.7 \times 10^{18}$ as given in (11.3).

Edge plasma turbulence and associated heat convection losses are found to increase with density. Large intermittent perturbations in density and temperature are reported in the SOL, called blobs or bursts, as the density limit is approached. Shortly after, the separatrix region as well as the plasma core are affected and an operational limit is set [80, 456].

Furthermore, for a given magnetic field B (the maximum value of B is restricted by technical reasons) the plasma pressure is limited due to MHD instabilities such as ideal kink or tearing mode instabilities. The ratio of the plasma pressure to the magnetic pressure

$$\beta = \frac{\langle p \rangle}{B^2/(2\mu_o)} \tag{11.5}$$

should be at least smaller than 0.1, but should be as high as possible—from an economical point of view. Beyond the stable region, a strong instability, called disruptive instability, occurs in usual operations. In tokamaks with elongated plasma cross-sections, β values up to 0.1 can be realized. The limit of the average β in tokamaks is found by MHD simulation codes to be

$$\beta^{\text{crit}}[\%] = \frac{g_\text{T} I_\text{p}[\text{MA}]}{a_\text{t}[\text{m}]\, B_\text{t}[\text{T}]} \tag{11.6}$$

with the so-called Troyon factor g_T=2–3.5 [14, 457].

When the radiation exceeds a certain fraction of the input power, a disruption of the plasma current occurs in tokamaks [14, 445]. This is generally assumed to be a consequence of the peripheral plasma region radiating to such an extent that the temperature profile collapses. Temperature collapses when the local radiation at the boundary exceeds the heat flux conducted to that radius. For a fixed impurity fraction of an ohmically heated plasma, the critical density is proportional to the plasma current, which in turn is proportional to B/R for a fixed q value. This behavior was derived from experimental data by Murakami et al. [458]. Though the balance between radiation and input power determines how much of the available parameter region is accessible, the deterioration of particle confinement is supposed to be the main driver behind the density limit [445]. Even more fueling is required to balance the losses due to deteriorated confinement leading to higher edge density and, subsequently, higher radiation, especially in the case of edge fueling. The current profile and MHD stability are altered to cause a disruption. With pellet injection, the density limit can be pushed upwards. There appears to be an absolute limit for all tokamaks given by [14]

$$n_\text{e}[1/\text{m}^3] \leq (1.2-2)\frac{10^{20}\, B[\text{T}]}{q_\text{s} R_\text{t}[\text{m}]} \tag{11.7}$$

where n is the average electron density and R_t is the major radius. When the value for NBI heated plasmas is higher, the value for ohmic heating is lower. The factor q_s is the safety factor

$$q_\text{s}(r) = \frac{r}{R}\frac{B_\text{toroidal}(r)}{B_\text{poloidal}(r)} \tag{11.8}$$

as a function of the radius. With $B_\text{poloidal} \approx \mu_o I_\text{p}/(2\pi a_\text{t})$ and using (11.8), the condition (11.7) is easily transformed to another familiar presentation of a tokamak density limit—the Greenwald limit [445, 459]

$$n_\text{e}[1/\text{m}^3] \leq \frac{10^{20}\, I_\text{p}[\text{MA}]}{\pi a_\text{t}^2[\text{m}]} \tag{11.9}$$

where a_t is the minor radius and I_p is the plasma current.

In addition, the Kruskal–Shafranov criterion demands that the safety factor q (see 11.8) should be larger than 1. This condition limits the value of the poloidal magnetic field B_{poloidal} to ensure magnetohydrodynamic plasma stability.

The β-limit (11.5) allows for a presentation in terms of the density using $p = 2n_{\text{e}}k_{\text{B}}T_{\text{e}}$ ($n_{\text{e}} = n_{\text{i}}$, $T_{\text{e}} = T_{\text{i}}$)

$$n_{\text{e}} \leq \beta^{\text{crit}} \frac{1}{4\mu_{\text{o}}k_{\text{B}}T_{\text{e}}} B^2 . \tag{11.10}$$

Multiplying (11.10) with the energy confinement time $\tau_{\text{E}} \simeq a^2/4\chi_{\perp}$, we obtain

$$n_{\text{e}} \cdot \tau_{\text{E}} \leq \beta^{\text{crit}} \frac{1}{4\mu_{\text{o}}k_{\text{B}}T_{\text{e}}} \frac{B^2 a_{\text{t}}^2}{4\chi_{\perp}} \tag{11.11}$$

a condition similar to the exhaust (3.19) and burn (3.26) criterions. The term $\chi_{\perp}$ is the anomalous energy diffusion coefficient, which describes the complex processes of radial energy transport within the context of a diffusion model.

The same procedure can be applied to condition (11.3)

$$n_{\text{e}} \cdot \tau_{\text{E}} \leq \frac{\epsilon_{\text{o}}}{m_{\text{e}}} \frac{B^2 a_{\text{t}}^2}{4\chi_{\perp}} . \tag{11.12}$$

Both conditions (11.11) and (11.12) limit the stable operational region of a fusion reactor even more as already done by the exhaust and burn criterions (see Sect. 3.2 and Fig. 3.2).

As seen in Fig. 11.2, low values of $\chi_{\perp} \leq 0.1$ are required, whereas under normal plasma operation values of about $\chi_{\perp} \approx 1\,\text{m}^2/\text{s}$ are measured in fusion experiments.

This problem can be solved by establishing so-called internal transport barriers (ITB), which have been observed in a wide range of tokamaks and helical systems. These barriers are produced by modification of the current profile applying auxiliary heating during the initial current ramp-up phase of tokamak discharges. They are characterized by low or negative magnetic shear and steep temperature and/or density gradients, corresponding to a local reduction of heat and/or particle transport (at constant heating power). In the experiments, the ion heat transport could be reduced down to the neoclassical values leading to high central ion temperatures. The much lower neoclassical electron heat conduction (in comparison to the ion heat conduction) is not reached experimentally even for very good electron confinement [460].

Combining (11.11) and (11.12), the intersection point of the two lines in Fig. (11.2) is given by ($c^2 = 1/(\mu_{\text{o}}\epsilon_{\text{o}})$)

$$k_{\text{B}}T_{\text{e}} = \beta^{\text{crit}} \frac{m_{\text{e}}c^2}{4} . \tag{11.13}$$

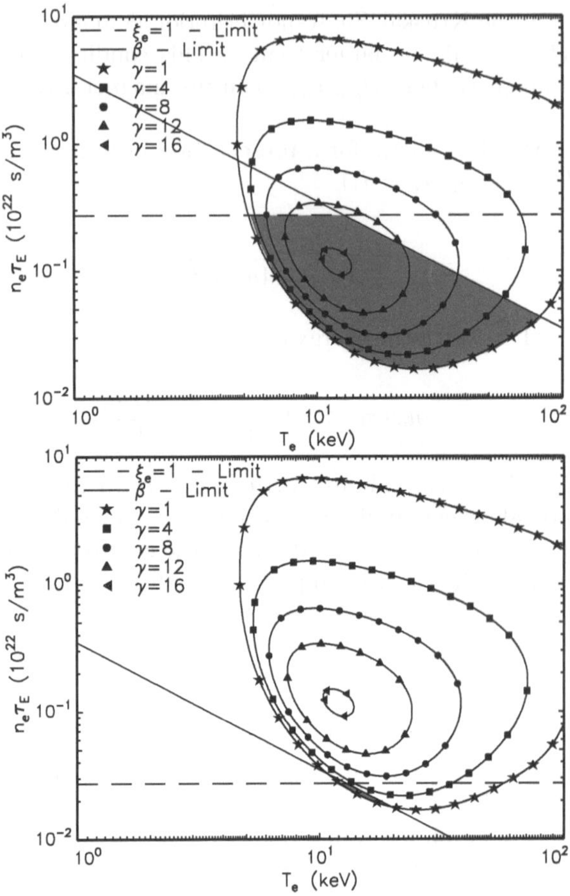

Fig. 11.2. Operation space in a future fusion reactor (shaded area) with helium as the only impurity corresponding to (3.19, 3.26, 11.11, 11.12) for $\chi_\perp = 0.1\,\mathrm{m^2/s}$ (upper) and $\chi_\perp = 1\,\mathrm{m^2/s}$ (lower), $\beta = 0.05$ and $B = 5.3\,\mathrm{T}$ (other parameters as in Fig. 3.2). To ensure a sufficient operation space, internal transport barriers (ITB) in the plasma core should help to reduce the cross-field transport

Using (11.11), the triple product can be presented in the form

$$n_\mathrm{e} \cdot k_\mathrm{B} T_\mathrm{e} \cdot \tau_\mathrm{E} = \frac{\beta^\mathrm{crit} B^2}{16\mu_\mathrm{o}} \frac{a_\mathrm{t}^2}{\chi_\perp} \,. \tag{11.14}$$

Since β can hardly exceed a value of 0.1 and the magnetic field is restricted to values below 10–15 T[1], either the transport across the magnetic field should be

[1] As a side remark, the world record for strongest magnetic field (22.5 T) has been achieved with metallic superconductor wire material by using Nb_3Al wire material and quick-heating and cooling methods by the Hitachi Corporation

reduced, or the size of the machine should be enlarged in order to obtain high triple products. This explains why the history of fusion science is the history of ever-increasing experimental devices. Increasing the size of the device is the simplest but most expensive way to obtain high triple products. Advanced scenarios, which are currently under development, could help to reduce the cross-field transport, i.e., the decrease of $\chi_\perp$, to reach the desired values.

For ITER parameters [18] ($B_\mathrm{T} = 5.3\,\mathrm{T}$, $a_\mathrm{T} = 2\,\mathrm{m}$, and assuming $\chi_\perp = 0.2\,\mathrm{m}^2/\mathrm{s}$, and $\beta^\mathrm{crit} = 0.05$) the triple products will be equal to $1.4 \times 10^6\,\mathrm{kg/(m\,s)}$ or 1.4 MPa s according to (11.14). As seen in Fig. 11.2, such parameters ensure burning conditions at $T = 10\,\mathrm{keV}$ for a value of $\gamma = 16$ and smaller. However, assuming a larger energy diffusion coefficient of $\chi_\perp = 1\,\mathrm{m}^2/\mathrm{s}$ yields only a triple product of 0.28 MPa·s, and burning cannot be sustained. It is therefore essential to reduce the radial energy transport, for example, by establishing transport barriers (see Sect. 9.3.6).

12

Material Operation Limits

Realizing a fusion energy source on earth requires a material wall surrounding the burning plasma in order to shield the plasma from the atmosphere. The particle and energy load on the plasma-facing components depends on the discharge conditions especially at the plasma edge. Vice versa, the kind of material chosen and the resulting particle fluxes due to erosion and emission affect the plasma operation. This delicate connection is determined by several criteria and general considerations.

First of all, the flux density across the last closed magnetic flux surface (LCFS) in toroidal devices increases with increasing plasma radius a_{t} [461]

$$\Gamma_{\mathrm{part.,energy}} = \frac{n}{\tau_{\mathrm{part.,energy}}} \cdot \frac{\text{plasma volume}}{\text{plasma surface}} \propto \frac{\pi a_{\mathrm{t}}^2 2\pi R_{\mathrm{t}}}{2\pi a_{\mathrm{t}} 2\pi R_{\mathrm{t}}} = \frac{a_{\mathrm{t}}}{2} \tag{12.1}$$

for constant confinement times τ and plasma density n, where a_{t} is the minor and R_{t} the major radius in toroidal geometry. However, using the presentation $\tau \simeq a_{\mathrm{t}}^2/4\chi_{\perp}$, i.e., assuming the same transport coefficient χ for small and large devices, yields the more favorable result $\Gamma \propto (1/a_{\mathrm{t}})$ that the load decreases with increasing plasma dimension (note: irrespective of the major radius R_{t}). In fact, the transport coefficient across the magnetic field is expected to be determined by microscale turbulence processes. It should not be connected to the size of the device.

The bombardment of materials by energetic particles (ions, electrons, neutral atoms, molecules) causes, in particular, the release of atoms from the material as the result of different erosion mechanisms. These atoms are emitted from the surface into the adjacent plasma region, where they are ionized and contribute there to the plasma impurities. With the erosion yield Y, which is the number of emitted atoms per incoming particle,

$$Y = \frac{\Gamma_{\mathrm{emission}}}{\Gamma_{\mathrm{influx}}} \tag{12.2}$$

and the flux density of eroded particles is given by

$$\Gamma_{\text{ero}} = \Gamma_{\text{influx}} Y \tag{12.3}$$

where Γ_{influx} is the flux density of particles striking the material surface.

In areas with grazing incidence of the magnetic field lines, the plasma ions (hydrogen, deuterium or tritium, and partly helium) mainly contribute to this flux, since the ions are moving along the field lines up to the intersecting surface. In this case, the erosion fluxes are comparatively high, but nevertheless the impurity concentration in the plasma center can be kept sufficiently low by arranging these areas far away from the plasma separatrix. In addition, a strong plasma flow toward these surface elements is required to slow down the ionized impurities and drive them back finally to their place of emission, where they stick with a high probability. In such divertor geometries, only a small fraction of ions are able to leave the divertor plasma region into the plasma center [59, 60].

In addition, the effect of prompt redeposition [397, 398] reduces the fluxes of particles leaving the surfaces right from the beginning. In contrast to the high plasma density near the divertor plates, the plasma density at the wall is rather low since the density profile in the scrape-off layer (SOL), i.e., outside the LCFS, is characterized by a short decay length λ_{n} of about 1–2 cm. Thus, emitted particles from the wall can travel a rather long distance up to the separatrix, where they are ionized by electron impact ionization, since $\lambda_{\text{iz}} \propto 1/n_{\text{e}}$. Being already close to the separatrix, the formed impurity ions have a chance to cross it and reach the plasma core by diffusion across the magnetic field. Solving the impurity problem in the divertor region as described above, the wall surfaces became, surprisingly, the main source of impurities in the plasma core [196, 401]. Despite the fact that under normal operation nearly no ions with their high erosion ability reach the wall elements. The wall can only be eroded by neutrals. Under ideal divertor operation, one would expect that the full recycling, i.e., ion impact (H^+) $\rightarrow$ molecule emission (H_2) $\rightarrow$ dissociation (H) $\rightarrow$ ionization or charge exchange (H^+), and consequently the occurrence of neutrals, should be restricted to the divertor plasma region. However in the experiments, a partial transparency of the divertor plasma for neutrals can hardly be avoided and not all of the neutrals can be pumped off. Furthermore, in steady state operation the sink/pump action of the divertor—as good or as bad as it works—should be compensated by gas puffing (and/or pellet injection) into the main plasma.

Besides erosion, the mechanical structure of fusion reactors are subjected to damage by the impact of neutrons. The resulting displacements of the lattice atoms to interstitial positions lead to vacancies, and, subsequently, to the production of voids, which are filled by helium and hydrogen implanted or produced by nuclear transmutations (Sect. 12.5).

12.1 Erosion Flux into the Plasma

For particle balance, the radial outflux of plasma ions should be equal to the influx of neutrals from the surfaces of the wall, including limiter and divertor plates

$$\begin{aligned} \Gamma_{\text{ions,out}} &= \Gamma_{\text{neutrals,in}} \\ \frac{n_{\text{H}^+}\,V_{\text{pl}}}{\tau_{\text{p}}\,S_{\text{pl}}} &= \frac{1}{2}\,n_{\text{H}_2} v_{\text{H}_2} = n_{\text{H}}\,v_{\text{H}}\;. \end{aligned} \tag{12.4}$$

Actually, it is sufficient to ensure balance of the particle fluxes integrated over the surface. The molecules H_2 emitted with thermal energies ($< 0.05\,\text{eV}$) from surfaces or injected through gas valves undergo predominantly Franck–Condon dissociation processes producing atoms with energies of $\simeq 3\,\text{eV}$. Therefore, the atomic density n_{H} in the SOL is smaller than the density of the molecules n_{H_2} by the factor $v_{H_2}/(2\,v_{\text{H}}) \approx 0.06$. Less than half of the produced atoms become ionized by electron impact ionization, while the majority transfers their electrons to the plasma ions in charge-exchange processes. Due to this charge exchange, the plasma ions are transformed to energetic neutrals with energies of about the ion temperature at the LCFS ($T_{\text{i}} \simeq 20$–$200\,\text{eV}$). These particles, no longer confined by the magnetic field, indeed have the ability to sputter the wall material, since their energy well exceeds the threshold energy for physical sputtering. In the case of carbon as wall material, already chemical erosion occurs at thermal energies (see below).

Summarizing, the flux density of particles sputtered from the wall can be estimated as

$$\Gamma_{\text{ero}} = \frac{n_{\text{e}}\,V_{\text{pl}}}{\tau_{\text{p}}\,S_{\text{pl}}}\,Y_{\text{H}^\text{o}}\,f_{\text{cx}} \tag{12.5}$$

introducing the fraction of charge exchange f_{cx} in the SOL in the case of gas puffing, or the charge-exchange efficiency in the plasma center in the case of deep pellet injection. Y_{H^o} is the sputtering yield due to hydrogen neutrals and $n_{\text{e}} \simeq n_{\text{H}^+}$.

12.2 Impurity Density in the Plasma Core

The central density of impurities, $n_{\text{imp}} = f_{\text{imp}}\,n_{\text{e}}$ [impurity ions/m^3], is directly related to a uniform influx rate of impurity atoms, $\Phi^{\text{o}} = \Gamma_{\text{ero}}\,S_{\text{pl}}$ [impurity neutrals/s] with the surface of the plasma S_{pl}. All neutrals are assumed to be ionized at a distance λ_{iz} inside the last closed flux surface. Neglecting neoclassical effects (in particular inward drift), the ion impurity transport is governed by an anomalously high diffusion perpendicular to the magnetic field lines. The corresponding diffusion equation for a certain ionization stage is then under steady state condition [462]

$$\frac{\partial n_{\text{imp}}^q}{\partial t} = \text{div}(D_\perp \,\text{grad}\, n_{\text{imp}}^q)$$
$$+ n_\text{e}\left[n_{\text{imp}}^{q-1} S_{q-1} + n_{\text{imp}}^{q+1} \alpha_{q+1} - n_{\text{imp}}^q (S_q + \alpha_q)\right] = 0 \qquad (12.6)$$

where n_{imp}^q is the density of the impurity ionization state q; S_q and α_q are the respective ionization and recombination coefficients. $D_\perp$ is the diffusion coefficient assumed to be equal for all ionization stages. The influx of neutral impurities at a velocity v penetrating into the plasma is reduced by ionization, and enhanced by recombination [462]

$$\text{div}(n_{\text{imp}}^{q=0}\, v) = -n_\text{e} n_{\text{imp}}^{q=1} S_1 + n_\text{e} n_{\text{imp}}^{q=2} \alpha_1 \,. \qquad (12.7)$$

The sum of (12.6) for all ionization stages together with (12.7) gives the steady state condition that the influx of neutrals should be balanced by diffusion losses $D_\perp(\partial n_{\text{imp}}/\partial r) = n_{\text{imp}}^{q=0}\, v = \Gamma_{\text{ero}}$ with the total ion impurity density $n_{\text{imp}} = \sum_{q=1}^{q_{\max}} n_{\text{imp}}^q$. Integration yields

$$n_{\text{tot}}(r) = \frac{v}{D_\perp} \int_0^r n_{\text{imp}}^{q=0}\, \text{d}r + n_{\text{imp}}(0) \,. \qquad (12.8)$$

With zero ion density at the plasma edge, $n_{\text{imp}}(a_\text{t}) = 0$, the central impurity density becomes simply

$$n_{\text{imp}}(0) = -\frac{v}{D_\perp} \int_0^r n_{\text{imp}}^{q=0}\, \text{d}r \simeq \frac{v\, n_{\text{imp}}^{q=0}\, \lambda_{\text{iz}}}{D_\perp} = \frac{\Gamma_{\text{ero}}\, \lambda_{\text{iz}}}{D_\perp} = \frac{\Phi^\text{o}\, \lambda_{\text{iz}}}{S_{\text{pl}} D_\perp} \qquad (12.9)$$

since the impurity neutrals are quickly ionized after crossing the separatrix at $r = a_\text{t}$, and therefore occur only in a thin shell $[a_\text{t} - \lambda_{\text{iz}}, a_\text{t}]$ near the plasma edge. In the "generation region" characterized by the thickness of about λ_{iz}, the impurity density increases to its maximum value n_{imp}, which is constant in the central region $[0, a_\text{t} - \lambda_{\text{iz}}]$. This flat profile is gradually filled up in reaching steady state. If the cross-field diffusion coefficient is very small in the edge region, then it takes some time to reach equilibrium.

The simple relation (12.9) obtained by Engelhardt and Feneberg predicts surprisingly well the experimental observations. A number of refinements have been made, for example, by implying a weak inward pinch velocity in order to describe the often observed peaked density profiles, or relaxing the hard boundary condition, which assumes an infinitely strong sink strength in the SOL (see in [2, 318, 463–465]).

It is worth mentioning that the diffusion coefficient at the edge—regardless of a possible strong radial dependence—determines the central impurity density according to (12.8), because only at the plasma edge is the density of neutral impurities, i.e., the source, different from zero.

Out of the scope of this model is the neoclassical effect of impurity accumulation in the core plasma, which gives a radial impurity profile proportional to the qth power of the plasma density [73]. This transport effect becomes very important when using high-Z elements as material for the wall [397].

12.3 Impurity Criterion

According to (12.9), the impurity concentration is related to the influx of impurities emitted from the wall. This flux has been estimated in Sect. 12.1 and can be directly used here. Similar derivations in a frame of a zero-dimensional model have been presented in [217, 376].

Combining (12.5) with (12.9) gives

$$n_{\mathrm{imp}}(r=0) = f_{\mathrm{imp}}\, n_{\mathrm{e}} = \frac{\Gamma_{\mathrm{ero}}\, \lambda_{\mathrm{iz}}}{D_{\perp}} = \frac{n_{\mathrm{e}}\, V_{\mathrm{pl}}\, Y_{\mathrm{H^o}}\, f_{\mathrm{cx}}\, \lambda_{\mathrm{iz}}}{D_{\perp} \tau_{\mathrm{p}}\, S_{\mathrm{pl}}} \tag{12.10}$$

and substituting $\tau_{\mathrm{p}} \simeq a_{\mathrm{t}}^2/(4D_{\perp})$

$$f_{\mathrm{imp}} = Y_{\mathrm{H^o}}\, f_{\mathrm{cx}}\, \frac{2\lambda_{\mathrm{iz}}}{a_{\mathrm{t}}} \tag{12.11}$$

with $V_{\mathrm{pl}} = 2\pi^2 a_{\mathrm{t}}^2 R_{\mathrm{t}}$ and $S_{\mathrm{pl}} = 4\pi^2 a_{\mathrm{t}} R_{\mathrm{t}}$. By evaluating τ_{p}, the recycling coefficient R_{cyc} does not appear, since the recycling process of the plasma ions itself is the matter of consideration here, and not enclosed in a global analysis as done for the He balance (see Sect. 3.2). Here, it is essential to count the number of erosion events at the wall and not the number of particles in a certain volume.

For a burning fusion plasma, the ratio $f_{\mathrm{imp}}/f_{\mathrm{crit}}$ (see Sect. 9.4, (9.102))

$$\gamma_{\mathrm{imp}} = \frac{f_{\mathrm{imp}}}{f_{\mathrm{crit}}} = Y_{\mathrm{H^o}}\, f_{\mathrm{cx}}\, \frac{2\lambda_{\mathrm{iz}}}{a_{\mathrm{t}}} \tag{12.12}$$

$$\times \frac{2\langle\sigma v\rangle_{\mathrm{DT}} E_{\alpha}(1-2f_{\mathrm{He}})q_{\mathrm{imp}} + 4c_{\mathrm{br}}\sqrt{k_{\mathrm{B}}T}(q_{\mathrm{imp}}^2 - q_{\mathrm{imp}}) + 4L_{\mathrm{imp}}}{\langle\sigma v\rangle_{\mathrm{DT}} E_{\alpha}(1-2f_{\mathrm{He}})^2 - 4c_{\mathrm{br}}\sqrt{k_{\mathrm{B}}T}(1+2f_{\mathrm{He}})} \le 1$$

should be kept well below unity. Remembering that (9.102) represents the most optimistic prediction. Reducing the sputtering by energetic charge-exchange neutrals, establishing conditions for near-plate ionization, and local redeposition, or simply enlarging the plasma dimension—these are the ways to achieve the impurity criterion $\gamma_{\mathrm{imp}} \le 1$.

12.4 Lifetime of Wall Elements

The erosion and redeposition behavior of different materials has been investigated in the past by specially designed probes exposed to the edge plasma of fusion experiments [173, 378, 379, 399, 466–475]. Two examples of experimental setups in such studies are shown in Figs. 12.1 and 12.2.

With an ion flux toward the plates of about

$$\Gamma_{\mathrm{ion}} \simeq c_{\mathrm{w}}\, n_{\mathrm{e}} = \sqrt{\frac{k_{\mathrm{B}}(T_{\mathrm{e}}+T_{\mathrm{i}})}{m_{\mathrm{i}}}}\, n_{\mathrm{e}}^{\mathrm{LCFS}} \simeq 8\cdot 10^{24}\, 1/(\mathrm{m}^2\mathrm{s}) \tag{12.13}$$

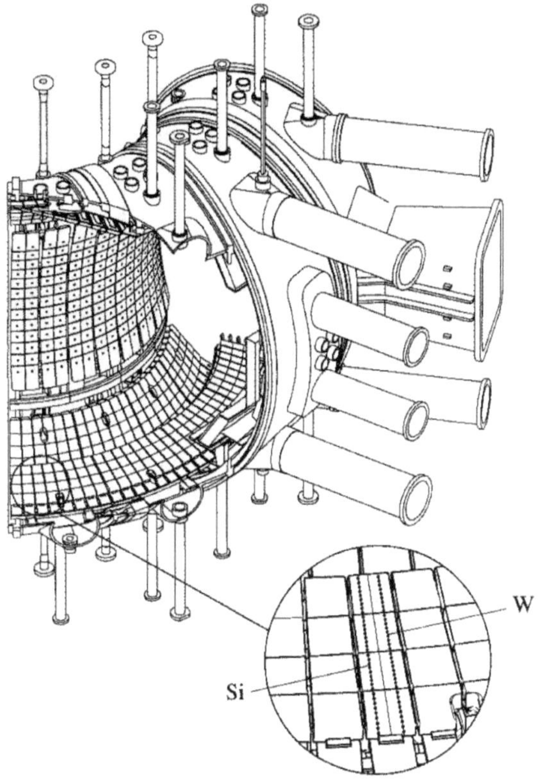

Fig. 12.1. The erosion and redeposition of silicon and tungsten evaporated onto the divertor plates of ASDEX-Upgrade has been measured and compared with simulation results in [173]. A redeposition of tungsten has been found close to the original marker spots, because of prompt redeposition during the first gyration. The eroded silicon atoms are transported far away from their place of origin. The observed erosion of the markers could only be explained by considering the protection action of deposited carbon impurities

along the magnetic field lines (with typical parameters $n_e = 2.5 \times 10^{20}\,1/\text{m}^3$, $T_e = T_i = 10\,\text{eV}$, $m_i = m_D$) and an average sputtering yield of about $Y = 0.01$, the eroded thickness of a carbon plate would be

$$\Delta_{\text{ero}} = \frac{Y\,\Gamma_{ion}\,t_{\text{discharge}}}{n_{\text{carbon}}} \simeq 0.7\,\text{mm} \tag{12.14}$$

after a discharge of 1000 s, where n_{carbon} is the atomic density of graphite, $n_{\text{carbon}} = 11.3 \times 10^{28}\,1/\text{m}^3$. A 1-cm-thick plate would already "disappear" after 15 discharges. Fortunately, tilting the magnetic field lines with respect to the plates gives a reduction factor of $\sin\alpha \approx \alpha = 0.02$–$0.04$ resulting in a 500 discharge campaign. With about 10 discharges per day, new plates have to be installed every 2 months. Obviously, an additional reduction effect is

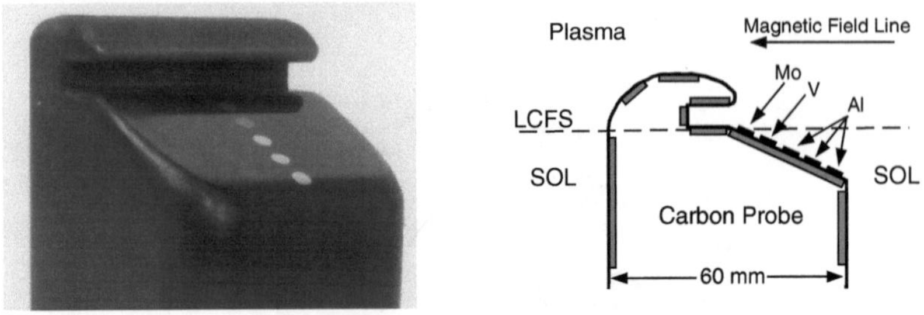

Fig. 12.2. Plasma-induced erosion and redeposition has been measured by surface layer analysis of a carbon probe with evaporated markers before and after exposure in the boundary plasma of the TEXTOR tokamak. Well-defined surface deposits of different materials were evaporated as dots of 3 mm diameter with thicknesses of 220 Å for Mo, 200 Å for V, and 500 Å for the three Al markers. Small probes (shaded areas) have been cut out of the probe after exposure for surface analysis. The hope to retrap most of the atoms eroded from the markers in the special slit has been fulfilled only for Mo atoms. Their redeposition in the slit came most likely from directly impinging sputtered neutrals. It turned out that the plasma parameters above the probe were too low to initiate sufficient ionization right in front of the slit. Most of the sputtered atoms have been ionized above the probe head. A deeper exposure of the probe into the plasma seemed desirable, but threatened a possible destruction of the probe [379]

urgently required. The gross erosion has to be minimized by redeposition in order to achieve an acceptable level of net erosion.

Sputtered atoms mostly leave the surface as neutrals. After ionization, their movement is governed by the magnetic and electric fields and by collisions with the plasma ions. If the ionization length is comparable with the gyration radius, prompt redeposition becomes possible and occurs with a certain probability P_{pr}. In other cases, most of the impurity ions are pushed back onto the surface due to friction with the upstreaming plasma flow (Sect.12.4.1). In the ideal case, i.e., zero net erosion, the ions should be pushed directly back to their place of origin, where they started as neutrals.

Net erosion is obtained by balancing the gross erosion due to all plasma ions i (including the material impurities) with redeposition of the sputtered material

$$\Gamma_{\mathrm{ero}}^{\mathrm{net}} = \Gamma_{\mathrm{ero}}^{\mathrm{gross}} - \Gamma_{\mathrm{redep}} \qquad (12.15)$$

where the fluxes of eroded $\Gamma_{\mathrm{ero}}^{\mathrm{gross}}$ and redeposited particles Γ_{redep} are given by the infinite series (Fig. 12.3)

$$\begin{aligned} \Gamma_{\mathrm{ero}}^{\mathrm{gross}} &= \Gamma_{\mathrm{e}} \sum_i Y_i f_i \Big(1 + P_{\mathrm{redep}} [Y_{\mathrm{self}} + 1 - s] \\ &\quad + (P_{\mathrm{redep}} [Y_{\mathrm{self}} + 1 - s])^2 + ... \Big) \\ &= \frac{\Gamma_{\mathrm{e}} \sum_i Y_i f_i}{1 - P_{\mathrm{redep}} [Y_{\mathrm{self}} + 1 - s]} \end{aligned} \qquad (12.16)$$

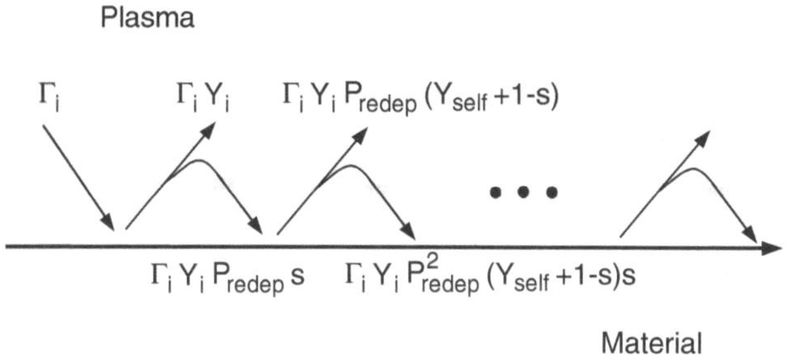

Fig. 12.3. Infinite erosion and redeposition cycle in the zero-dimensional model

and

$$\begin{aligned}\Gamma_{\text{redep}} &= s\,\Gamma_{\text{e}}P_{\text{redep}}\sum_i Y_i f_i\Big(1 + P_{\text{redep}}[Y_{\text{self}} + 1 - s] \\ &\quad +(P_{\text{redep}}[Y_{\text{self}} + 1 - s])^2 + ...\Big) \\ &= \frac{s\,\Gamma_{\text{e}}P_{\text{redep}}\sum_i Y_i f_i}{1 - P_{\text{redep}}[Y_{\text{self}} + 1 - s]}\,, \qquad (12.17)\end{aligned}$$

which converge as long as $P_{\text{redep}}[Y_{\text{self}} + 1 - s] < 1$. The probability of prompt redeposition is included here in P_{redep}, i.e., $P_{\text{pr}} \leq P_{\text{redep}}$. The concentrations of the plasma ions f_i are given with respect to the electron density according to charge neutrality $\sum_i f_i q_i = 1$. The electron flux density is $\Gamma_{\text{e}} = n_{\text{e}}c_{\text{s}}\sin\alpha$ (note that $\Gamma_{\text{e}} = \Gamma_{\text{i}}$ under floating conditions (Chap. 7)). Self-sputtering of the material characterized by the sputtering coefficient Y_{self} initiates a repetitive cycle of erosion and redeposition represented by the series (12.16) and (12.17). The redeposited material impurities stick to the surface with the probability s. Using (12.16) and (12.17) in (12.15) yields

$$\Gamma_{\text{ero}}^{\text{net}} = \Gamma_{\text{e}}\,\frac{(1 - sP_{\text{redep}})\sum_i Y_i f_i}{1 - P_{\text{redep}}[Y_{\text{self}} + 1 - s]} = \Gamma_{\text{e}}\,Y_{\text{eff}}\,. \qquad (12.18)$$

Introducing the effective sputtering yield Y_{eff} considers the effects of self-sputtering, redeposition, and sticking. As seen from (12.18), the effective sputtering can be very small, if the sticking coefficient s as well as the redeposition probability P_{pr} are close to unity. A large self-sputtering yield Y_{self} can lead to excessive sputtering. If $P_{\text{redep}}[Y_{\text{self}} + 1 - s] \geq 1$, then runaway erosion occurs. In the case of $s = 1$ and $Y_{\text{self}}{=}0$, the relation (12.18) simply reduces to $\Gamma_{\text{ero}}^{\text{net}} = \Gamma_{\text{e}}(1{-}P_{\text{redep}})\sum_i Y_i f_i$. With zero sticking and negligible self-sputtering, the effective sputtering yield is equal to $Y_{\text{eff}} = \sum_i Y_i f_i/(1 - P_{\text{redep}})$.

Surface erosion is a critical issue for the lifetime of the divertor components in future fusion experiments such as ITER. It was shown in [476] that the new divertor concept [56], where heat is dissipated mainly by volume radiation and charge exchange, does not lessen this problem, but even aggravates it by reducing the favorable effect of redeposition.

12.4.1 Simple Geometrical Model of Redeposition

Exposing a solid probe with a surface area of $l_x \cdot l_y$ into the edge plasma of fusion experiments (Fig. 12.4), the surface concentration of each element $n_s^k(x,y,t)$ at a certain point is changed by erosion processes and by redeposition according to

$$\begin{aligned}\frac{\partial n_s^k(x,y,t)}{\partial t} &= \Gamma^k(x,y,t)(1-R_N^k) \\ &\quad - \sum_i \Gamma^i(x,y,t) Y_{i\to k}(E_o, \Omega_o, n_s^k(x,y,t)) \\ &\quad + \int_0^{l_x}\int_0^{l_y} \frac{dx'dy'}{\Delta S} \sum_i \int_E \int_\Omega \int_0^\infty \frac{1}{\lambda_{iz}^k(r)} \exp\left(-\int_o^r \frac{dr'}{\lambda_{iz}^k(r')}\right) \\ &\quad \times \Gamma^i(x',y',t) \frac{\partial^2 Y_{i\to k}}{\partial\Omega\partial E}(E_o, \Omega_o, n_s^k(x',y',t); E, \Omega)\, dr d\Omega dE\end{aligned} \tag{12.19}$$

where $\Gamma^i(x,y,t)$ are the plasma ion flux densities of species i bombarding the surface, R_N^k is the particle reflection coefficient, $Y_{i\to k}(E_o, \Omega_o, n_s^k)$ are the sputtering yields of the element k under bombardment with ions i, $\partial^2 Y_{i\to k}/(\partial\Omega\partial E)$ represent the energy and angular distributions of sputtered particles, which depend on the incident energy E_o and the angle of incidence $\Omega_o = (\theta_o, \varphi_o)$, and $dS = dxdy$ denotes the small surface element around the point of deposition (x,y) [378]. The term

$$\frac{1}{\lambda_{iz}^k(r)} \exp\left(-\int_o^r \frac{dr'}{\lambda_{iz}^k(r')}\right) dr \tag{12.20}$$

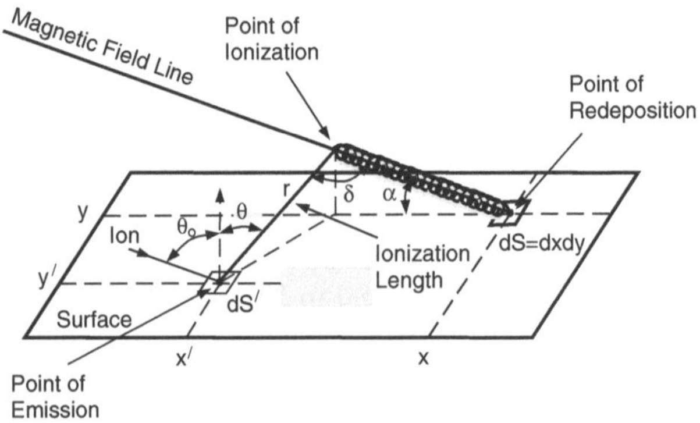

Fig. 12.4. Model geometry

is the probability for the sputtered neutrals moving along the direction $\Omega = (\theta, \varphi)$ to be ionized in the interval $[r, r + \mathrm{d}r]$ (Fig. 12.4). The distance R from the point of sputtering (x', y') to the point of deposition (x, y) is determined by the length of flight before ionization r, the angle θ, the azimuthal angle φ, and the angle α between the surface and the magnetic field lines

$$r = R\sqrt{\tan^2\alpha/(\sin^2\theta\tan^2\alpha + \cos^2\theta + \sin 2\theta\cos\varphi\tan\alpha)} \tag{12.21}$$

with $R = \sqrt{(x'-x)^2 + (y'-y)^2}$ (Fig. 12.4). The mean ionization length λ_{iz} depends on the local plasma parameter above the surface and the velocity of the sputtered atoms.

The first term on the right-hand side in (12.19) describes the deposition of plasma impurities with a sticking probability $(1 - R_{\mathrm{N}})$. The erosion of the surface element k by the incoming flux of all ion species i is given by the second term. The third term represents the redeposition of atoms which have been sputtered, ionized in the plasma, and returned to the surface. Note that the sputtering yields for the different species depend on the actual surface concentrations.

The sputtered atoms move along straight trajectories until they are ionized in the plasma. After ionization, they precess around the magnetic field lines, while retaining their initial parallel velocity $v_{\|\mathrm{o}} = -\sqrt{2E/M}\cos\delta$ (Fig. 12.4). Those ions moving away from the surface will be slowed down by the upstreaming plasma flow. They may finally be pushed and accelerated towards the surface. The change of parallel velocity is given—in the simplest case—by collisional friction according to

$$\frac{\mathrm{d}v_{\|}}{\mathrm{d}t} = \nu(v_{\mathrm{pl}} - v_{\|}) \tag{12.22}$$

with the collisions frequency ν and the velocity of the plasma flow v_{pl}. Cross-field transport can be taken into account approximately by introducing a diffusion coefficient $D_{\perp}$ such that an impurity is replaced by a step $\sqrt{4D_{\perp}\Delta t}$ in time Δt.

After reaching the sheath, the ions increase their velocity due the acceleration in the electric sheath. They impinge on the surface with an energy of $(\phi_{\mathrm{w}} \simeq -3k_{\mathrm{B}}T_{\mathrm{e}}/e)$

$$E_{\mathrm{o}} = M\frac{v_{\|}^2}{2} - \phi_{\mathrm{w}}Q \simeq M\frac{v_{\|}^2}{2} + 3k_{\mathrm{B}}T_{\mathrm{e}}q \tag{12.23}$$

where M is the mass of the ion, Q its charge, and q its charge state.

Such geometrical models, where the point of deposition is determined by the point of emission (and the direction and energy of emission), have been widely used in erosion/redeposition studies (for example in [477–481]) as well as with the REDEP code [421, 476, 482], which has been applied to predict the erosion/redeposition behavior in ITER.

To be more accurate in describing the impurity transport in the edge plasma, Monte Carlo codes such as WBC and ERO have been developed. This presents an opportunity to consider a large number of different reactions—not just ionization. For example, several hundred reactions are important by modeling the transport of hydrocarbons C_xH_y. The assumption of guiding center motion fails near the surface. The point of deposition is shifted due to the electric drift in the sheath (Sect. 12.4.2). The analysis of such effects requires the correct solution of the equation of motion (Sect. 9.3.9).

12.4.2 Net Erosion at Divertor Plates

For a toroidally symmetric tokamak, the large transport in the toroidal and poloidal directions is irrelevant and can be ignored. The question, whether net erosion or net deposition dominates at the plates, depends on the radial transport. The distance in the radial direction Δy from the point of emission to the point of deposition is mainly determined by the magnetic field geometry $\Delta y_\theta = \lambda_{\mathrm{iz}}/\tan\theta$, where λ_{iz} is the ionization length and θ being the inclination angle of the target plate with respect to the magnetic flux surface (Fig. 12.5). As shown in [173, 427, 483], an additional transport effect has to be taken into account: the $\boldsymbol{E} \times \boldsymbol{B}$ motion of the impurity ions in the electric sheath especially in the case of nearly parallel inclination of the magnetic field lines (Fig. 12.5), since the electric field in the sheath is directed toward the surface and the magnetic field parallel to it. It is noteworthy that this motion is not the "classical" drift motion of the gyro-center of the particles in crossed electric and magnetic fields, but rather characterized by strongly distorted trajectories of the impurity ions, where hardly a gyro-center can be defined, due to the inhomogeneity of the electric field. As discussed in Sect. 7.5, the sheath thickness is on the order of the gyro-radius of the hydrogen plasma ions in the case of nearly parallel magnetic field incidence, while the gyro-radii of the impurity ions are larger. Therefore, the gradient of the electric field in the sheath is too strong to allow a regular gyro-motion of the ions.

The described drift effect nevertheless leads to a drift-like shift of the redeposition point—to the point D_1 instead of point D_2 (Fig. 12.5).

Assuming that the electron density (n_{e}) and temperature (T_{e}) profiles are constant along the field lines and decay exponentially from the separatrix across them

$$
\begin{aligned}
n_{\mathrm{e}}(z,y) &= n_{\mathrm{e}}^{\mathrm{o}} \exp[-(y \sin\theta + z \cos\theta)/\lambda_{\mathrm{n}}] \\
T_{\mathrm{e}}(z,y) &= T_{\mathrm{e}}^{\mathrm{o}} \exp[-(y \sin\theta + z \cos\theta)/\lambda_{\mathrm{n}}]
\end{aligned}
\qquad (12.24)
$$

where z is the distance above, y the distance along the plate, $n_{\mathrm{e}}^{\mathrm{o}}$ and $T_{\mathrm{e}}^{\mathrm{o}}$ are the values at the separatrix (the plane defined by $y\sin\theta + z\cos\theta = 0$), and λ_{n} is the decay length (Fig. 12.5). Note that in the "private region" (from Fig. 4.3) ($y\sin\theta + z\cos\theta < 0$) $\lambda_{\mathrm{n}} = \lambda_{\mathrm{n,l}}$ becomes negative. In the outboard region of the separatrix ($y\sin\theta + z\cos\theta > 0$) $\lambda_{\mathrm{n}} = \lambda_{\mathrm{n,r}}$ is constant and positive, but

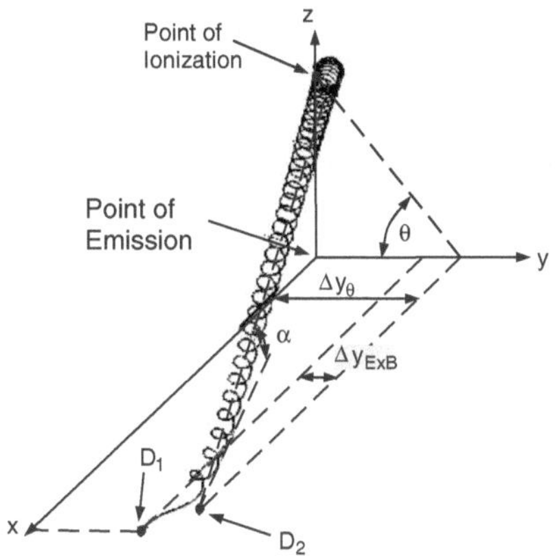

Fig. 12.5. Trajectory of a sputtered atom. The magnetic field is directed toward the viewer. The variable x is the toroidal direction and y the radial direction. The xy-plane is the divertor plate. In the electric sheath right above the surface, the ion drifts to the left and impinges at point D_1. Without this drift, the particle would be deposited at point D_2

$\lambda_{\rm n,r}$ may be different from $|\lambda_{\rm n,l}|$. The plasma parameters have their highest values at the separatrix; hence, the maximum gross erosion is expected at the point where the separatrix intersects the divertor plate. Thus a sample profile of gross erosion $\Gamma_{\rm ero}(y)$ can be represented as shown in Fig. 12.6. In the simplest case, i.e., assuming a constant (and small) radial transport step Δy, no self-sputtering, and all sputtered atoms return to the target, the deposition profile $\Gamma_{\rm dep}(y)$ is defined as $\Gamma_{\rm dep}(y) = -\Gamma_{\rm ero}(y - \Delta y)$. The resulting net erosion profile $\Gamma_{\rm ero/dep}$ can be calculated by

$$\begin{aligned}\Gamma_{\rm ero/dep}(y) &= \Gamma_{\rm ero}(y) + \Gamma_{\rm dep}(y) \\ &= \Gamma_{\rm ero} - \Gamma_{\rm ero}(y - \Delta y) \approx \frac{\partial \Gamma_{\rm ero}}{\partial y}\Delta y \ .\end{aligned} \tag{12.25}$$

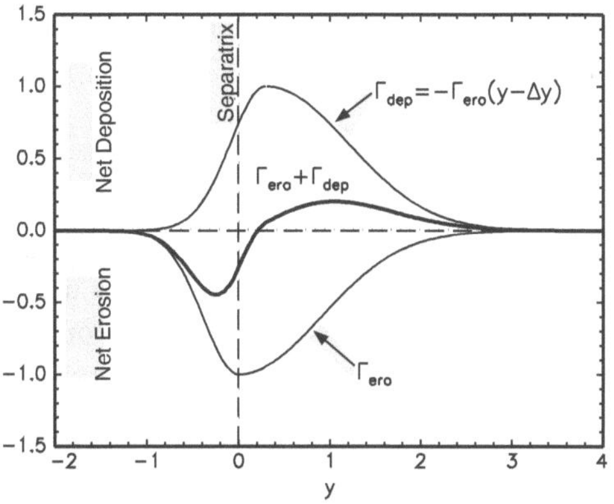

Fig. 12.6. Sample erosion/deposition profile across the divertor plate

Experiments show that the profiles of net erosion along the divertor plates are indeed of this type [473]. Additional confirmation is obtained by calculation with the REDEP code by Brooks [421, 482]. The material which is mainly eroded in the separatrix region, i.e., the region with the highest plasma parameters, is transported outward into regions with reduced plasma load and deposited there.

A straightforward way to reduce the net erosion is to decrease the electron temperature. This might not be possible to the desired extent and is, furthermore, only successful in suppressing physical sputtering. In contrast, chemical erosion does not show such a significant dependence on T_e and a decrease of the electron temperature leads rather to increasing ionization lengths and, subsequently, to a reduction of the favorable redeposition effect.

Another way is to minimize the transport length Δy. This is possible using the $\boldsymbol{E} \times \boldsymbol{B}$ drift effect (note: without changing the plasma parameters). For this, the relation

$$\Delta y = \Delta y_\theta + \Delta y_{E\times B} \rightarrow 0 \tag{12.26}$$

must be satisfied. The term $\Delta y_{E\times B}$ can only be obtained numerically. For this purpose, Monte Carlo codes such as the ERO code [378] or WBC code [424], where the ion transport equation is solved in detail, should be used.

For a more quantitative analysis, the amount of eroded/deposited material has to be determined for each point i at the divertor plate by the balance equation

$$\Gamma^i_{\mathrm{ero/dep}} = -(\Gamma^i_{\mathrm{pl}}\, Y^i_{\mathrm{pl}} + \Gamma^i_{\mathrm{c}}\, Y^i_{\mathrm{self}})\,(1 - s\, P^i_{\mathrm{pr}}) + s\, \Gamma^i_{\mathrm{c}} \tag{12.27}$$

where $\Gamma_{\mathrm{ero/dep}}$ is the flux density of eroded/deposited particles, Γ_{pl} denotes the flux densities of the plasma ions (fuel and impurity ions, but not the material impurities), Y_{pl} is the sputtering yield due to plasma ions, Y_{self} is the sputtering yield caused by the eroded and then deposited target atoms, and s is the sticking coefficient. The probability of prompt redeposition P_{pr} is determined by the ratio of the ionization length λ_{iz} to the gyro-radius ρ according to (9.76). The self-sputtering by promptly redeposited target atoms is neglected in (12.27).

The amount of material originating from other areas of the divertor plate i' and then deposited, mainly due to the friction with the upstreaming plasma flow at point i, is denoted by Γ^i_{c}

$$\Gamma^i_{\mathrm{c}} = \sum_{\forall i' \neq i} (\Gamma^{i'}_{\mathrm{pl}}\, Y^{i'}_{\mathrm{pl}} + \Gamma^{i'}_{\mathrm{c}}\, Y^{i'}_{\mathrm{self}})\,(1 - s\, P^{i'}_{\mathrm{pr}}) \tag{12.28}$$

in (12.27). In [483], it is shown how Γ_{c} and the following $\Gamma_{\mathrm{ero/dep}}$ can be determined for each point along the divertor plate with the aim of obtaining the radial profile of erosion or deposition. It turned out in the analysis that the main parameter is the displacement $\Delta y = \Delta y_\theta + \Delta y_{E\times B}$. Using the normalized displacement γ_{ero},

$$\gamma_{\mathrm{ero}} = \frac{\Delta y}{|\lambda_{\mathrm{n,l}}|} = \frac{\Delta y_\theta + \Delta y_{E\times B}}{|\lambda_{\mathrm{n,l}}|}\,, \tag{12.29}$$

the ratio of net to gross erosion is found to be proportional to $\tanh(C\,\gamma_{\rm ero})$, where C is a fitting coefficient [483].

The peak net erosion at the point where the separatrix strikes the divertor plate, given as the thickness of the eroded material layer, can be estimated by

$$\Delta_{\rm ero}^{\rm divertor} = \frac{t_{\rm exp}\,\Gamma_{\rm pl}\,Y_{\rm pl}\,(1 - s\,P_{\rm pr})}{n_{\rm o}\,(1 - P_{\rm pr}[Y_{\rm self} + 1 - s])}\,\tanh(C\,\gamma_{\rm ero}) \qquad (12.30)$$

where $t_{\rm exp}$ is plasma exposure time and $n_{\rm o}$ the atomic density of the target material. For a divertor target made of carbon, $C \approx 3.5$. Note that the contribution from prompt redeposition is distinguished in (12.30) from the contribution of redeposition due to the upstreaming plasma flow, which is considered by the $\tanh(C\,\gamma_{\rm ero})$ term (compare with (12.18)).

With (12.30), one can estimate the peak erosion at the divertor plates in a quite realistic way, namely taking into account the total erosion by the plasma ions, the self-sputtering by the target atoms, the local redeposition, i.e., prompt redeposition, as well as the redeposition due to friction with the plasma ions. The criterion of zero net erosion (12.26) reads then as $\gamma_{ero} = 0$, and can be used to optimize the magnetic field angles α and θ.

The effect of additional erosion processes can also be calculated by means of (12.30) as far as the corresponding sputtering yields and typical radial transport steps Δy are known. Target material not retained in the divertor and/or material, which eroded from wall surfaces in the main chamber, can be considered by adding a deposited thickness $\Gamma_{\rm dep}\,t_{\rm exp}/n_{\rm o}$ in (12.30), which reduces of course the net erosion. Under detached plasma conditions, the divertor can turn from a net erosion into a net deposition zone [484].

The way to reduce the peak net erosion at the divertor plates is to minimize the radial transport step. The displacements Δy_θ and $\Delta y_{E\times B}$ should have opposite signs, which can be achieved by an appropriate arrangement of the divertor plates with respect to the magnetic field structure [483]. For technical reasons, this may not always be possible for both inner and outer divertor plates. The magnetic field angles θ and α can be optimized according to (12.26). Interestingly, by simply changing the magnetic field direction the peak erosion may consequently increase by a substantial factor.

It should be noted that the mechanical and thermal properties of the redeposited material can be different from the base material mainly due to the large content of deuterium within it [485]. Thus, even under the condition of zero net erosion, damage of the divertor plates cannot be fully avoided.

12.4.3 Net Erosion at Wall Plates

Particles emitted from the wall have only a small chance to return to the place of emission after they become ionized in the edge plasma. They may diffuse into the central plasma or out of the confined plasma region, therefore, losing the *memory* of where they came from. In fact, zero local redeposition would not immediately cause a problem with respect to net erosion, as the resulting deposition onto the wall surfaces would be as uniform as erosion. Unfortunately, deposition of eroded wall particles occurs predominantly at the divertor plates. The impurity ions in the scrape-off layer are led parallel along the magnetic field lines usually much faster into the divertor region as they can move across the magnetic field to the wall. Hence, the net erosion at the wall is nearly equal to the gross erosion at the wall (see (12.5)). The amount eroded is given by

$$\Delta_{\rm ero}^{\rm wall} = \frac{Y\,\Gamma_{\rm H^\circ}\,t_{\rm discharge}}{n_{\rm carbon}} = \frac{Y_{\rm H^\circ} n_{\rm e}\,V_{\rm pl}\,t_{\rm discharge}}{\tau_{\rm p}\,S_{\rm pl} n_{\rm carbon}}\,f_{\rm cx}$$
$$= Y_{\rm H^\circ}\,n_{\rm e}\,t_{\rm discharge}\,f_{\rm cx}\,\frac{2D_\perp}{n_{\rm carbon}\,a_{\rm t}} \tag{12.31}$$

with $\tau_{\rm p} \simeq a_{\rm t}^2/(4D_\perp)$, $V_{\rm pl} = 2\pi^2 a_{\rm t}^2 R_{\rm t}$ and $S_{\rm pl} = 4\pi^2 a_{\rm t} R_{\rm t}$. Taking typical values of $Y_{\rm H^\circ} = 0.02$, $n_{\rm e} = 2\times10^{20}\,1/{\rm m}^3$, $t_{\rm discharge} = 1000\,{\rm s}$, $f_{\rm cx} = 0.5$, $D_\perp = 1\,{\rm m}^2/{\rm s}$, $n_{\rm carbon} = 11.3\times10^{28}\,1/{\rm m}^3$, and $a_{\rm t} = 0.5\,{\rm m}$, relation (12.31) gives $\Delta_{\rm ero}^{\rm wall} \simeq 0.08\,{\rm mm}$ per discharge.

12.5 Neutron Irradiation

ITER will be the first fusion device in which the degradation of properties due to neutron damage is a prime concern. The neutron flux will set off a variety of effects such as hardening, embrittlement, change of volume like void swelling, shape changes like irradiation creep and growth, activation, tritium permeation, and retention. Resulting failure processes are fracture (strength, ductility), creep rupture, fatigue corrosion, and oxidation. Especially at low irradiation temperatures, the thermal conductivity degrades due to atom displacements from positions in the lattice to interstitial positions. Large dislocation loops and defect clusters are observed. The lifetime of electrical and electronic parts will be reduced to rather short times.

Neutrons produce two types of damage which can significantly alter the properties of structural materials: (1) transfer of kinetic energy from the neutrons to the atoms in the lattice that leads to vacancies and interstitials, and (2) nuclear reactions resulting in transmutations that alter the composition. The changes in microstructure and microchemistry, which result from processes of diffusion, agglomeration, and annihilation of these point defects and new transmuted elements, can significantly alter the mechanical, physical, and chemical properties of materials. In a fusion reactor, neutrons with energies up to 14.1 MeV appear, compared to a maximum energy of about 4 MeV in the neutron spectrum of fission. It has been confirmed that no new type of damage is created by 14 MeV neutrons [486]. The displacement damage (displacements per atom (dpa)) can be calculated using the neutron energy spectrum by simulating the primary recoil spectra, damage energy deposition, and displacement cascade formation. In principle, a description of defect evolution combined with dislocation theory and micromechanical models can predict the effect of neutrons [486] to some degree, but the development of new materials depends on experiments using an adequate neutron source, which is not available at present.

The dpa is the average number of times an atom has been knocked from its original lattice position. It is used to compare irradiation doses. For ITER, the integrated neutron dose is still limited and will not exceed values of 1 dpa for the components of the first wall [487]. However, in a fusion reactor the tremendous number of several hundred dpa are expected [486]. For a fluence of about 10^{25} 14 MeV neutrons per square meter approximately 10 dpa are expected. For comparison, already at < 0.01 dpa, the thermal conductivity of graphite starts to degrade, while mechanical properties are not yet affected.

Due to nuclear transmutations helium and hydrogen are produced deep in the bulk material. Especially helium, accumulating in the voids of the lattice, can build-up very high pressures producing swelling of the volume. Helium has a very low solubility in the lattice and, even at very low concentration, cavities (bubbles) are formed in which the surface tension forces are approximately balanced by the internal gas pressure. With increasing neutron damage the tritium retention and, subsequently, the inventory increase until saturation is reached at 0.1–1 dpa [488].

Metallic materials dramatically change their mechanical properties under neutron irradiation, for example, their flow stresses increase, and their ductility decreases. Both effects depend strongly on the irradiation temperature, but the saturation temperature is rather insensitive to the temperature [488]. Swelling is observed in austenitic steel above 400 degrees centigrade. Fatigue, fatigue-crack growth, and fracture toughness are also influenced by neutron irradiation.

Low activation materials should be used, since activation is an intrinsic feature of the DT fusion accompanied by the production of energetic neutrons. Depending on the composition of the structural materials, decay times up to several thousand years are possible. This should clearly be avoided in order to reduce the problems with radioactive waste after shutdown of the reactor. Whether fusion is accepted by the public as a clean and viable energy source depends on the careful selection of materials and alloys, in which the induced radioactivity decays at least partly within a few hours or days to substantially lower values. For example, N, Nb, and Mo produce radioactivity with very long half-lives. In Fe-Ni-Cr austenitic steels the Ni content should be reduced by replacing it with Mn. Mo and Nb have to be eliminated as an alloying element, e.g., in high-temperature martensitic steels [486]. Nevertheless, any maintenance work in the reactor will require remote handling due to the background radiation level. Human access to the device will not be possible shortly after operation is stopped due to the large decay times of most of the used materials.

Under pulse operation with induced high thermal stresses, the first wall components are exposed to fatigue conditions. The neutrons will induce radiation creep in the metallic materials that can drastically reduce the lifetime. The radiation creep–fatigue interaction is a major concern for the expected lifetime of the first wall materials [437].

13

Choice of Materials

It is the long and often contradictory list of requirements that makes an adequate choice of materials for fusion application so difficult. Plasma-facing and structural materials are to be distinguished. There are requirements dictated by plasma performance (e.g., the necessity to minimize impurity contamination and the radiation losses of the confined plasma), requirements of engineering integrity, of component lifetime (e.g., to withstand thermal stresses, to have acceptable erosion), and requirements of safety (minimize tritium and radioactive dust inventories). A good plasma-facing material should have the following properties:

- Low sputter yield (physical and chemical)
- Low thermal erosion during off–normal heat loads such as disruptions and ELMs
- Low atomic number
- Compatibility with hydrogen isotopes (low chemical interaction rate, low trapping probability, low permeability)
- Low chemical reactivity with steam and oxygen in the case of loss of vacuum, or loss of coolant
- Good out gassing behavior
- Low tritium retention, and therefore low T inventory
- High melting point
- Low vapor pressure
- High thermal conductivity
- Low thermal expansion
- High strength
- Low elastic modules
- High fracture stress
- High thermal shock resistance
- Low activation by neutrons, low afterheat,

- Low neutron–induced degradation of material properties (such as the thermal conductivity)
- Resistance to fatigue damage
- Good mechanical properties, producibility and handling, low cost

The first experiments at the beginning of fusion research suffered, besides other problems, from high radiation levels in the plasma center due to middle-Z elements such as iron and nickel when using a vessel made of steel. The elements from the center of the periodic table with their good mechanical properties have in general rather high erosion yields due to low sublimation energies, and, to make it worse, a high ability to radiate in the plasma. Only by covering the wall, especially the limiters, by low-Z materials such as carbon, was further progress in fusion research possible. High-Z elements feature a very low erosion yield but their radiation capability is quite threatening. Thus, the higher tolerable impurity concentration seems to favor wall materials of low atomic number (see Sects. 9.4 and 12.3). Having different conditions at different positions at the wall, for example, high heat and particle fluxes at limiter and divertor plates, and less demanding conditions with fluxes of charge-exchange neutrals onto baffles and wall elements, one can try to optimize the plasma-facing materials for the specific regions (Fig. 13.1).

For commercial power stations, it will no longer be possible to find adequate solutions among existing materials and alloys. It will be necessary to develop new materials for these demanding applications and to characterize them before construction. Only if this problem is tackled soon and with substantial effort can one hope to finish the development in time [489]. In fact, no single alloy has properties that meet all the listed requirements. In particular, none of the existing materials for the wall has been shown to resist to fluences up to 200 dpa (displacements per atom) without drastic changes of its properties. In addition, one should be aware that pulsed operation, probably inevitable in tokamaks, strongly reduce the lifetime of all constructions and materials.

Not only is the choice of the materials directly facing the plasma critical, but the target plates design should also demonstrate distinguished features such as optimized heat removal capability based on an effective bonding of the armor material to the actively cooled metal structure.

The reference design for the ITER divertor plate consists of 5 mm to 1 cm thick graphite fiber composite tiles brazed to a water-cooled metallic heat sink. Good thermal contact between the graphite tiles and the heat sink material is crucial. The quality requirements for uniform brazing of the graphite tiles to the heat sink are extremely high. Poor thermal contact over a small region due to poor braze could lead to very large temperature gradients and potentially to melting of the heat sink, and to rupturing of the water cooling tubes embedded in the heat sink. The engineering margin of such designs is very small and the hazard that initially small regions of bad thermal contact between the brazed graphite and the heat sink could enlarge in size during operation is substantial.

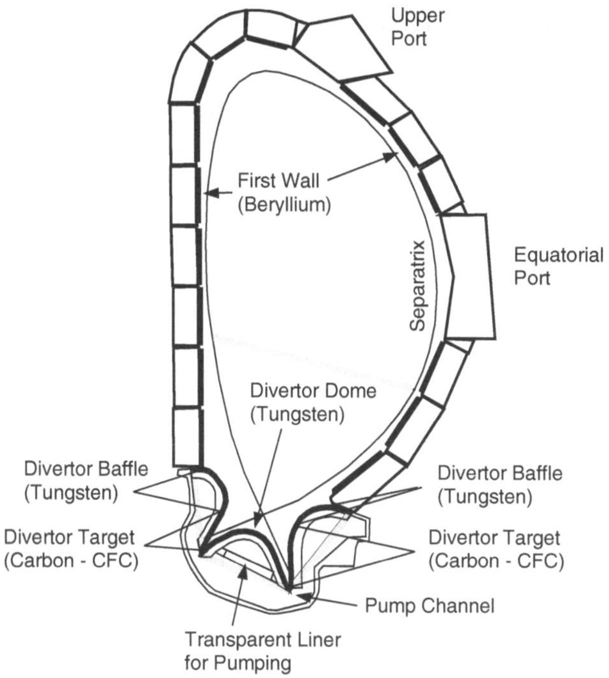

Fig. 13.1. Cross-section of ITER with plasma-facing components including the first wall, the V-shaped divertor slots with the divertor targets, baffles, and the divertor dome [490]

In general, surface roughness, which is especially critical for carbon fiber composites, is an issue of concern. Most of the processes associated with plasma–material interaction are affected. Each inhomogeneity on the surface represents a leading edge and is subjected to high heat load leading to melting or sublimation. The optimistic anticipation that the plasma will smooth the surface by eroding the peaks is not justified. On the contrary, an even more pronounced peak structure of the material is formed. Small amounts of metal impurities may protect small areas from erosion, whereas the surrounding areas are sputtered. In the case of carbon fiber composites (CFC), the material of the carbon matrix is usually more easily sputtered than the carbon fibers.

Besides the wall plates, other components in the vessel such as mirrors for diagnostics purposes should also withstand the high heat and particle fluxes. In the case of mirrors, often a sharp drop of reflectance in certain wavelength regions have been observed due to transformation of the surface layer, formation of oxide, hydroxide, and hydrocarbon layers [491, 492]. Vacuum annealing at elevated temperatures could result in partial restoration of the reflectance.

13.1 Candidates of Materials

The list of candidates out of the periodic table is rather short, at least concerning the plasma-facing materials. Metals with intermediate atomic numbers, such as copper or iron, are generally not favored because of high sputtering leading to plasma contamination. The observed breathing phenomenon in LHD is an impressive example [493, 494]). Similarly, the bad experience with metal walls in the earlier JET [495] is not encouraging. At the same time, their thermophysical properties at high temperatures are insufficient [351]. Vanadium, tantalum, and titanium have suitable properties, but are subjected to exothermic reactions with hydrogen. Low-Z elements such as beryllium and carbon and high-Z elements such as tungsten are left (Table 13.1).

Beryllium, used on the first wall as surface layer material, has shown good performance in JET. Owing to its large getter capability of oxygen and strong hydrogenic pumping, it allowed for easy density control, particularly for neutral beam heating, and also for the achievement of peaked density profiles [496, 497]. It is non-reactive with hydrogenic isotopes and a good thermal conductor. However, its low melting temperature (1560 K) and

Table 13.1. Thermophysical properties of different materials (thermal conductivity k in W/(mK), specific heat c_p in J/(kgK), density ρ in kg/m^3, melting temperature T_m in °C) as taken from the ITER Material Properties Handbook. The surface temperatures T are given in °C. The ranges of validity are also indicated (in brackets)

Property	Relation
	Thermal properties of beryllium
k $(20 < T < 1000)$	$189.9 - 0.27T + 2.54 \times 10^{-4}T^2$ $-1.01 \times 10^{-7}T^3$
c_p $(20 < T < 1200)$	$1741.8 + 3.34T - 3.11 \times 10^{-3}T^2$ $+1.27 \times 10^{-6}T^3$
ρ $(20 < T < 1200)$	$1823 - 6.9 \times 10^{-2}T - 1.51 \times 10^{-5}T^2$
T_m	1283
	Thermal properties of tungsten
k $(20 < T < 3500)$	$174.93 - 0.11T + 5.01 \times 10^{-5}T^2$ $-7.83 \times 10^{-9}T^3$
c_p $(20 < T < 3000)$	$128.31 + 3.28 \times 10^{-2}T$ $-3.41 \times 10^{-6}T^2$
ρ $(20 < T < 1500)$	$19302.7 - 0.24T - 2.24 \times 10^{-8}T^2$
T_m	3410
	Thermal properties of CFC (N31)
k $(20 < T < 1000)$	$46.4 + 0.85T - 1.49 \times 10^{-3}T^2$ $+1.21 \times 10^{-6}T^3 - 5.04 \times 10^{-10}T^4$ $+1.04 \times 10^{-13}T^5 - 8.35 \times 10^{-18}T^6$
c_p $(20 < T < 1200)$	$868 + 1.87T - 1.26 \times 10^{-3}T^2$ $+3.96 \times 10^{-7}T^3 - 4.48 \times 10^{-11}T^4$
ρ $(20 < T < 1200)$	1923
$T_\mathrm{sublimation}$	3367

the relatively high physical sputtering yield [498]—also with respect to self-sputtering [499]—prevent its application as divertor material. At the first wall in future fusion devices, the heat load will reach values that cannot be handled by components made from Be. Safety concerns are the potential toxicity in manufacturing, the possible production of explosive hydrogen-oxygen mixtures in the case of water steam impact, and the potential production of toxic dust during plasma exposure. It can be provided as bulk material based on powder metallurgy or deposited as layers of several hundred μm by plasma spraying techniques.

Extensive experience of using carbon is available, since the progress of fusion research in the last 30 years is accompanied by the use of carbon as a plasma-facing material. Carbon has excellent thermophysical properties, shows no melting, has high thermal shock resistance, and preserves its shape even under extreme temperature excursions. Carbon can withstand, to some degree, off-normal operation, e.g., disruptions, ELMs, plasma excursions, and is, more or less, the only material for high heat flux components. Once emitted due to erosion processes, carbon serves as an effective radiator in the edge plasma. The weak point of carbon is chemical erosion by oxygen and by hydrogen. Even at low energies (a few eV) and at room temperature, carbon is eroded by the formation of volatile hydrocarbons. Thick layers of hydrogenated carbon films are produced on all plasma-facing components, but also on components in remote areas. These layers, which tend to flake off, contain large amounts of hydrogen isotopes. The ability to collect tritium could prevent the use of carbon in future fusion reactors, if no effective cleaning techniques are developed in the meantime. The variety of carbon-based materials developed over the last few years is impressive. CFCs for example, offer properties well beyond the capability of metals with respect to thermal conductivity, heat load resistance, and strength. These highly advanced materials produced in extensive technological procedures have found their application in various areas of modern industry.

The use of high-Z materials has been strictly avoided over the years after the first bad experience at the beginning of fusion research. A tiny concentration in the plasma center is sufficient to extinguish the plasma by radiation (see Sect. 9.4). The development of plasma scenarios with a rather cold edge brought materials such as tungsten under consideration again [397, 500–502]. Starting with the exposure of small probes made of high-Z materials in different fusion devices [399, 400, 503–505] ASDEX-Upgrade is operating now with wall and divertor plates covered almost completely by tungsten without degradation of plasma performance [506–509]. Having plasma temperatures in front of the plate in the eV range, the erosion yield is suppressed due to the high threshold energies for physical sputtering. No notable erosion is observed up to a plasma temperature of about 20 eV. Concentration of oxygen in the plasma should be kept below 1% in order to avoid significant erosion [510, 511]. The probability of prompt redeposition (Sect. 9.3.3) for tungsten is close to unity. Tungsten shows low tritium retention after saturation is reached. The problems are high heat load and off-normal events such as disruptions and ELMs, which cause significant melting of tungsten and a possible loss of the melt layer [508]. Effective mitigation of disruption would be necessary in fusion reactors based on the tokamak concept. Another issue of concern is the often observed phenomenon of high-Z accumulation [73, 411], which is especially critical for stellarators.

Solid tungsten has reduced heat load capability [512]. Special brush structures with superior thermomechanical behavior [513] have been developed to solve this problem, but at the expense of leading edges. A promising concept is the preparation of tungsten layers plasma-sprayed on CFC materials [514,515]. As in the production of CFC materials the structure, the morphology and, subsequently, the properties, depend sensitively on the applied technological procedures. The choice of the interlayer between tungsten and CFC and its formation is a process of detailed optimization. The tungsten materials developed so far behave poorly under neutron irradiation [487]. Irradiation embrittlement occurs at a rather low dose over a wide range of irradiation temperatures [516].

13.1.1 Discussion of Plasma-Facing Materials

The erosion of Be and C is comparable. Without chemical sputtering, carbon would behave much better than Be. Tungsten shows only negligible erosion at low plasma temperatures, i.e., at low ion impact energies. The minimum target thickness of about 2 mm is determined by structural and lifetime arguments, while the maximum thickness of about 1 cm is dictated by effective heat removal.

The lifetime of C is limited by chemical erosion. Tungsten suffers from loss of plasma control, disruptions, and ELMs, since melt layer losses would thin the targets significantly in the case of high heat loads. Beryllium cannot at all stand the conditions at the divertor targets.

Obviously, the available materials should be applied in the different regions of plasma–wall interaction in accordance to their specific features.

In ITER, beryllium, being inadequate for the divertor, is foreseen for the first wall, tungsten will be used for the less-intense regions of plasma contact in the divertor, i.e., for the baffles and the vertical targets [18,490] (Fig. 13.1). Carbon-based materials find their application at the divertor targets near the strike points, i.e., at the places of high heat load. It is still under discussion, whether ITER should initially start with a tungsten wall, i.e., tungsten plasma-sprayed on a bulk material, replacing beryllium. The decision depends on the plasma performance in ASDEX-Upgrade, which operates now with an almost full tungsten coating [475].

However, the use of different materials in one device will certainly lead to a material mixture due to erosion, transport, and redeposition processes in a way that the aspired optimization might be lost after some plasma operation time. The degree of mixing as observed in fusion experiments reached such levels that occasionally a "new" element, the "tokamakium", has been claimed [517]. The composition, behavior, and properties of mixed materials are still very uncertain.

With carbon in the device, one has to deal with the large amount of tritium codeposited and thus collected at nearly all material surfaces. Unfortunately, codeposition does not show saturation in contrast to implantation of tritium. Adequate recovery methods have to be developed in order to keep the tritium inventory in the device at the low level desired.

13.1.2 Construction Materials

Other than the plasma-facing components, the construction and blanket materials [518, 519]. Blankets in future fusion reactors are located right behind the first wall and have several functions such as tritium breeding and recovery as well as radiation shielding. But its main function is conversion of the kinetic energies of the neutrons into heat, which can be used to produce electric energy as in a conventional heat and power plant. For the blanket, the options of water-cooled ceramic breeders, water-cooled solid PbLi, Li-aqueous salt, and helium-cooled ceramic breeders were considered [520]. At first option, ceramic breeders were selected with low-pressure, low-temperature water as coolant, based on safety considerations. The leading ceramic breeders are Li_2O, Li_4SiO_4, Li_2ZrO_3, and $LiAlO_2$ [521]. Annealed austenitic stainless steel will be used for structural material, however, the structural loads will be relatively low. Furthermore, a large amount of beryllium as a neutron multiplier will be used to enhance the tritium breeding ratio.

Copper alloys, niobium alloys, or molybdenum alloys can be used for heat sink applications. Four types of construction materials are under investigation: austenitic stainless steels, martensitic stainless steels, vanadium alloys, and fiber reinforced ceramic composites [522].

The characterizing elements in austenitic steel are Cr and Ni, or Cr and Mn. In order to assure a faster radioactive decay, elements like Ni and Mo producing long-lived isotopes have been replaced by manganese and carbon as austenite stabilizers. Martensitic steels contain a rather high percentage of Cr (9–13%) and of C or Mn to stabilize the structure of the martensitic phase, but their ferromagnetic properties are of concern.

Austenitic stainless steels will be employed in ITER, but their high swelling rate, high susceptibility to He embrittlement, and low heat conductivity make them unsuitable for fusion reactors. Ferritic/martensitic steels have been successfully used in fast-breeder reactors up to a damage level of 100 dpa. A Japanese R&D program is developing reduced activation ferritic steels by replacing several alloy elements. It is hoped that the effect of ferromagnetism on plasma operation is sufficiently small and would be restricted to the rather beneficial effect of field ripple reduction [523].

At temperatures over 800 K, austenitic and martensitic steels are inferior to vanadium alloys concerning mechanical properties. Vanadium alloys have favorable physical properties and show low activation. The limitations that arise from the use of vanadium alloys at elevated temperatures are embrittlement resulting from the pick-up of interstitial C, N, and O impurities, the poor oxidation resistance, and the ability to retain high concentrations of hydrogen isotopes. During production, fabrication, and welding the atmosphere should be controlled to avoid picking up interstitial elements [524].

Reinforcing a ceramic matrix with ceramic fibers as the second-phase material increases the overall strength in comparison with monolithic ceramic, which has low toughness strength resulting in low reliability. The fibers with their high tensile strength relieve the matrix of load. High-temperature properties, low density, low thermal expansion, and low neutron activation are the favorable features of ceramic composites, while hermiticity, radiation stability, and joining require further developments [524].

In general, composite materials offer tremendous potential as structural materials in fusion applications. New components with improved performances are continuously put into operation as new fibers and new manufacturing processes are developed. For example, SiC fiber reinforced copper used as a heat sink material combines high thermal conductivity and mechanical strength, allowing for the operation of the divertor at higher surface temperatures [525].

13.2 Alternative Concepts and Innovative Ideas

As discussed above, the choice of a solid material for the plasma-facing components in a fusion reactor is quite limited. In the past, several innovative concepts have been proposed: self-renewable protecting layers, liquid materials, solid/liquid combinations, and moving parts.

The use of liquid materials has been proposed by many material experts. This concept has been discussed not only with respect to magnetic confinement [526], but also concerning inertial confinement fusion [33]. Lithium is the favored material for such applications. It is foreseen as a liquid curtain to protect the wall, and as a forced flow with rather high speed (up to 10 m/s) to cover the divertor targets. The advantage of being able to always offer a fresh surface to the plasma cannot compensate the main problem of such schemes—MHD instabilities. A conducting material in strong magnetic fields reacts sensitively to small disturbances, leading to flow destruction, and, finally, would contaminate the plasma significantly. In addition, higher sputtering yields for the liquid phase have been observed in comparison to the same solid materials [527–529].

In another concept, the lithium on the surface of an appropriate material is supplied by diffusion and segregation [530,531]. The permanent loss of the thin surface layer due to erosion is therefore balanced by these temperature-controlled processes, which continuously pump the lithium out of the bulk material onto the top. Unfortunately, the erosion is by far too large under normal conditions.

As an improvement, capillary porous materials, e.g., vanadium, filled with lithium are proposed [532–534], but these materials also behave poorly under high heat load.

In situ coating with Be, B (diborane flushing B_2D_6) or injection of carborane ($C_2B_{10}H_{12}$ [535]) or Li (aerosol injection) can cover three functions: (1) reduction of oxygen, (2) lowering of hydrogen recycling, and (3) reduced erosion of the bulk material, but only in the case that the thin layers are maintained during longer operation [443]. The idea is to use the effect of redeposition. Particles eroded from the layer will be transported back to the target. The question remains whether such a sensitive balance of erosion and redeposition could be established. Otherwise, a substantial amount of additional material must be frequently introduced in order to repair the film, and most of the film material will be deposited elsewhere in the device.

With the aim of avoiding a frequent replacement of eroded and/or destroyed target modules, the concept of a moving belt has been suggested [536]. However, such a belt should be very long, since a refreshment/regeneration of used target elements can be performed presumably only outside the device. The technical realization of moving parts under high vacuum conditions with sufficient heat transfer to cooling structures in toroidal geometry seems infeasible. In general, the possibility to ensure an effective cooling of moving structures is an insurmountable task.

13.3 Open Questions

There are still major gaps in our present understanding [537]. Among them are: (1) the lack of material data and rate coefficients, (2) the carbon–tritium cycle, (3) the tritium/carbon layer removal, (4) the dust problem, (5) the behavior of mixed materials, and (6) the effects of neutron irradiation.

Most of the complex problems in fusion research cannot be handled without extensive numerical calculations, but the relevance of code simulations is closely connected to the availability of reliable databases. This concerns material data including the thermophysical properties of alloys and mixed materials in a wide range of surface temperature [370] and rate coefficients for the overwhelming variety of possible reactions in the plasma, e.g., the break-up reactions of hydrocarbon molecules C_xH_y. Especially for low plasma temperatures around 1 eV rate coefficients are missing or are very uncertain. Many complex particle–material interactions such as the sticking of hydrocarbons or the sputtering are parameterized in order to be able to include such effects in larger impurity transport calculations. The next step should be done here by providing, for example, sticking coefficients as functions of the incoming energy and surface temperature, and providing sputtering yields including synergistic effects owing to the simultaneous bombardment with different ion species [538]. Such a detailed description for the large number of important processes can be obtained by dedicated computer simulations, e.g., molecular dynamic simulations, and experiments addressed to the specific problem under consideration. Studies of hydrocarbon modeling with the best knowledge of existing reaction coefficients and sticking models are presented in [210, 539, 540] and [541].

The question of tritium inventory is mainly a question of carbon erosion, transport, and deposition, since most of the tritium will be found in amorphous hydrogenated carbon layers. The generation and transport cycle of carbon still bares a lot of critical issues. Firstly, the yield and composition of emitted hydrocarbons need to be further investigated, especially at very low impact energies and low surface temperatures. Secondly, several hundred reactions of the hydrocarbon molecules with electrons and protons are partly listed with very uncertain rate coefficients. The transport of hydrocarbon ions in the edge plasma is a highly complex problem. And finally, the processes at the surface such as sticking and transformation, i.e., the formation of other molecules as the impinging species, are affected by many parameters such as the impact energy, the involved species, the surface composition, and the bond structure.

Despite the ongoing progress in understanding of the main processes involved in the carbon cycle, it seems impossible to fully avoid any codeposition of fuel ions such as deuterium and tritium. The development of efficient tritium removal techniques, for example using oxygen, is thus urgently needed in order to keep carbon on the list of promising candidates.

In general, ventilation with oxygen should be avoided otherwise it becomes difficult to re-establish plasma operation and to reduce the large impurity concentration in the start-up phase. Cleaning the hydrogen discharges offers only small removal rates. The additional possibility of a high-temperature bake at 900 K to desorb the tritium thermally out of the layers is technically hard to realize for the full device. If codeposition cannot be mitigated, only the mechanical removal of whole target plates and modules out of the vessel for external treatment is left as the most expensive procedure.

The so-called scavenger technique [150] by using nitrogen in the plasma with the aim of forcing the formation of stable and volatile nitrogen–carbon species such as CN, HCN, C_2N_2 to prevent and reduce film formation could be a promising tool, and should be further investigated.

Radiative plasma termination is proposed [542] to recover tritium from plasma-deposited layers in ITER. A short and highly energetic radiation pulse should be triggered by a massive impurity injection that terminates the plasma discharge. The plasma-facing surfaces are heated and release the stored tritium into the vessel, where it can be pumped out and recovered.

Especially for carbon-based materials, the formation of a thick deposited layers with a high hydrogen content (so-called amorphous C:H layers) poses a critical issue with respect to the large amount of trapped tritium. In addition, these layer are highly stressed and flake off occasionally [363, 543].

The degradation of properties due to neutron irradiation is the major concern. Up to now, only limited tests have been performed. A specialized 14 MeV neutron generator to test materials for fusion is still not available. Many of the experimental results have been obtained by simulating the effects of a fusion energy spectrum through implantation of He or doping with B to approximate the effective dpa of He occurring in fusion. Fission reactors may offer a solution, but the neutron energy spectrum of these reactors is much softer than the spectrum of the fusion neutrons.

Different types of neutron sources have been analyzed in the 1990s, concluding that accelerator driven D-Li sources are most suitable. As the result, the concept of such a device has been elaborated. The International Fusion Materials Irradiation Facility (IFMIF) is an accelerator-based D-Li stripping neutron source for production of high-energy neutrons. Two deuteron beams strike a liquid lithium target and produce neutrons with a peak around 14–16 MeV [544]. There are also proposals to use the European Spallation Source (ESS) for testing of fusion materials for DEMO instead or in parallel to IFMIF. This spallation neutron source is driven by a proton linear accelerator with a beam energy of 1.33 GeV and a beam power of 10 MW [545]. The neutron spectrum of a spallation source has a long high-energy tail, but only the stripping source IFMIF is able to provide the required high neutron flux around 14 MeV. In addition, the recoil energy spectra in the ESS are softer than those in IFMIF and DEMO [546]. The present time scale foresees the construction start of IFMIF in 2009 or 2010 and the start of operation in 2016 or 2017 with one accelerator line, and in 2019 or 2020 with two accelerator lines [518].

14

Summary and Outlook

Enclosing a plasma simply in a vessel results in extremely large energy and particle fluxes on the wall. These losses are reduced considerably by confining the plasma within a magnetic field. However, the magnetic confinement of a hot plasma is not perfect. Particles can still leave the plasma and hit the surfaces. The wall components act as a sink for energy and particles, and at the same time they are a source of particles. Once the wall surfaces are saturated, the recycling of particles comes close to the possible maximum: for every two incoming ions (D^+), one particle, usually a neutral molecule (D_2) with thermal energy, is released from the surface. The electrons hitting the wall take part in the recombination process of the incoming ions. Electrons can leave the solid, if the temperature of the surface is sufficiently high, or by processes such as secondary electron emission and field emission. Most of the incoming energy, however, is absorbed by the solid material. Only a fraction of backscattered particles are able to bring a considerable amount of energy back into the plasma. Processes such as diffusion, trapping in the material, and the release of gas atoms depend on many parameters including the history of the solid wall and its temperature.

The energy and particle transfer from the plasma to the wall is governed by the electric sheath, which is established just near the surface to ensure zero charge transfer to the solid. The solid walls are negatively charged with respect to the plasma in order to repel the fast electrons and to attract the ions. It was shown, via a new derivation of the Bohm criterion, that the potential differences in the sheath and in the presheath obey an energy minimum criterion. The ions are accelerated in the electric field of the sheath and may cause, after impact, significant erosion.

The history of emitted target atoms in fusion experiments based on the concept of magnetic confinement can be described within the cycle of erosion, transport, and deposition (Fig. 14.1). After a target atom is emitted, for example, by physical sputtering, chemical erosion, or thermal sublimation, it is going to be ionized at a position not far away from the surface. Once ionized, electric and magnetic forces are exerted on the particle, now called an

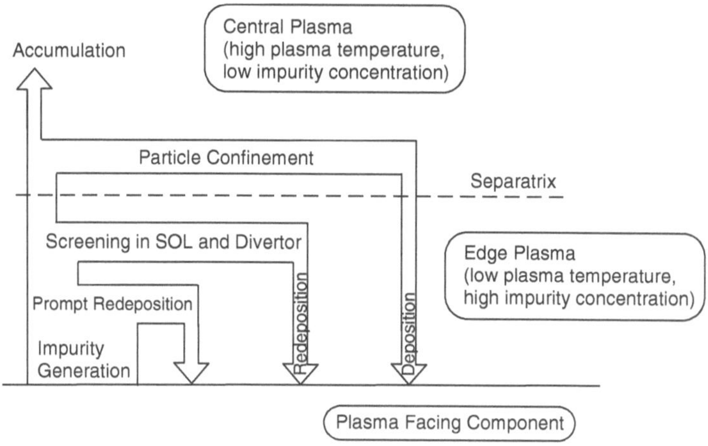

Fig. 14.1. The impurity cycle in fusion experiments. The pumping out of volatile impurities is not included here

impurity ion. If the ionization length is smaller than its gyro-radius, then it has a good chance to be promptly redeposited very close to the place of emission. Otherwise, it changes momentum and energy in collisions with plasma ions and other impurity ions. It may be subjected, besides multiple ionization and recombination, to a large number of different reactions in the plasma, in particular charge exchange. As long as the impurity ion stays in the plasma of the scrape-off layer, it will be efficiently transported along the magnetic field lines until it strikes a divertor or limiter plate. This process is called redeposition.

A small fraction of the impurities is able to penetrate deeply into the plasma core by crossing the separatrix. Now, the transport parallel to the magnetic field lines as well as transport in the poloidal direction (diamagnetic flows) do not play a role and only the cross-field transport matters; it is characterized by anomalously high diffusion coefficients. After a while, the particle confinement time, each impurity crosses the separatrix again, now from the plasma core into the SOL, and is deposited at one of the wall components. This process is called deposition and is characterized by the loss of information about the origin, where the particle originally came from.

For a burning fusion plasma, it is essential to keep the plasma in the central region hot enough to maintain the fusion reactions. In order to reduce the large cross-field transport of energy, so-called internal transport barriers have to be established. On the other hand, the impurity ions lead to dilution of the fuel and strong radiation losses. Hence, their concentration should be kept as small in a fusion reactor. Transient effects such as sawteeth in the core and ELMs in the edge plasma, used as "plasma cleaners", may be helpful for efficient impurity and He ash control. ELM control by pellet injection is a topic of large relevance.

Impurity ions impinging on the surface of a wall component sticks with a certain probability. It may cause sputtering as well as defects in the material. In the case of hydrocarbons, the interaction process with the surface involves not only sticking but also transformation. Due to the impact of one certain hydrocarbon, one or more hydrocarbon radicals of a different type may be emitted.

The deposition of impurities determines the magnitude of the net effects, whether an area is dominated by net erosion or net deposition. The gross effects are usually much larger than the net ones. The erosion at the divertor plates in future fusion experiments such as ITER has to be reduced by redeposition in order to achieve an acceptable level of net erosion. The favorable effect of redeposition can be enhanced exploiting the $\boldsymbol{E} \times \boldsymbol{B}$ drift motion of the impurity ions in the electric sheath. By a suitable choice of the magnetic field direction and geometry the peak net erosion at the divertor plates can be minimized. A simple relation has been derived which allows for estimating the peak net erosion concerning all important effects: erosion by the plasma ions, self-sputtering, prompt redeposition, and redeposition due to friction with the plasma streaming towards the plates.

Along with erosion and deposition processes, surface modification occurs. The material composition changes as well as the surface profile, and the material roughness. Hence, erosion depending sensitively on the surface composition is affected too. Deposited impurities such as C or Be protect targets made from other materials efficiently by formation of layers on their surfaces.

The build-up of layers on the surface of wall components, even at very remote areas, is inherently connected to the tritium retention problem. Tritium ions arrive simultaneously with the impurities at the surface and are codeposited there. The concentration of radioactive tritium in such layers can reach high values, and the amount of collected tritium, being proportional to the layer thickness, can exceed the amount implanted (as well as the amount of tritium trapped deeply in the bulk material) by several orders of magnitude. The process of codeposition works very efficiently for carbon impurities, but has been also observed for Be and W deposition. Techniques for efficient tritium removal and recovering should be further developed in order to cope with the tritium retention problem.

The theoretical description of the involved processes will more and more rely upon computer simulations. They are based on reliable data on atomic and molecular processes, and material properties. It is not (yet) possible to follow a large number of representative particles in their motion through the whole plasma and further in the solid material. Though such a comprehensive analysis is today out of the reach of existing computer systems, computer simulations nevertheless provide a valuable tool to test theoretical models and assumptions. Such simulations act as a bridge between the different scales in space and time. In plasma–material interaction studies these scales cover several orders of magnitude: beginning from the fast scattering and sputtering processes up to the rather slow processes involved in changing the surface composition. Simulations dedicated to certain processes are usually restricted to a narrow range of space and time. Analyzing the results from such simulations, theoretical models can be derived, which provide simplifying assumptions (scaling rules) to be used in simulations of processes characterized by other scales.

Computer simulations should be used with great care, and the validation against experimental results is an indispensable part of the numerical analysis. Laboratory experiments should provide as much detailed knowledge as possible. The more parameters can be measured the less are left in the analysis, which otherwise could be used incorrectly as fitting parameters. Only by collecting the information obtained in different laboratory experiments can a successful analysis of the *big* fusion experiments, where all processes take place at the same time, be possible. Often, the output of computer simulations appears as incomprehensible as the experiments themselves. Only theoretical analysis can provide then the desired understanding by formulating the right questions. New results, new knowledge, and a better understanding of the real world can hardly be obtained without the trinity of experiment, theory, and simulation (Fig. 14.2).

The choice of plasma-facing materials is always a compromise, and is determined by a large number of different criteria and operation limits. In particular, a new plasma density limit (11.3) as well as a new impurity criterion (12.12) have been proposed. Unfortunately, this choice is based only on a very limited list of possible candidates. While for the next step devices such as ITER some compromise solutions can be found, there is at present no material available for a commercial fusion reactor. Either the heat loads are too large (for Be and W), or the build-up of a large tritium inventory by codeposition (in the case of C) prevents the use of materials under consideration.

The plasma-exposed materials in fusion experiments suffer from major erosion and destruction events. The emitted material is distributed via the plasma to all neighboring surfaces and deposited there. After some discharges, the surface composition shows significant concentration of all materials present

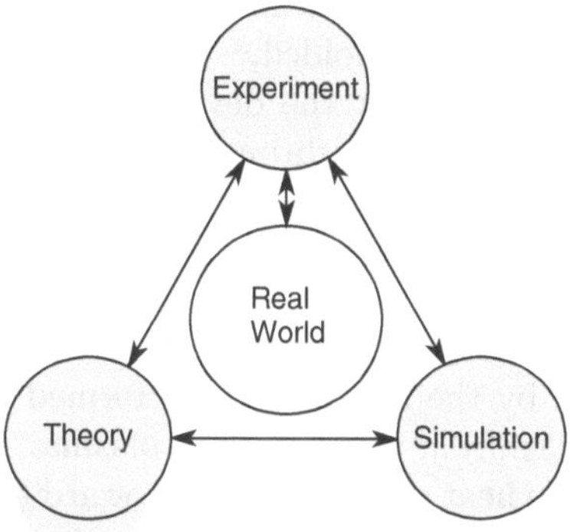

Fig. 14.2. The trinity of experiment, theory, and simulation

in the plasma chamber. Having more than one material in the device, will lead to a material mix and, thus, to a possible loss of the desired properties. The processes of material mixing need further investigations. It is expected that ITER will give the answers to this issue of concern.

Is it perhaps possible to use only one material for both the divertor and the first wall? Beryllium is not suitable for the divertor plates. It melts already at modest heat loads. Tungsten could be used, in principle, also for the strike zones at the divertor plate, but will survive only a limited number of major disruption events and ELMs. Tungsten can be used also as wall material in the main chamber, but should be protected from energetic charge-exchange neutrals. These neutrals have energies larger than the threshold energy for physical sputtering. The tungsten atoms emitted from the wall can penetrate deeply into the plasma and cause the serious problem of high-Z radiation losses. Carbon would be the ideal material applicable for all surfaces. It withstands high thermal loads and radiates, as desired, only in the plasma edge, but it is susceptible to chemical erosion. The threshold energy for chemical sputtering is rather low and the yields quite significant even at room temperature. Most of the tritium inventory would be collected in amorphous carbon–tritium layers.

The use of carbon as the only material would be possible if efficient tritium removal techniques could be developed, or if the formation of thick layers could be prevented. The use of tungsten as the only material appears possible if disruptions and ELMs could be mitigated. In addition, the recycling process should be localized in the divertor in order to prevent charge-exchange processes occurring outside the divertor. If tungsten is eroded in the divertor, redeposition and divertor retention reduce the outflux of tungsten into the main chamber. Tungsten should not be used as bulk material but rather as a thick plasma-sprayed coating on CFC material.

The use of CFC material covered by tungsten could be an option for a fusion reactor based on the stellarator concept. As mentioned, disruptions do not occur in stellarator configurations, and the possibility of continuous operation is an essential advantage from the engineering point of view. It must be shown whether high-Z accumulation as predicted by neoclassical transport theory is pertinent or can be avoided.

One may speculate whether the development of a new material system, where Li is continuously pressed through a CFC material with a rather loose composite of fibers (i.e., with reduced matrix material), can be successful. Under regular conditions, a very thin film of Li has to keep up on top of the target controlled by the pressure at the rear side of the material and the material temperature. Large thermal loads caused by disruptions or ELMs have to be absorbed by the Li vapor cloud formed immediately after impact of the plasma particles from the disrupted plasma. The CFC material should efficiently conduct the heat along the fibers toward the heat sink of the cooling system installed beneath the targets. The chemical erosion of graphite would be significantly reduced, and lithium would assist in effective wall pumping.

Once eroded, both materials—as low-Z elements—radiate almost completely in the plasma edge, where a significant conversion of the outflowing power into radiation is required. It may be possible to use the outstanding properties of both materials, while compensating for their handicaps.

In general, a reliable solution of the material problem can only be found in the joint development of new materials and suitable plasma scenarios. Fifteen years ago, the designed heat load at the strike point of the divertor plates reached values of 25–30 MW/m^2. Soon it became clear that there was no technical concept of target plates in sight that could handle more than 10 MW/m^2 under steady state operation. As a consequence, the new divertor concept using all available processes to spread the generated power over larger areas was developed resulting in much lower values of the power flux density onto the targets.

The development of adequate plasma solutions is a major task. A robust solution has to be found considering both aspects: the plasma conditions and operation scenarios on one side and the design of the plasma-facing components on the other. Establishing transport barriers inside the separatrix could help to reduce the power flux into the scrape-off layer, while reaching burn conditions in the core. The fluxes of particles and energy out of the confined region have to be distributed over larger areas to reduce the local power loads.

Despite all experience in plasma control that will be gathered during the next few years, one concern remains—the high flux of neutrons. Today, there is no solid material available which is able to withstand the expected fluence leading to several hundred displacements per atom without significant degradation. In particular, advanced materials such as carbon fiber enforced composites suffer from neutron bombardment and lose their advantageous properties such as heat conduction by disintegration of their optimized and complex structure. Future research must concentrate on this issue. If no satisfying solution can be found, there is still the pragmatic, but rather expensive alternative of regular replacements of divertor and wall elements.

Appendix A

A.1 Some Important Relations and Parameters

Stefan–Boltzmann Law:

$$E = \sigma_{\mathrm{SB}} T^4$$
$$\sigma_{\mathrm{SB}} = \frac{2\pi^5 k_{\mathrm{B}}^4}{15c^2h^3} = \frac{C_1}{15}\left(\frac{\pi}{C_2}\right)^4 = 5.668 \times 10^{-8} \left[\frac{\mathrm{W}}{\mathrm{m^2K^4}}\right] \tag{A.1}$$

where $C_1 = 3.7418 \times 10^{-16}\,\mathrm{Wm^2}$ and $C_2 = 1.4388 \times 10^{-2}\,\mathrm{mK}$.

Thermal Diffusivity:

$$a = \frac{k}{\rho \cdot c_{\mathrm{p}}} \left[\frac{\mathrm{m^2}}{\mathrm{s}}\right] \tag{A.2}$$

with $[k]$=W/(mK), $[c_{\mathrm{p}}]$=J/(kgK), and $[\rho]$=kg/m^3.

Coefficient ζ:

$$\zeta = \frac{2}{\sqrt{\pi\,\rho\, c_{\mathrm{p}}\, k}} \left[\frac{\mathrm{Km^2}}{\mathrm{Ws^{1/2}}}\right] \tag{A.3}$$

used in

$$T_{\mathrm{s}} = T_{\mathrm{so}} + P\,\zeta\,\sqrt{t} \tag{A.4}$$

to calculate the rise of surface temperature due to a heat load of $P\,[\mathrm{W/m^2}]$ onto a half-infinite solid during the time t. T_{o} is the initial surface temperature at $t = 0$. For most materials, $\zeta \simeq (0.5\text{–}1.0)\times 10^{-4}$ in a wide range of surface temperatures.

Electron Plasma Frequency:

$$\omega_{\mathrm{pe}} = \sqrt{\frac{n_{\mathrm{e}} e^2}{m_{\mathrm{e}} \epsilon_{\mathrm{o}}}} \text{ [rad/s]} \tag{A.5}$$

Ion Plasma Frequency:

$$\omega_{\mathrm{pi}} = \sqrt{\frac{n_{\mathrm{i}} q^2 e^2}{m_{\mathrm{i}} \epsilon_{\mathrm{o}}}} \text{ [rad/s]} \tag{A.6}$$

Debye Length:

$$\lambda_{\mathrm{D}} = \sqrt{\frac{\epsilon_{\mathrm{o}} k_{\mathrm{B}} T}{n_{\mathrm{e}} e^2}} \text{ [m]} \tag{A.7}$$

Number of Particles Inside a Debye Sphere:

$$N_{\mathrm{D}} = \frac{4}{3}\pi \lambda_{\mathrm{D}}^3 n_{\mathrm{pl}} = \frac{4\pi}{3}\left(\frac{\epsilon_{\mathrm{o}} k_{\mathrm{B}} T}{n_{pl}^{1/3} e^2}\right)^{3/2} \text{ [particles]} \tag{A.8}$$

Electron Cyclotron Frequency:

$$\omega_{\mathrm{ce}} = \frac{eB}{m_{\mathrm{e}}} \text{ [rad/s]} \tag{A.9}$$

Ion Cyclotron Frequency:

$$\omega_{\mathrm{ci}} = \frac{qeB}{m_{\mathrm{i}}} \text{ [rad/s]} \tag{A.10}$$

Electron Gyro-Radius:

$$\rho_{\mathrm{ce}} = \frac{v_{\perp \mathrm{e}}}{\omega_{\mathrm{ce}}} = \frac{m_{\mathrm{e}} v_{\perp \mathrm{e}}}{eB} \text{ [m]} \tag{A.11}$$

Ion Gyro-Radius:

$$\rho_{\text{ci}} = \frac{v_{\perp\text{i}}}{\omega_{\text{ci}}} = \frac{m_{\text{i}} v_{\perp\text{i}}}{qeB} \text{ [m]} \tag{A.12}$$

Redeposition Parameter:

$$\gamma_p = \frac{\lambda_{\text{iz}}}{\rho} \frac{v_{\perp\text{i}}}{\omega_{ci}} = \frac{m_{\text{i}} v_{\perp\text{i}}}{qeB} \text{ [m]} \tag{A.13}$$

Magnetization Parameter (Dielectric Susceptibility):

$$\xi_{\text{e,i}} = \frac{\rho_{\text{e,i}}^2}{\lambda_{\text{D}}^2} = \frac{n_{\text{pl}} m_{\text{e,i}}}{\epsilon_{\text{o}} B^2} \tag{A.14}$$

Ion Sound Speed:

$$c_{\text{s}} = \sqrt{\frac{\gamma_{\text{e}} k_{\text{B}} T_{\text{e}} + \gamma_{\text{i}} k_{\text{B}} T_{\text{i}}}{m_{\text{i}}}} \text{ [m/s]} \tag{A.15}$$

where γ_{i} can be set equal to 3, since ions are subjected to one-dimensional compression, whereas electrons are fast enough to reach thermal equilibrium so that $\gamma_{\text{e}} = 1$. The sound speed depends on the electron temperature, since the arising electric field is proportional to this temperature and ion mass, due to the inertia of the fluid.

Particle Drifts:

In an electric field:

$$\boldsymbol{v}_{\text{E}} = (\boldsymbol{E} \times \boldsymbol{B})/B^2 \text{ [m/s]} \tag{A.16}$$

In gravity:

$$\boldsymbol{v}_{\text{g}} = m(\boldsymbol{g} \times \boldsymbol{B})/(qB^2) \text{ [m/s]} \tag{A.17}$$

Due to a force $\boldsymbol{F}$:

$$\boldsymbol{v}_{\text{F}} = (\boldsymbol{F} \times \boldsymbol{B})/(qB^2) \text{ [m/s]} \tag{A.18}$$

Due to a gradient of $\boldsymbol{B}$:

$$\boldsymbol{v}_{\text{G}} = -\frac{m v_{\perp}^2}{2qB} \frac{(\boldsymbol{\nabla} B) \times \boldsymbol{B}}{B^2} \text{ [m/s]} \tag{A.19}$$

Due to curvature of $\boldsymbol{B}$:

$$\boldsymbol{v}_{\mathrm{C}} = -\frac{mv_{\parallel}^2}{qB^3}\left[(\boldsymbol{B}\cdot\boldsymbol{\nabla})\frac{\boldsymbol{B}}{B}\right] \times \boldsymbol{B} \ [\mathrm{m/s}] \tag{A.20}$$

Due to polarization:

$$\boldsymbol{v}_{\mathrm{p}} = m\frac{\partial \boldsymbol{E}_{\perp}/\partial t}{qB^2} \ [\mathrm{m/s}] \tag{A.21}$$

Maxwellian Distribution:

The thermodynamical definition of temperature T is

$$3k_{\mathrm{B}}T/2 = m\langle v^2\rangle/2 = \frac{m}{2}\int_v f(v)\, v^2\, \mathrm{d}^3 v\ . \tag{A.22}$$

The number of particles having a velocity in the interval $[v, v+\mathrm{d}v]$ is

$$nf(v)\mathrm{d}^3 v = n\left(\frac{m}{2\pi k_{\mathrm{B}}T}\right)^{(3/2)} \exp\left(-\frac{mv^2}{2k_{\mathrm{B}}T}\right)\mathrm{d}^3 v \tag{A.23}$$

where n is the particle density.

The one-dimensional distribution is given by

$$\begin{aligned} g(v_x)\mathrm{d}v_x = &\left(\frac{m}{2\pi k_{\mathrm{B}}T}\right)^{3/2} \exp\left(-\frac{mv_x^2}{2k_{\mathrm{B}}T}\right)\mathrm{d}v_x \\ &\times \int_{-\infty}^{+\infty} \exp\left(-\frac{mv_y^2}{2k_{\mathrm{B}}T}\right)\mathrm{d}v_y \int_{-\infty}^{+\infty} \exp\left(-\frac{mv_z^2}{2k_{\mathrm{B}}T}\right)\mathrm{d}v_z\ . \end{aligned} \tag{A.24}$$

Since each integral equals $(2\pi \mathrm{k}T/m)^{1/2}$ one obtains

$$g(v_x)\mathrm{d}v_x = \left(\frac{m}{2\pi k_{\mathrm{B}}T}\right)^{1/2} \exp\left(-\frac{mv_x^2}{2k_{\mathrm{B}}T}\right)\mathrm{d}v_x \tag{A.25}$$

with $\int_{-\infty}^{+\infty} g(v_x)\mathrm{d}v_x = 1$. The same relation is valid for $g(v_y)$ and $g(v_z)$.

The variance is given by

$$\langle v_i^2\rangle = \int_{-\infty}^{+\infty} g(v_i)v_i^2\mathrm{d}v_i = \frac{k_{\mathrm{B}}T}{m} \tag{A.26}$$

resulting in $\sqrt{\langle v_i^2\rangle} = \sqrt{k_{\mathrm{B}}T/m}$ and $m\langle v_i^2\rangle/2 = k_{\mathrm{B}}T/2$. The velocity components are statistically independent on each other, i.e.,

$$f(\boldsymbol{v})\mathrm{d}\boldsymbol{v} = g(v_x)\mathrm{d}v_x \cdot g(v_y)\mathrm{d}v_y \cdot g(v_y)\mathrm{d}v_y\ . \tag{A.27}$$

The velocity distribution as a function of $v = |\boldsymbol{v}|$ is

$$F(v)\mathrm{d}v = \int\limits_{\theta}\int\limits_{\varphi} f(v)v^2 \sin\theta \mathrm{d}\theta \mathrm{d}\varphi \mathrm{d}v = 4\pi v^2 f(v)\mathrm{d}v \tag{A.28}$$

since $\mathrm{d}^3 v = v^2 \sin\theta \mathrm{d}\theta \mathrm{d}\varphi \mathrm{d}v$ and $4\pi v^2 \mathrm{d}v$ represents the volume of a sphere in the velocity space

$$F(v) = 4\pi \left(\frac{m}{2\pi \mathrm{k}T}\right)^{(3/2)} v^2 \exp\left(-\frac{mv^2}{2\mathrm{k}T}\right) . \tag{A.29}$$

The average velocity is shown to be

$$\langle v \rangle = \int\int\limits_{-\infty}^{+\infty}\int f(v)\, v\, \mathrm{d}v_x \mathrm{d}v_y \mathrm{d}v_z = \int\limits_{0}^{+\infty} F(v)\, v\, \mathrm{d}v = \left(\frac{8k_\mathrm{B}T}{\pi m}\right)^{1/2} . \tag{A.30}$$

The average square velocity is given by

$$\langle v^2 \rangle = \int\limits_{0}^{+\infty} F(v)v^2\, \mathrm{d}v = 3k_\mathrm{B}T/m . \tag{A.31}$$

The most probable velocity is then

$$v_\mathrm{mp} = \sqrt{2k_\mathrm{B}T/m} \tag{A.32}$$

satisfying the condition

$$\left.\frac{\mathrm{d}F(v)}{\mathrm{d}v}\right|_{v_\mathrm{mp}} = 0 . \tag{A.33}$$

The energy distribution ($E = mv^2/2$) can be presented as

$$g(E)\mathrm{d}E = \frac{2}{\sqrt{\pi}}\left(\frac{E}{(k_\mathrm{B}T)^3}\right)^{1/2} \exp\left(-\frac{E}{k_\mathrm{B}T}\right) \mathrm{d}E . \tag{A.34}$$

The particle flux density is given as

$$\begin{aligned} \Gamma_z &= n\int\limits_{0}^{\infty} f(v)v^3 \mathrm{d}v \int\limits_{0}^{\pi/2} \sin\theta\cos\theta\, \mathrm{d}\theta \int\limits_{0}^{2\pi} \mathrm{d}\varphi = n\,\pi \int\limits_{0}^{\infty} f(v)v^3 \mathrm{d}v \\ &= n(k_\mathrm{B}T/(2\pi m))^{1/2} = n\langle v \rangle/4 . \end{aligned} \tag{A.35}$$

Other often used distributions are

$$f(v = \sqrt{v_x^2 + v_y^2 + v_z^2})$$

$$= 4\pi \left(\frac{m}{2\pi k_B T}\right)^{(3/2)} v^2 \exp\left(-\frac{mv^2}{2k_B T}\right) \rightarrow \langle v \rangle = \sqrt{\frac{8k_B T}{\pi m}} \quad (A.36)$$

$$f(v = \sqrt{v_x^2 + v_y^2}) = \left(\frac{m}{k_B T}\right) v \exp\left(-\frac{mv^2}{2k_B T}\right) \rightarrow \langle v \rangle = \sqrt{\frac{\pi k_B T}{2m}} \quad (A.37)$$

$$f(v = |v_x|) = 2\left(\frac{m}{2\pi k_B T}\right)^{1/2} \exp\left(-\frac{mv^2}{2k_B T}\right) \rightarrow \langle v \rangle = \sqrt{\frac{2k_B T}{\pi m}} \quad (A.38)$$

with $\int_0^\infty v f(v) \mathrm{d}v = \langle v \rangle$.

Collision Frequencies:

The collision frequency for momentum transfer between a test particle a and field particles b with their charges q_a, q_b, their masses m_a, m_b, and their thermal velocities $v_{\mathrm{th},a}$, $v_{\mathrm{th},b}$, respectively, is given by

$$\nu_{ab} = \frac{16\sqrt{\pi}}{3m_a}\left(\frac{1}{m_a} + \frac{1}{m_b}\right) \frac{q_a^2 q_b^2 e^4 n_b \ln \Lambda}{(4\pi\epsilon_o)^2 (v_{\mathrm{th},a}^2 + v_{\mathrm{th},b}^2)^{3/2}} . \quad (A.39)$$

For electron and singly charge ions ($q_e = -1, q_i = 1$) the collision frequencies are

$$\nu_{ii} = \frac{1}{(4\pi\epsilon_o)^2} \frac{4}{3} \frac{\sqrt{\pi}\, e^4\, n_i \ln \Lambda}{\sqrt{m_i}\, (k_B T_i)^{3/2}} \quad (A.40)$$

$$\nu_{ee} = \frac{1}{(4\pi\epsilon_o)^2} \frac{4}{3} \frac{\sqrt{\pi}\, e^4\, n_e \ln \Lambda}{\sqrt{m_e}\, (k_B T_e)^{3/2}} \quad (A.41)$$

$$\nu_{ei} = \frac{1}{(4\pi\epsilon_o)^2} \frac{4}{3} \frac{\sqrt{2\pi}\, e^4\, n_i \ln \Lambda}{\sqrt{m_e}\, (k_B T_e)^{3/2}} \quad (A.42)$$

$$\nu_{ie} = \frac{1}{(4\pi\epsilon_o)^2} \frac{4}{3} \frac{\sqrt{2\pi m_e}\, e^4\, n_e \ln \Lambda}{m_i\, (k_B T_e)^{3/2}} \quad (A.43)$$

where $\nu_{ie} = (m_e/m_i)\,\nu_{ei}$, $\nu_{ii} = \nu_{ee}\sqrt{m_e/m_i}$, and $\nu_{ei} = \sqrt{2}\,\nu_{ee}$.

A.2 Simple Particle Mover

The equation of motion (9.99) is given in Cartesian coordinates by

$$
\begin{aligned}
M\frac{\mathrm{d}v_x}{\mathrm{d}t} &= qE_x + Qv_yB_z - Qv_zB_y \\
M\frac{\mathrm{d}v_y}{\mathrm{d}t} &= qE_y - Qv_xB_z + Qv_zB_x \\
M\frac{\mathrm{d}v_z}{\mathrm{d}t} &= qE_z + Qv_xB_y - Qv_yB_x
\end{aligned}
\tag{A.44}
$$

with the components of the electric field $\boldsymbol{E} = (E_x, E_y, E_z)$ and the magnetic field $\boldsymbol{B} = (B_x, B_y, B_z)$. Using an implicit scheme of second order accuracy [547] in substituting the derivatives by differences yields

$$
\frac{\mathrm{d}v_\mathrm{i}}{\mathrm{d}t} = \frac{v_\mathrm{i}^{n+1} - v_\mathrm{i}^n}{\Delta t} \quad \text{and} \quad v_\mathrm{i} = \frac{v_\mathrm{i}^{n+1} + v_\mathrm{i}^n}{2}
$$

for $i = (x, y, z)$. This scheme conserves the kinetic energy. The index n denotes the steps in time. The system of equations (A.44) can be expressed in matrix notation

$$
\begin{pmatrix} 1 & -C\,B_z & C\,B_y \\ C\,B_z & 1 & -C\,B_x \\ -C\,B_y & C\,B_x & 1 \end{pmatrix}
\begin{pmatrix} v_x^{n+1} \\ v_y^{n+1} \\ v_z^{n+1} \end{pmatrix}
=
\begin{pmatrix} v_x^n + 2C\,E_x + C\,B_z v_y^n - C\,B_y v_z^n = s_1 \\ v_y^n + 2C\,E_y - C\,B_z v_x^n + C\,B_x v_z^n = s_2 \\ v_z^n + 2C\,E_z + C\,B_y v_x^n - C\,B_x v_y^n = s_3 \end{pmatrix}
$$

with $C = Q\Delta t/(2M)$, where Q is the charge and M is the mass of the particle. The corresponding determinant is

$$
\Delta = 1 + C^2(B_x^2 + B_y^2 + B_z^2) = 1 + C^2B^2\,. \tag{A.45}
$$

The new components for the velocity after time step Δt are given by

$$
\begin{aligned}
v_x^{n+1} &= \frac{s_1(1 + C^2B_x^2) + s_2(C\,B_z + C^2B_xB_y) + s_3(C^2B_xB_z - C\,B_y)}{1 + C^2B^2} \\
v_y^{n+1} &= \frac{s_1(C^2B_xB_y - C\,B_z) + s_2(1 + C^2B_y^2) + s_3(C\,B_x + C^2B_zB_y)}{1 + C^2B^2} \\
v_z^{n+1} &= \frac{s_1(C^2B_zB_x + C\,B_y) + s_2(C^2B_yB_z - C\,B_x) + s_3(1 + C^2B_z^2)}{1 + C^2B^2}\,.
\end{aligned}
$$

The new positions are calculated as follows:

$$
\begin{aligned}
x^{n+1} &= x^n + (v_x^{n+1} + v_x^n)\Delta t/2 \\
y^{n+1} &= y^n + (v_y^{n+1} + v_y^n)\Delta t/2 \\
z^{n+1} &= z^n + (v_z^{n+1} + v_z^n)\Delta t/2\,.
\end{aligned}
\tag{A.46}
$$

An appropriate choice of the time step is necessary in order to reduce numerical errors.

A.3 Symbols

An attempt has been made to avoid double denotation of symbols throughout the text. For this purpose, it has been necessary to add subscripts in the case of commonly used symbols for certain physical parameters. The use of remaining double denotations (e.g., E for energy and electric field, ρ for gyro-radius and density) hopefully becomes clear from the context. The usage of Z for atomic numbers (nuclear charges) in solid state physics as well as for charge states in plasma physics especially leads to some confusion. In the following text, Z is used to denote atomic numbers, q is the charge state, and $Q = qe$ is the charge.

The following subscripts have been used more frequently: (e) electron, (eff) effective, (E) energy, (i) ion, (imp) impurity, (o) initial condition, (pl) plasma, (q) charge state, ($\perp$) perpendicular, ($\|$) parallel.

a	Thermal diffusivity
a_t	Minor radius in toroidal geometry
a_L	Lindhard screening length
a_o	Bohr radius
a_s	Screening length
A	Mass number in [amu]
$A_{k\to i}$	Transition probability
B	Magnetic field
c	Speed of light in vacuum
c_A, c_B	Concentration of species A and B
c_br	Bremsstrahlung coefficient
c_p	Heat capacity at constant pressure
c_s	Ion sound speed
C	Coefficient
C_d	Fitting parameter
d	Thickness
d_m	Dimension
D_B	Bohm diffusion coefficient
D	Diffusion coefficient
$D_\perp$	Cross-field diffusion coefficient
e	Elementary charge
E	Energy
E	Electric field
$E_\|$	Electric field parallel to $\boldsymbol{B}$
$E_\perp$	Electric field perpendicular to $\boldsymbol{B}$
E_em	Energy of emitted electrons
E_F	Fermi energy
E_i	Ion energy at sheath entrance
E_s	Heat of sublimation
E_th	Threshold energy for sputtering

E_{therm}	Thermal energy
E_{TF}	Thomas–Fermi energy
E_{Y}	Young modulus
E_{α}	Helium energy
f	Fraction of particles
f_{esc}	Escape probability
f_{G}	Gamov factor
f_{i}	Impurity concentration
f_{P}	Probability
f_{y}	Yamamura parameter
$f(v)$	Velocity distribution
$f(\boldsymbol{r}, \boldsymbol{v}, t)$	Distribution function in phase space
F	Force
F_{R}	Friction force
$F(v)$	Velocity distribution
g	Standard acceleration of gravity
g_{T}	Troyan factor
$g(v)$	Velocity distribution
$g(E)$	Energy distribution
G	Deposition distribution
h	Planck constant
i	Index
I	Electric current
I_{ion}	Ionization energy
I_{p}	Plasma current
j	Electric current density
j_{ce}	Electron emission current density
j_{e}	Electron current density
j_{i}	Ion current density
j_{s}	Thermionic current density
k	Heat conductivity
k_{B}	Boltzmann constant
k_{L}	Lindhard–Scharff coefficient
l	Length
l_{p}	Mean free path
l_{tr}	Transport length
L	Lagrange function
L_{c}	Connection length of wall elements
L_{i}	Impurity radiation function
L_{z}	Cooling rate for certain impurity and charge state
m	Mass
m_{e}	Electron mass
m_{i}	Ion mass
m_{p}	Exponent of the power potential
M_1	Mass of incident particle

M_2	Mass of target atom
M_φ	Angular momentum
n	Particle density
n_o	Atomic density in solids
n_e	Electron density
n_{pl}	Plasma density
n_q	Density of plasma impurities of charge state q
n_s	Surface concentration in particles/m^2
N	Number of particles
N_D	Number of particles in a Debye sphere
N_F	Number of Frenkel pairs
N_t	Number of scattering centers
N_Z	Number of quantum states
p	Pressure
p_x, p_y, p_z	Components of the momentum
P	Heat flux density
P_{alpha}	Helium fusion power
P_{brems}	Bremsstrahlung losses
P_e	Electron heat flux density
P_i	Ion heat flux density
P_{pr}	Probability of prompt redeposition
P_q	Radiation losses for a certain charge state
P_{redep}	Probability of redeposition
P_v	Power loss density
P_w	Total heat flux density
Pe	Peclet number
P_{rad}	Radiation power
q	Charge state of plasma particles
q_i	Charge state of impurity ions
q_s	Safety factor
Q	Charge
Q_t	Total charge
Q_E	Inner heat source
Q_{RES}	Fitting parameter for RES calculation
Q_y	Fitting parameter
r	Radius
r_o	Gas-kinetic radius
r_ρ	Normalized range
R	Reflection coefficient
R_{cyc}	Recycling coefficient
R_d	Depth
R_E	Energy reflection coefficient
R_N	Particle reflection coefficient
R_o	Total range

R_t	Major radius in toroidal geometry
$R(z)$	Depth distribution
s	Sticking coefficient
s_n	Nuclear stopping cross-section
S	Surface area
S_n	Stopping cross-section
S_q	Ionization rate coefficient
S_{wall}	Area of wall elements
t	Time
t_c	Self-collision time
t_d	Deflection time
t_{exp}	Exposure time
t_E	Energy exchange time
t_{pl}	Plasma oscillation time
t_s	Slowing down time
T	Temperature
T_e	Electron temperature
T_i	Ion temperature
T_m	Melting temperature
T_s	Surface temperature
u	Velocity in the center-of-mass system
U	Potential energy
U_o	Energy of a motionless electron in vacuum
$U(r)$	Interaction potential
v	Velocity
v_A	Alvén velocity
v_B	Bohr velocity
v_c	Velocity of the center of mass
v^s	Recession/deposition speed
$v_{\parallel}$	Parallel velocity
$v_{\perp}$	Transverse velocity
V	Volume
W_E	Plasma energy content
W_{th}	Thermal energy
W_T	Thermionic work function
x	Cartesian x-coordinate, length
y	Cartesian y-coordinate
Y	Sputtering yield
Y_{eff}	Effective sputtering yield
z	Cartesian z-coordinate, depth
Z	Nuclear charges, atomic number
Z_1	Atomic number of incident particles
Z_2	Atomic number of bulk atoms
Z_{eff}	Effective charge state

α	Angle
α-particle	Helium
α_e	Coefficient of electron temperature gradient
α_i	Coefficient of ion temperature gradient
α_T	Thermal expansion coefficient
α_q	Recombination rate coefficient
β	Ratio of plasma to magnetic pressure
β^*	Critical β value
γ	Ratio of particle to energy confinement time
γ_B	Branching ratio
γ_G	Gaunt factor
γ_k	Kinematic factor
γ_i, γ_e	Ratio of the specific heats, c_p/c_V
γ_{ero}	Ratio of displacement to density decay length
γ_{imp}	Ratio of f_{imp} to f_{crit}
γ_E	Energy transmission factor
γ_p	Ratio of gyro-radius to ionization length
γ_{rad}	Ratio of radiation to input power
γ_{RE}	Generation rate of runaway electrons
γ_σ	Material stress parameter
$\mathit{\Gamma}$	Flux density
$\mathit{\Gamma}_{coupl}$	Coupling constant
δ	Emission coefficient
$\mathit{\Delta}$	Distance, displacement, thickness
Δ	Increment
ϵ_o	Permittivity of vacuum
ϵ	Exhaust efficiency
ε	Reduced energy
ε_g	Grayness coefficient
$\bar{\varepsilon}$	Average energy losses
ζ	Material constant
η	Resistivity
η_v	Viscosity
θ	Angle, poloidal angle
ι	Rotational transform
κ	Elongation
λ_D	Debye length
λ_{iz}	Ionization length
λ_n	Plasma density decay length
λ_{SOL}	Plasma decay length in the SOL
$\lambda_{n,l}$, $\lambda_{n,r}$	Plasma decay length in the divertor
$\Lambda(T)$	Integral thermal conductivity

$\ln \Lambda$	Coulomb logarithm
μ	Reduced mass
μ_b	Mobility
μ_o	Permeability of vacuum
ν	Collision frequency
ν_P	Poisson ratio
ξ	Magnetization parameter
ρ	Density, charge density
ρ_p	Collision parameter, impact parameter
ρ_{ce}	Electron gyro-radius
ρ_{ci}	Ion gyro-radius
σ	Cross-section
σ^*	Scattering parameter
σ_R^*	Backscattering parameter
σ_c	Plasma conductivity
σ_{SB}	Constant in Stefan–Boltzmann law
σ_{tr}	Transport cross section
σ_{ts}	Thermal stress
σ_y	Yield strength
$\langle \sigma v \rangle$	Rate coefficient
τ	Characteristic time
τ_p	Particle confinement time
τ_E	Energy confinement time
τ_{iz}	Ionization time
τ^*	Effective confinement time
φ	Angle, azimuthal angle
ϕ	Electric potential
ϕ_{fl}	Floating potential
ϕ_s	Potential at sheath entrance
ϕ_w	Wall potential
Φ	Error function
Φ^o	Particle flux
χ	Dielectric susceptibility
$\chi_\perp$	Anomalous energy transport coefficient
ω	Angular frequency
ω_{ce}	Electron cyclotron frequency
ω_{ci}	Ion cyclotron frequency
ω_{pe}	Electron plasma frequency
ω_{pi}	Ion plasma frequency
Ω	Solid angle

A.4 Abbreviations

Alcator C-Mod	Tokamak at MIT, Boston (US) (*http://www.psfc.mit.edu/research/alcator*)
ASDEX-Upgrade	Axially Symmetric Divertor Experiment: tokamak at IPP, Garching (Germany) (*http://www.ipp.mpg.de/ippcms/eng/pr/forschung/asdex*)
A*THERMAL-S	Simulation code: time-dependent heat conduction [337]
B2-Eirene	Simulation code: 2D MHD coupled with a Monte Carlo transport code for neutrals [430]
CM-system	Center-of-mass system
CFC	Carbon fiber composites
CHS	Torsatron at NIFS, Nagoya (Japan): Compact Helical System (*http://rd-w3server.nifs.ac.jp/chs*)
CIC	Cloud-in-cell method
CPU	Central processing unit
CX	Charge exchange
dpa	Number of displacements per atom
DEMO	Name for a prototype fusion reactor demonstrating power production
DT	Deuterium–tritium mixture
ECR	Electron cyclotron resonance
ECRH	Electron cyclotron resonance heating: heating method using microwave radiation close to the resonance gyro-frequency of plasma electrons
EDA	Engineering design activities of ITER
ELM	Edge localized mode: relaxation instability of steep edge plasma profiles in the H-mode
erf	error function
ERO	Simulation code: erosion and impurity transport using Monte Carlo techniques [548]
ESS	European Spallation Source
ESEE	Electron impact secondary electron emission
FOREV	Simulation code: material erosion by large heat loads [338]
HEIGHTS	Simulation code: two-dimensional radiation–magnetohydrodynamic model [335]
high-Z	Materials such as molybdenum and tungsten with large atomic numbers
HL-1M	Tokamak at SWIP, Chengdu (China) (*http://www.swip.ac.cn*)
HT-7	Hefei Tokamak-7: superconducting tokamak at Institute of Plasma Physics, Hefei (China) (*http://english.cas.ac.cn*)
HTS	High-temperature superconductor
H-mode	Plasma regime of improved confinement characterized by steep gradients of the plasma parameters at the plasma edge

ICRF	Ion cyclotron resonance frequency
ICRH	Ion cyclotron resonance heating: heating method using radio frequencies close to the resonance gyro-frequency of plasma ions
IFMIF	International Fusion Materials Irradiation Facility
IPP	Max Planck Institute for Plasma Physics at Garching and Greifswald (Germany) (*http://www.ipp.mpg.de)*
ISEE	Ion impact secondary electron emission
ITB	Internal transport barrier
ITER	International Thermonuclear Experimental Reactor (ITER meaning "the way" in Latin): international project of a prototype fusion reactor based on the tokamak design (*http://www.iter.org*)
ITER–FEAT	Fusion Energy Advanced Tokamak
JET	Joint European Torus: up to now the world's largest tokamak experiment at Culham (UK) (*http://www.jet.efda.org*)
LHD	Large Helical Device: superconducting torsatron-type experiment at NIFS, Toki (Japan) (*http://www.lhd.nifs.ac.jp/en*)
LCFS	Last closed flux surface: boundary between the core plasma, where the magnetic field lines are closed, and the scrape-off layer
L-mode	The normal confinement mode in tokamaks
L-system	Laboratory system
low-Z	Materials such as beryllium, lithium and carbon with small atomic numbers
MARFE	Multi-faceted radiation from the edge: toroidally symmetric, but poloidally localized, radiation instability
MARLOWE	Simulation code: particle transport in crystal matter [271]
MC	Monte Carlo method: use of random numbers
MD	Molecular dynamics: many-particle interaction
MHD	Magnetohydrodynamics: theory of conducting fluids or plasmas in a magnetic field
NBI	Neutral beam injection for heating and refueling
NGP	Nearest grid point method
PKA	Primary knock-on atom
PIC	Particle-in-cell: kinetic simulation of plasmas with self-consistently obtained electromagnetic fields
PSI	Plasma surface interaction
PSI-1/2	Plasma generator: linear DC-arc discharge device at the Humboldt University, Berlin (Germany) (*http://plasma.physik.hu-berlin.de/psi/psi.html*)
Q-machine	Quiescent machine: generating plasmas with no electric field gradients and no instabilities by using thermionic electron emission and contact ionization for ion production

RBS	Rutherford backscattering technique for surface analysis
REDEP	Simulation code: erosion and redeposition [482]
RES	Radiation-enhanced sublimation
RND	Random number in the interval $[0, 1]$
Sawteeth	Periodic relaxation oscillations of the central plasma Parameters localized in a region roughly within the $q_s = 1$ surface
SEE	Emission of secondary electrons
SKA	Secondary knock-on atom
SOL	Scrape-off layer: outer region of the plasma where the field lines intersect wall components (e.g., limiters or divertor plates)
TEXTOR	Tokamak Experiment for Technology Oriented Research: tokamak at IPP, Jülich (Germany) (*http://www.fz-juelich.de/ipp/textor*)
Tore-Supra	Superconducting tokamak at CEA, Cadarache (France) (*http://www-fusion-magnetique.cea.fr*)
TFTR	Tokamak Fusion Test Reactor: tokamak at PPPL, Princeton (US) (*http://www.pppl.gov/projects/pages/tftr.html*)
TRIM	Simulation code: particle transport in matter [266]
TRIM-SP	Simulation code: sputtering of solids and alloys [549]
TZM	Alloy of molybdenum with 0.5% Ti, 0.1% Zr
W7-X	New superconducting stellarator under construction at Greifswald (Germany) based on advanced design criteria (*http://www.ipp.mpg.de/ippcms/eng/pr/forschung/w7x*)
WBC	Simulation code: impurity transport in the plasma edge [424]
X-point	Point where two magnetic flux surfaces appear to cross (e.g., between magnetic islands or in the divertor region of tokamaks)

A.5 Fundamental Physical Constants

Table A.1. Physical constants used throughout the text

Constant, symbol	Value	Unit
Speed of light (vac.), c	2.99792458×10^{8}	$\mathrm{m\ s^{-1}}$
Permittivity of vacuum, ϵ_{o}	$8.854187817 \times 10^{-12}$	$\mathrm{C\ m^{-1}V^{-1}}$
Permeability of vacuum, μ_{o}	$4\pi \times 10^{-7}$	$\mathrm{T\ m\ A^{-1}}$
Elementary charge, e	$1.602176462(63) \times 10^{-19}$	C
Electron mass, m_{e}	$9.10938188(72) \times 10^{-31}$	kg
Electron volt, eV	$1.602176462(63) \times 10^{-19}$	J
Proton mass, m_{p}	$1.67262158(13) \times 10^{-27}$	kg
Proton–electron mass ratio, m_p/m_{e}	1836.1526675(39)	1
Atomic mass unit, amu	$1.66053873(13) \times 10^{-27}$	kg
Boltzmann constant, k_{B}	$1.3806503(24) \times 10^{-23}$	$\mathrm{J\ K^{-1}}$
Planck constant, h	$6.62606876(52) \times 10^{-34}$	J s
Planck constant over 2π, $\hbar = h/2\pi$	$1.054571596(82) \times 10^{-34}$	J s
Molar gas constant, R	8.314472(15)	$\mathrm{J\ K^{-1}\ mol^{-1}}$
Avogadro constant, N_{A}	$6.02214199(47) \times 10^{23}$	$\mathrm{mol^{-1}}$
Gravitational constant, G	$6.673(10) \times 10^{-11}$	$\mathrm{m^{3}\ kg^{-1}\ s^{-2}}$
Stefan–Boltzmann constant, σ	$5.670400(40) \times 10^{-8}$	$\mathrm{W\ m^{-2}\ K^{-4}}$
Bohr radius, a_{o}	$5.291772083(19) \times 10^{-11}$	m
Classical electron radius, r_{e}	$2.817940285(31) \times 10^{-15}$	m
Fine-structure constant, α	$7.297352533(27) \times 10^{-3}$	
Inverse fine-structure constant, α^{-1}	137.03599976(50)	
Rydberg constant, R_{∞}	$1.0973731568549(83) \times 10^{7}$	$\mathrm{m^{-1}}$
Electron magn. moment, μ_{e}	$-9.28476362(37) \times 10^{-24}$	$\mathrm{J\ T^{-1}}$
Proton magn. moment, μ_{p}	$1.410606633(58) \times 10^{-26}$	$\mathrm{J\ T^{-1}}$

The numbers in the brackets indicate the deviation of the last digit given, for example, 6.6260755(4) means 6.6260755 ± 0.0000004. These data are taken from the CODAS compilation [550,551]. Additional information can be found in [552–555].

Table A.2. Important conversion factors

Length	Energy	Pressure
1 µm = 10^{-6} m	1 cal = 4.19 J	1 Pa = 9.87 $\times 10^{-6}$ atm
1 nm = 10^{-9} m	k=8.617 $\times 10^{-5}$ eV/K	1 Pa =10^{-5} bar
1 Å=10^{-10} m	1 eV ≡ 11605 K	1 Torr =133.$\bar{3}$ Pa
1 pm = 10^{-12} m	1 rydberg=13.61 eV	1 bar = 750 Torr

A.6 Physical Properties of Elements

In the following table a number of physical properties of elements up to uranium are summarized. They are

- Z: atomic number
- A: mass number
- n_o: atomic density, $n_o = \rho/(A \cdot \text{amu})$
- ρ: density
- c_p: heat capacity
- k: heat conductivity
- W: work function
- I_{ion}: ionization energy
- E_s: sublimation energy

The values given are valid for normal conditions and could serve for estimations and quick comparison. One should be aware of their particularly strong dependence on temperature and structure since some elements such as carbon are used to produce materials with apparently different properties, among them fine-grain graphite and carbon fiber composites (CFC).

Table A.3 Physical properties of elements

Z		A (amu)	$n_o, \times 10^{28}$ (1/m^3)	$\rho, \times 10^3$ (kg/m^3)	c_p (J/(kgK))	k (W/(mK))	W (eV)	I_{ion} (eV)	E_s (eV)
1	H	1.00794	–	–	–	–	–	13.6	–
2	He	4.0026	–	–	–	–	–	24.59	–
3	Li	6.941	4.633	0.534	3582	84.7	2.4	5.39	1.67
4	Be	9.01218	12.362	1.85	1825	200	3.9	9.32	3.38
5	B	10.811	12.929	2.32	–	–	4.5	8.3	5.73
6	C	12.0107	11.3	≈ 1.8	≈ 800	≈ 150	4.7	11.26	7.42
7	N	14.0067	–	–	–	–	–	14.53	–
8	O	15.9994	–	–	–	–	–	13.62	–
9	F	18.9984	–	–	–	–	–	17.42	–
10	Ne	20.1797	–	–	–	–	–	21.56	–
11	Na	22.9898	2.541	0.97	1228	141	2.3	5.14	–
12	Mg	24.3051	4.311	1.74	1023	156	3.6	7.65	–
13	Al	26.9815	6.026	2.7	897	237	4.2	5.99	3.36
14	Si	28.0855	–	–	–	–	4.8	8.15	4.7
15	P	30.9738	–	–	–	–	–	10.49	–
16	S	32.067	–	–	–	–	–	10.36	–
17	Cl	35.4527	–	–	–	–	–	12.97	–
18	Ar	39.948	–	–	–	–	–	15.76	–
19	K	39.0983	1.371	0.89	757	102.4	2.2	4.34	–
20	Ca	40.078	2.314	1.54	647	200	2.6	6.11	–
21	Sc	44.956	4.005	2.99	568	15.8	3.3	6.56	–
22	Ti	47.867	5.674	4.51	523	21.9	4.0	6.83	4.89
23	V	50.9415	7.093	6.0	489	30.7	4.1	6.75	5.33
24	Cr	51.9962	8.281	7.15	449	93.7	4.6	6.77	4.12
25	Mn	54.938	8.002	7.3	479	7.82	3.8	7.43	–
26	Fe	55.845	8.485	7.87	449	80.2	4.3	7.9	4.34
27	Co	58.9332	9.054	8.86	421	100	4.4	7.88	4.43
28	Ni	58.6934	9.132	8.9	444	90.7	4.5	7.64	4.46
29	Cu	63.546	8.491	8.96	385	401	4.5	7.73	3.52
30	Zn	65.39	6.576	7.14	388	116	4.2	9.39	–

Table A.3 continued

Z	A (amu)	n_o, $\times 10^{28}$ $(1/m^3)$	ρ, $\times 10^3$ (kg/m^3)	c_p (J/(kgK))	k (W/(mK))	W (eV)	I_{ion} (eV)	E_s (eV)
31 Ga	69.723	5.105	5.91	371	40.6	4.0	6.0	2.82
32 Ge	72.61	–	–	–	–	4.8	7.9	3.88
33 As	74.921	–	–	–	–	5.1	9.79	–
34 Se	78.96	–	–	–	–	4.7	9.75	–
35 Br	79.904	–	–	–	–	–	11.81	–
36 Kr	83.80	–	–	–	–	–	14.0	–
37 Rb	85.4678	1.078	1.53	363	58.2	2.2	4.18	–
38 Sr	87.62	1.814	2.64	301	35.3	2.3	5.69	–
39 Y	88.9059	3.028	4.47	298	17.2	3.3	6.22	–
40 Zr	91.224	4.304	6.52	278	22.7	3.9	6.63	6.33
41 Nb	92.9064	5.555	8.57	265	53.7	4.0	6.76	7.59
42 Mo	95.94	6.403	10.2	251	138	4.3	7.09	6.83
43 Tc	[98]	–	–	–	–	4.4	7.28	–
44 Ru	101.07	7.21	12.1	238	117	4.6	7.36	–
45 Rh	102.9055	7.257	12.4	243	150	4.7	7.46	–
46 Pd	106.42	6.791	12.0	246	71.8	4.8	8.34	3.91
47 Ag	107.8682	5.862	10.5	235	429	4.3	7.58	2.97
48 Cd	112.412	4.655	8.69	232	96.8	4.1	8.99	–
49 In	114.818	3.834	7.31	233	81.6	3.8	5.79	2.49
50 Sn	118.711	3.683	7.26	228	66.6	4.4	7.34	–
51 Sb	121.760	3.304	6.68	207	24.3	4.1	8.61	–
52 Te	127.60		–	–	–	4.7	9.01	–
53 I	126.9045	–	–	–	–	6.8	10.45	–
54 Xe	131.29	–	–	–	–	–	12.13	–
55 Cs	132.9054	0.875	1.93	242	35.9	1.8	3.89	–
56 Ba	137.327	1.587	3.62	204	18.4	2.5	5.21	–
57 La	138.906	2.666	6.15	195	13.4	3.3	5.58	–
58 Ce	140.116	2.91	6.77	192	11.3	2.7	5.54	–
59 Pr	140.908	2.893	6.77	193	12.5	2.7	5.46	–
60 Nd	144.24	2.927	7.01	190	16.5	3.2	5.52	–

Table A.3 continued

Z	A (amu)	n_o, $\times 10^{28}$ (1/m^3)	ρ, $\times 10^3$ (kg/m^3)	c_p (J/(kgK))	k (W/(mK))	W (eV)	I_{ion} (eV)	E_s (eV)
61 Pm	[145]	–	–	–	–	3.1	5.58	–
62 Sm	150.36	3.012	7.52	197	13.3	2.7	5.64	–
63 Eu	151.964	2.077	5.24	182	13.9	2.5	5.67	–
64 Gd	157.25	3.025	7.9	236	10.5	3.1	6.15	–
65 Tb	158.925	3.119	8.23	182	11.1	3.1	5.86	–
66 Dy	162.50	3.169	8.55	170	10.7	3.2	5.94	–
67 Ho	164.93	3.213	8.8	165	16.2	3.2	6.02	–
68 Er	167.26	3.266	9.07	168	14.5	3.2	6.11	–
69 Tm	168.934	3.322	9.32	160	16.9	3.1	6.18	–
70 Yb	173.04	2.401	6.9	155	38.5	2.6	6.25	–
71 Lu	174.967	3.387	9.84	154	16.4	3.1	5.43	–
72 Hf	178.49	4.487	13.3	144	23	3.5	6.83	–
73 Ta	180.948	5.458	16.4	140	57.5	4.1	7.55	–
74 W	183.84	6.322	19.3	132	174	4.5	7.86	8.68
75 Re	186.207	6.727	20.8	137	47.9	5.0	7.83	–
76 Os	190.23	7.151	22.59	130	87.6	4.7	8.44	–
77 Ir	192.217	7.049	22.5	131	147	4.7	8.97	–
78 Pt	195.078	6.637	21.5	133	71.6	5.3	8.96	5.86
79 Au	196.967	5.901	19.3	129	317	4.7	9.23	3.8
80 Hg	200.59	4.062	13.53	140	8.34	4.5	10.44	–
81 Tl	204.383	3.477	11.8	129	46.1	3.7	6.11	–
82 Pb	207.2	3.284	11.3	129	35.3	4.0	7.42	2.03
83 Bi	208.98	2.821	9.79	122	7.87	4.4	7.29	–
84 Po	[209]	2.651	9.2	–	–	4.7	8.42	–
85 At	[210]	–	–	–	–	–	–	–
86 Rn	[222]	–	–	–	–	–	10.75	–
87 Fr	[223]	–	–	–	–	1.6	4.07	–
88 Ra	[226]	1.332	5	–	–	3.2	5.28	–
89 Ac	[227]	2.653	10	–	–	2.7	5.17	–
90 Th	232.038	3.037	11.7	113	54	3.5	6.31	–
91 Pa	231.036	4.014	15.4	–	–	3.3	5.89	–
92 U	238.029	4.832	19.1	116	27.6	3.4	6.19	5.42

References

1. D.E. Post, R. Behrisch, (Eds.), *Physics of Plasma–Wall Interaction in Controlled Fusion* (Plenum Press, New York, 1986)
2. P.C. Stangeby, G.M. McCracken, Nucl. Fusion **30**(7), 1225 (1990)
3. R.K. Janev, H.W. Drawin, (Eds.), *Atomic and Plasma–Material Interaction Processes in Controlled Thermonuklear Fusion* (Elsevier, Amsterdam, 1993)
4. W.O. Hofer, J. Roth, (Eds.), *Physical Processes of the Interaction of Fusion Plasmas with Solids* (Academic Press, San Diego, 1996)
5. G. Federici, C.H. Skinner, J.N. Brooks, J.P. Coad, C. Grisolia, A.A. Haasz, A. Hassanein, V. Philipps, C.S. Pitcher, J. Roth, W.R. Wampler, D.G. Whyte, Nucl. Fusion **41**(12R), 1967 (2001)
6. R. Janev, (Ed.), *Atomic and Plasma–Material Interaction Data for Fusion* (Supplement to Nuclear Fusion, Volume 1, 1991)
7. R. Janev, (Ed.), *Atomic and Plasma–Material Interaction Data for Fusion* (Supplement to Nuclear Fusion, Volume 2, 1992)
8. R. Janev, (Ed.), *Atomic and Plasma–Material Interaction Data for Fusion* (Supplement to Nuclear Fusion, Volume 3, 1992)
9. R. Janev, (Ed.), *Atomic and Plasma–Material Interaction Data for Fusion* (Supplement to Nuclear Fusion, Volume 4, 1993)
10. R.S. Pease, J. Nucl. Mater. **191–194**, 7 (1992)
11. R.J. Bickerton, Plasma Phys. Control. Fusion **35**, B3 (1993)
12. J. Raeder, I. Cook, F.H. Morgenstern, E. Salpietro, R. Bünde, E. Ebert, Safety and Enviromental Assessment of Fusion Power (SEAFP). Tech. Rep. EUR-FUBRU XII-217/95, European Commission, Directorate General XII, Fusion Programme, Brussels (1995)
13. G.H. Miley, Tech. Rep. 00-2218-17, University of Illinois (1974)
14. K. Miyamoto, *Fundamentals of Plasma Physics* (Iwanami Book Service Center, Tokyo, 1997)
15. H.S. Bosch, G.M. Hale, Nucl. Fusion **32**, 611 (2002)
16. G. Gamov, Zeitschrift für Physik **51**, 204 (1928)
17. U. S. Geological Survey, http://minerals.usgs.gov
18. R. Aymar, P. Barabaschi, Y. Shimomura, the ITER team, Plasma Phys. Control. Fusion **44**, 519 (2002)
19. F.F. Chen, *Plasma Physics and Controlled Fusion Vol.1: Plasma Physics* (Plenum Press, New York, London, 1984)

20. J.D. Lawson, Proc. Phys. Soc. **B 70**, 6 (1957)
21. J.L. Cecchi, J. Nucl. Mater. **93 & 94**, 28 (1980)
22. R. Behrisch, V. Prozesky, Nucl. Fusion **30**(10), 2166 (1990)
23. M. Bornatici, F. Engelmann, Physica Scripta **63**, 93 (2001)
24. D. Reiter, G.H. Wolf, H. Kever, J. Nucl. Mater. **176 & 177**, 756 (1990)
25. U. Samm, M.Z. Tokar, B. Unterberg, J. Nucl. Mater. **241–243**, 827 (1997)
26. W.P. West, S.L. Allen, N.H. Brooks, D.A. Buchenauer, T.N. Carlstrom, J.W. Cuthbertson, E.J. Doyle, T.E. Evans, M.E. Fenstermacher, D.N. Hill, A.W. Hyatt, R.C. Isler, G.L. Kackson, R. Jong, C.C. Klepper, C.J. Lasnier, A.W. Leonard, M.A. Mahdavi, R. Maingi, G.R. McKee, W.H. Meyer, R.A. Moyer, D.G. Nilson, T.W. Petrie, G.D. Porter, T.L. Rhodes, M.J. Schaffer, R.D. Stambaugh, D.M. Thomas, S. Tugarinov, M.R. Wade, J.G. Watkins, D.G. Whyte, R.D. Wood, Plasma Phys. Control. Fusion **39**, A295 (1997)
27. H.S. Bosch, D. Coster, R. Dux, C. Fuchs, G. Haas, A. Herrmann, S. Hirsch, A. Kallenbach, J. Neuhauser, R. Schneider, J. Schweinzer, M. Weinlich, A.U. team, N. team, J. Nucl. Mater. **241–243**, 82 (1997)
28. H.S. Bosch, D. Coster, R. Dux, G. Haas, A. Kallenbach, M. Kaufmann, K. Lackner, J. Neuhauser, S. de Pena Hempel, W. Poschenrieder, R. Schneider, A.U. team, N. team, I. team, E. team, Plasma Phys. Control. Fusion **39**, 1771 (1997)
29. R. Preuss, E. Ascasibar, A. Dinklage, V. Dose, J.H. Harris, A. Kus, S. Okamura, F. Sano, U. Stroth, J. Talmadge, H. Yamada, 32nd EPS Plasma Physics Conference, Tarragona, 2005 (to be published)
30. J.J. Duderstadt, G.A. Moses, *Inertial Confinement Fusion* (John Wiley and Sons, New York, 1982)
31. J.D. Lindl, *Inertial Confinement Fusion: The Quest for Ignition and Energy Gain using Indirect Drive* (Springer Verlag, 1997)
32. S. Pfalzner, *An Introduction to Inertial Confinement Fusion* (IoP, 2004)
33. A.R. Raffray, D. Haynes, F. Najmabadi, J. Nucl. Mater. **313–316**, 23 (2003)
34. Physicsweb, http://www.physicsweb.org/articles/news/9/9/2/1
35. J.A. Bittencourt, *Fundamentals of Plasma Physics* (Pergamon Press, Oxford, New York, Beijing, 1986)
36. R.J. Goldston, P.H. Rutherford, *Introduction to Plasma Physics* (IOP Publishing Ltd, London, 1995)
37. H. Wobig, Plasma Phys. Control. Fusion **35**, 903 (1993)
38. G. Grieger, W. Lotz, P. Merkel, J. Nührenberg, J. Sapper, H. Wobig, the W7-X team, R. Burhenn, V. Erckmann, U. Gasparino, L. Giannone, H.J. Hartfuss, R. Jaenicke, G. Kühner, H. Ringler, A. Weller, F. Wagner, the W7-AS team, Phys. Fluids **B 4**(7), 2081 (1992)
39. G. Grieger, C.D. Beidler, H. Maassberg, E. Harmeyer, F. Herrnegger, J. Junker, J. Kisslinger, W. Lotz, P. Merkel, J. Nührenberg, F. Rau, J. Sapper, A. Schlüter, F. Sardei, H. Wobig, Physics and Engineering Studies for Wendelstein 7-X. Tech. Rep. IAEA-CN-53/G-I-6, IPP, Garching (1994)
40. J. Geiger, Equilibrium and Stability of Toroidally Confined Plasmas. Tech. Rep. IPP 5-49, IPP, Garching (1992)
41. E. Strumberger, Nucl. Fusion **32**(5), 737 (1992)
42. E. Strumberger, Contrib. to Plasma Phys. **34**(2/3), 120 (1994)
43. E. Strumberger, Nucl. Fusion **36**(7), 891 (1996)

44. D. Hildebrandt, H. Wolff, J. Roth, R. Brakel, P. Grigull, F. Rau, F. Sardei, E. Taglauer, the W7-AS team, the NBI group, the ECRH group, J. Nucl. Mater. **196–198**, 436 (1992)
45. J. Nührenberg, E. Strumberger, Contrib. to Plasma Phys. **32**(3/4), 204 (1992)
46. P.C. Stangeby, J. Phys. D: Appl. Phys. **18**, 1547 (1985)
47. B. Labombard, A.A. Grossman, R.W. Conn, J. Nucl. Mater. **176 & 177**, 548 (1990)
48. P.C. Stangeby, C.S. Pitscher, J.D. Elder, Nucl. Fusion **32**(12), 2079 (1992)
49. U. Samm, J. Boedo, G. Bertschinger, K.H. Dippel, H. Euringer, K.H. Finken, D. Gray, D. Hillis, A. Pospieszczyk, D. Reiter, M. Tokar, B. Unterberg, J. Nucl. Mater. **196–198**, 633 (1992)
50. U. Samm, G. Bertschinger, P. Bogen, J.D. Hey, E. Hintz, L. Könen, Y.T. Lie, A. Pospieszczyk, D. Rusbüldt, R.P. Schorn, B. Schweer, M. Tokar, B. Unterberg, Plasma Phys. Control. Fusion **35**, B167 (1993)
51. J. Neuhauser, R. Wunderlich, J. Nucl. Mater. **145–147**, 877 (1987)
52. J. Neuhauser, Plasma Phys. Control. Fusion **34**(13), 2015 (1992)
53. K. Lackner, R. Schneider, Fusion Engineering and Design **22**, 107 (1993)
54. D.F. Düchs, G. Haas, D. Pfirsch, H. Vernickel, J. Nucl. Mater. **53**, 102 (1974)
55. P.C. Stangeby, Nucl. Fusion **33**(11), 1695 (1993)
56. G. Janeschitz, K. Borrass, G. Federici, Y. Igikhanov, A. Kukushkin, H.D. Pacher, G.W. Pacher, M. Sugihara, J. Nucl. Mater. **220–222**, 73 (1995)
57. G. Janeschitz, G. Fussmann, J. Hofmann, L.B. Ran, H.R. Yang, J. Roth, E. Taglauer, the ASDEX team, the NI team, the ICRH team, J. Nucl. Mater. **162–164**, 624 (1989)
58. P.C. Stangeby, J.D. Elder, D. Reiter, G.P. Maddison, Contrib. to Plasma Phys. **34**(2/3), 306 (1994)
59. P.C. Stangeby, Nucl. Fusion **35**, 1391 (1995)
60. P.C. Stangeby, J.D. Elder, J. Nucl. Mater. **230–222**, 193 (1995)
61. K. Borrass, R. Schneider, R. Farengo, Nucl. Fusion **37**, 523 (1997)
62. D. Naujoks, Contrib. to Plasma Phys. **41**, 375 (2001)
63. D. Naujoks, Contrib. to Plasma Phys. **42**, 356 (2002)
64. H. Ikezi, Phys. Fluids **29**(6), 1764 (1986)
65. T. Trottenberg, A. Melzer, A. Piel, Plasma Sources SCi. Technol. **4**, 450 (1995)
66. C.K. Goertz, Reviews of Geophysics **27**(2), 271 (1989)
67. M. Horanyi, C.K. Goertz, The Astrophysical Journal **361**, 155 (1990)
68. A. Melzer, A. Homann, A. Piel, Phys. Rev. **(E) 53**(3), 2757 (1996)
69. A. Melzer, V.A. Schweigert, I.V. Schweigert, A. Homann, S. Peters, A. Piel, Phys. Rev. **(E) 54**(1), R46 (1996)
70. L. Spitzer, R. Härm, Phys. Rev. **89**, 977 (1953)
71. G. Schmidt, *Physics of High Temperature Plasmas. An Introduction* (Academic Press, New York and London, 1966)
72. G. Fussmann, Teilchentransport in magnetisch eingeschlossenen Plasmen. Tech. Rep. IPP 1/273, IPP, Garching (1992)
73. G. Fussmann, A.R. Field, A. Kallenbach, K. Krieger, K.H. Steuer, the ASDEX-team, Plasma Phys. Control. Fusion **33**, 1677 (1991)
74. T.E. Stringer, The Non-Ambipolarity of Neoclassical Transport and its Consequences. Tech. Rep., JET, Oxford (1993)
75. H. Park, E. Heo, W. Horton, D. Choi, Phys. Plasmas **4**(9), 3273 (1997)

76. D. Bohm, E.H. Burhop, H.S. Massey, *The Characteristics of Electrical Discharges in Magnetic Fields, Eds. A. Guthrie and R. K. Wakerling* (McGraw–Hill, New York, 1949)
77. L. Spitzer, Letters to the Editor **00**, 659 (1960)
78. B.D. Scott, New Journal of Physics **4**, 52.1 (2002)
79. O. Grulke, T. Klinger, New Journal of Physics **4**, 67.1 (2002)
80. J.L. Terry, N.P. Basse, I. Cziegler, M. Greenwald, O. Grulke, B. LaBombard, S.J. Zweben, E.M. Edlund, J.W. Hughes, L. Lin, Y. Lin, M. Porkolab, M. Sampsell, B. Veto, S.J. Wukitch, Nucl. Fusion **45**, 1321 (2005)
81. A.J. Wootton, B.A. Carreras, H. Matsumoto, K. McGuire, W.A. Peebles, C.P. Ritz, Phys. Fluids B **2**(12), 2879 (1990)
82. H.Y.W. Tsui, A.J. Wooton, J.D. Bell, R.D. Bengsten, D. Diebold, J.H. Harris, Phys. Fluids B **B5**, 2491 (1993)
83. K.H. Burrell, Phys. Plasmas **4**(5), 1499 (1997)
84. J.B. Taylor, Plasma Phys. Control. Fusion **39**, A1 (1997)
85. P.C. Liewer, Nucl. Fusion **25**, 543 (1985)
86. W. Horton, B. Hu, J.Q. Dong, P. Zhu, New Journal of Physics **4**, 14.1 (2002)
87. V. Naulin, New Journal of Physics **4**, 28.1 (2002)
88. J. Schou, P. Borgesen, O. Ellegaard, H. Sorensen, Phys. Rev. **B 34**(1), 93 (1986)
89. A.L. Boers, Nucl. Instr. and Meth. **B4**, 98 (1984)
90. J.B. Sanders, H.E. Roosendaal, F. Vitalis, Radiation Effects **38**, 201 (1978)
91. J.F. Ziegler, J.M. Manoyan, Nucl. Instr. and Meth. **B35**, 215 (1988)
92. J. Lindhard, M. Scharff, Phys. Rev. **124**(1), 128 (1961)
93. O.B. Firsov, Sov. Tech. Phys. **5**, 1517 (1959)
94. D.R. Hartree, Proc. Cambridge Philos. Soc. **24**, 88 (1928)
95. V. Fock, Z. Phys. **61**, 126 (1930)
96. J.C. Slater, Phys. Rev. **81**, 385 (1951)
97. M. Gryzinski, Phys. Rev. **138**, A336 (1965)
98. M. Gryzinski, Phys. Rev. **138**, A305 (1965)
99. M. Gryzinski, Phys. Rev. **138**, A322 (1965)
100. E. Gerjuoy, Phys. Rev. **148**, 54 (1966)
101. P. Sigmund, Phys. Rev. **A 26**(5), 2497 (1982)
102. N. Bohr, Kgl. Danske Videnskab. Selskab, Mat.-fys. Medd. **18**, 1 (1948)
103. N. Bohr, J. Lindhard, Kgl. Danske Videnskab. Selskab, Mat.-fys. Medd. **28**, 1 (1954)
104. J.F. Ziegler, J.P. Biersack, U. Littmark, *The Stopping and Range of Ions in Solids, Vol. 1* (Pergamon, New York, 1984)
105. W.E. Lamb, Phys. Rev. **58**, 696 (1940)
106. W. Brandt, M. Kitagawa, Phys. Rev. **B 25**(9), 5631 (1982)
107. J. Lindhard, M. Scharff, H.E. Schiott, Kgl. Danske Videnskab. Selskab, Mat.-fys. Medd. **33**(14), 1 (1963)
108. J. Lindhard, V. Nielsen, M. Scharff, P.V. Thomson, Kgl. Danske Videnskab. Selskab, Mat.-fys. Medd. **36**(10), 1 (1968)
109. W. Eckstein, *Computer Simulation of Ion–Solid Interactions.* Springer Series in Materials Science (Springer–Verlag, Berlin, Heidelberg, New York, 1991)
110. M.I. Rasanov, I.S. Tilinin, *Surface Analysis by Backscattering* (Energoatomisdat, Moscow, 1985)
111. P. Sigmund, Rev. Roum. Phys. **17**(9), 1079 (1972)

112. Z. Luo, S. Wang, Phys. Rev. **36**(4), 1885 (1987)
113. V.C. Remisovich, D.B. Rogoskin, M.I. Rasanov, *Range Fluctuations of Charged Particles* (Energoatomisdat, Moscow, 1988)
114. W. Eckstein, Calculated Trapping Curves of D in C and Si. Tech. Rep. IPP 9/33, IPP, Garching (1980)
115. O.B. Firsov, Sov. Phys.-JETP **36**, 1076 (1959)
116. O.S. Oen, M.T. Robinson, Nucl. Instr. and Meth. **132**, 647 (1976)
117. E. Fermi, E. Teller, Phys. Rev. **72**, 399 (1947)
118. R.H. Ritchie, Nucl. Instr. and Meth. **198**, 81 (1982)
119. J.J. Thomson, *Conduction of Electricity Through Gases* (University Press, Cambridge, 1903)
120. C.G. Darwin, Phil. Mag. **23**, 901 (1912)
121. H.A. Bethe, Ann. Phys. (Leipzig) **5**, 325 (1930)
122. F. Bloch, Ann. Phys. (Leipzig) **16**, 285 (1933)
123. F. Bloch, Z. Phys. **81**, 363 (1933)
124. Particle Data Group, Eur. Phys. J. C **3**, 1 (1998)
125. S.P. Ahlen, Rev. Mod. Phys. **52**, 121 (1980)
126. J.F. Ziegler, *Handbook of Stopping Cross–Sections for Energetic Ions in all Elements* (Pergamon, New York, 1980)
127. M. Gryzinski, Phys. Rev. **107**, 1471 (1957)
128. M. Gryzinski, Phys. Rev. **115**, 374 (1959)
129. D. Semrad, P. Bauer, Nucl. Instr. and Meth. **B12**, 24 (1985)
130. R. Wedell, Nucl. Instr. and Meth. **B12**, 17 (1985)
131. I. Nagy, J. Laszlo, J. Giber, Z. Phys. A - Atoms and Nuclei **321**, 221 (1985)
132. C. Varelas, J.P. Biersack, Nucl. Instr. and Meth. **79**, 213 (1970)
133. N.Q. Lam, G.K. Leaf, H. Wiedersich, J. Nucl. Mater. **88**, 289 (1980)
134. H. Wiedersich, Nucl. Instr. and Meth. **B7/8**, 1 (1985)
135. R. Kelly, Bombardment–Induced Compositional Change with Alloys, Oxides, Oxysalts and Halides: I. The Role of Surface Binding Energy. Tech. Rep., IBM, N. Y. U. S. A. (1989)
136. R. Kelly, Nucl. Instr. and Meth. **B39**, 43 (1989)
137. R. Kelly, Material Science and Engineering **A115**, 11 (1989)
138. H. Wiedersich, J. Nucl. Mater. **206**, 121 (1993)
139. G.H. Kinchin, R.S. Pease, Rep. Prog. Phys. **18**, 1 (1955)
140. J. Yu, W.F. Sommer, J.N. Bradbury, W.V. Green, M. Victoria, J. Nucl. Mater. **227**, 266 (1996)
141. L.E. Rehn, P.R. Okamoto, Nucl. Instr. and Meth. **B39**, 104 (1989)
142. W. Möller, PIDAT – A Computer Program for Implant Diffusion and Trapping. Tech. Rep. IPP 9/44, IPP, Garching (1983)
143. W. Möller, J. Nucl. Mater. **162–164**, 138 (1989)
144. B.M.U. Scherzer, Nucl. Instr. and Meth. **B45**, 57 (1990)
145. R.A. Causey, J. Nucl. Mater. **162–164**, 151 (1989)
146. A.A. Haasz, P. Frenzen, J.W. Davis, S. Chiu, C.S. Pitcher, J. Appl. Phys. **77**(1), 66 (1995)
147. W. Möller, J. Nucl. Mater. **162–164**, 138 (1989)
148. D. Hildebrandt, M. Akbi, B. Jüttner, W. Schneider, J. Nucl. Mater. **226–269**, 532 (1999)
149. M. Mayer, R. Behrisch, H. Plank, J. Roth, G. Dollinger, C.M. Frey, J. Nucl. Mater. **230**, 67 (1996)

150. F.L. Tabares, V. Rohde, the ASDEX-Upgrade Team, Plasma Phys. Control. Fusion **46**, B381 (2004)
151. J.W. Davis, A.A. Haasz, J. Nucl. Mater. **266–269**, 478 (1999)
152. V. Philipps, H.G. Esser, J. von Seggern, H. Reimer, M. Freisinger, E. Vietzke, P. Wienhold, J. Nucl. Mater. **266–269**, 386 (1999)
153. W. Jacob, B. Landkammer, C.H. Wu, J. Nucl. Mater. **266–269**, 552 (1999)
154. R.T. Nachtrieb, B.L. LaBombard, J.L. Terry, J.C. Reardon, W.L. Rowan, W.R. Wampler, J. Nucl. Mater. **266–269**, 896 (1999)
155. R. Behrisch, W. Eckstein, Nucl. Instr. and Meth. **B 82**, 255 (1993)
156. P. Sigmund, Phys. Rev. **184**, 383 (1969)
157. P. Sigmund, Nucl. Instr. and Meth. **B27**, 1 (1987)
158. G. Falcone, P. Sigmund, Appl. Phys. **25**, 307 (1981)
159. R. Behrisch, (Ed.), *Sputtering by Particle Bombardment I, II.* Topics Appl. Phys., Vols. 47 and 52 (Springer–Verlag, Berlin, Heidelberg, 1981, 1983)
160. R. Behrisch, G. Maderlechner, B.M.U. Scherzer, Appl. Phys. **18**, 391 (1979)
161. J. Bohdansky, J. Roth, H.L. Bay, J. Appl. Phys. **51**, 2861 (1980)
162. J. Bohdansky, C.L. Chen, W. Eckstein, J. Roth, J. Nucl. Mater. **103 & 104**, 339 (1981)
163. J. Bohdansky, Nucl. Instr. and Meth. **B2**, 587 (1984)
164. C. Garcia-Rosales, W. Eckstein, J. Roth, J. Nucl. Mater. **218**, 8 (1994)
165. W. Eckstein, C. Garcia-Rosales, J. Roth, Nucl. Instr. and Meth. **B 83**, 95 (1993)
166. C. Garcia-Rosales, J. Nucl. Mater. **211**, 202 (1994)
167. Y. Yamamura, J. Bohdansky, Vacuum **35**(12), 561 (1985)
168. W. Eckstein, C. Garcia-Rosales, J. Roth, W. Ottenberger, Sputtering Data. Tech. Rep. IPP 9/82, IPP, Garching (1993)
169. Y. Yamamura, Y. Itikawa, N. Itoh, Tech. Rep. IPPJ-AM-26, Nagoya University, Nagoya (1983)
170. W. Eckstein, Nucl. Instr. and Meth. **B18**, 344 (1987)
171. V.A. Abramov, Y.I. Igitkhanov, V.I. Pistunovich, V.A. Pozharov, J. Nucl. Mater. **162–164**, 462 (1989)
172. D. Naujoks, W. Eckstein, J. Nucl. Mater. **230**(6), 93 (1996)
173. D. Naujoks, J. Roth, K. Krieger, G. Lieder, M. Laux, J. Nucl. Mater. **210**, 43 (1994)
174. C.A. Ordonez, W.D. Booth, R. Carrera, M.E. Oakes, J. Nucl. Mater. **185**, 130 (1991)
175. W. Eckstein, J. Roth, Nucl. Instr. and Meth. **B53**, 279 (1991)
176. D. Naujoks, W. Eckstein, J. Nucl. Mater. **220–222**, 993 (1995)
177. H. Maer, K. Schmid, W. Eckstein, J. Nucl. Mater. **337–339**, 480 (2005)
178. E. Vietzke, K. Flaskamp, V. Philipps, J. Nucl. Mater. **128–129**, 545 (1984)
179. E. Vietzke, K. Flaskamp, V. Philipps, J. Nucl. Mater. **145–147**, 443 (1987)
180. E. Vietzke, V. Philipps, K. Flaskamp, P. Koidl, C. Wild, Surface and Coatings Technology **47**, 156 (1991)
181. J. Küppers, Surf. Science Reports **22**, 249 (1995)
182. M. Wittmann, J. Küppers, Surf. Science Reports **22**, 249 (1995)
183. A. Horn, A. Schenk, J. Biener, Chemical Physics Letters **231**, 193 (1994)
184. C. Hopf, Chemical Erosion and Ion–Induced Growth of Armorphous Hydrocarbon Films. Tech. Rep. IPP 9-134, Max-Planck-Institut für Plasmaphysik, Garching (2003)

185. C. Hopf, A. von Keudell, W. Jacob, Nucl. Fusion **42**, L27 (2002)
186. C. Hopf, A. von Keudell, W. Jacob, J. Appl. Physics **94**(4), 2373 (2003)
187. J. Roth, J. Nucl. Mater. **266–269**, 51 (1999)
188. E. Vietzke, J. Nucl. Mater. **290–293**, 158 (2001)
189. J. Roth, C. Garcia-Rosales, Nucl. Fusion **36**, 1647 (1996)
190. C. Garcia-Rosales, M. Balden, J. Nucl. Mater. **290–293**, 173 (2001)
191. M. Balden, C. Garcia-Rosales, R. Behrisch, J. Nucl. Mater. **290–293**, 52 (2001)
192. H. Yagi, H. Toyoda, H. Sugai, J. Nucl. Mater. **313–316**, 284 (2003)
193. J. Roth, J. Nucl. Mater. **145–147**, 87 (1987)
194. D.M. Goebel, J. Bohdansky, R.W. Conn, Y. Hirooka, B. LaBombard, W.K. Leung, R.E. Nygren, J. Roth, G.R. Tynan, Nucl. Fusion **28**(6), 1041 (1988)
195. J.W. Davis, T. Haasz, J. Nucl. Mater. **241–243**, 37 (1997)
196. A. Kallenbach, A. Bard, A. Carlson, R. Dux, Physica Scripta **T81**, 43 (1999)
197. H. Grote, W. Bohmeyer, P. Kornejew, H.D. Reiner, G. Fussmann, R. Schlögl, G. Weinberg, C.H. Wu, J. Nucl. Mater. **266–269**, 1059 (1999)
198. J. Roth, R. Preuss, W. Bohmeyer, S. Brezinsek, A. Cambe, E. Casarotto, R. Doerner, E. Gauthier, G. Federici, S. Higashijima, J. Hogan, A. Kallenbach, A. Kirschner, H. Hubo, J.M. Layet, T. Nakano, V. Philipps, A. Pospieszczyk, R. Pugno, R. Ruggieri, B. Schweer, G. Sergienko, M. Stamp, Nucl. Fusion **44**, L21 (2004)
199. J. Roth, A. Kirschner, W. Bohmeyer, S. Brezinsek, A. Cambe, E. Casarotto, R. Doerner, E. Gauthier, G. Federici, S. Higashijima, J. Hogan, A. Kallenbach, H. Kubo, J.M. Layet, T. Nakano, V. Philipps, A. Pospieszczyk, R. Preuss, R. Pugno, R. Ruggieri, B. Schweer, G. Sergienko, M. Stamp, J. Nucl. Mater. **337–339**, 970 (2005)
200. J. Roth, C. Garcia-Rosales, R. Behrisch, W. Eckstein, J. Nucl. Mater. **191–194**, 45 (1992)
201. A.Y.K. Chen, J.W.D. an A. A. Haasz, J. Nucl. Mater. **290–293**, 61 (2001)
202. C. Hopf, W. Jacob, A. von Keudell, J. of Appl. Physics **97**, 1 (2005)
203. A.V. Krasheninnikov, E. Salonen, K. Nordlund, J. Keinonen, C.H. Wu, Contrib. to Plasma Phys. **42**(2-4), 451 (2002)
204. E. Salonen, K. Nordlund, J. Keinonen, C.H. Wu, Physical Review B **63, 195415**, 1 (2001)
205. E. Salonen, K. Nordlund, J. Keinonen, C.H. Wu, J. Nucl. Mater. **313–316**, 404 (2003)
206. E. Salonen, Physica Scripta **111**, 133 (2004)
207. P. Träskelin, E. Salonen, K. Nordlund, A.V. Krasheninnikov, J. Keinonen, C.H. Wu, J. Nucl. Mater. **313–316**, 52 (2003)
208. J.N. Brooks, Transport Calculations of Chemically Sputtered Carbon near a Plasma Divertor Surface. Tech. Rep. ANL/FPP/TM-259, Argonne National Laboratory, Argonne, Illinois 60439, U. S. A. (1992)
209. A. Kirschner, J.N. Brooks, V. Philipps, J. Nucl. Mater. **313–316**, 444 (2003)
210. D. Naujoks, D. Coster, H. Kastelewicz, R. Schneider, J. Nucl. Mater. **266–269**, 360 (1999)
211. J. Roth, J. Bohdansky, J.B. Roberto, J. Nucl. Mater. **128 & 129**, 534 (1984)
212. A.A. Haasz, J.W. Davies, J. Nucl. Mater. **224**, 141 (1995)
213. R.E. Nygren, J. Bohdansky, A. Pospieszczyk, R. Lehmer, Y. Ra, R.W. Conn, R. Dorner, W.K. Leung, L. Schmitz, J. Nucl. Mater. **176 & 177**, 445 (1990)
214. P. Franzen, J. Nucl. Mater. **228**, 1 (1996)

215. R. Behrisch, J. Nucl. Mater. **93 & 94**, 498 (1980)
216. M.Z. Tokar, A.V. Nedospasov, A.V. Yarochkin, Nucl. Fusion **32**(1), 15 (1992)
217. B.M.U. Scherzer, J. Vac. Sci. Technol. **13**(1), 420 (1976)
218. Y. Ueda, K. Tobita, Y. Katoh, J. Nucl. Mater. **313–316**, 32 (2003)
219. M.Y. Ye, H. Kanehara, S. Fukuta, J. Nucl. Mater. **313–316**, 72 (2003)
220. K. Tokunaga, M.J. Baldwin, R.P. Doerner, N. Noda, Y. Kubota, N. Yoshida, T. Sogabe, T. Kato, B. Schedler, J. Nucl. Mater. **337–339**, 887 (2005)
221. D. Nishijima, M.Y. Ye, N. Ohno, S. Takamura, J. Nucl. Mater. **313–316**, 97 (2003)
222. C.A. Moyer, K. Orvek, Surf. Science **114**, 295 (1982)
223. A. Arnau, Surf. Sci. Rep. **27**, 117 (1997)
224. T. Neidhart, F. Pichler, F. Aumayr, H. Winter, M. Schmid, P. Varga, Phys. Rev. Lett. **74**, 5280 (1995)
225. F. Aumayr, H. Winter, Phil. Trans. R. Soc. Lond. A **362**, 77 (2004)
226. G. Hayderer, S. Cernusca, M. Schmid, P. Varga, H. Winter, F. Aumayr, D. Niemann, V. Hoffmann, N. Stolterfoht, C. Lemell, L. Wirtz, J. Burgdörfer, Phys. Rev. Lett. **86**(16), 3530 (2001)
227. K. Baudin, Nucl. Instr. and Meth. B **96**, 341 (1994)
228. H. Dammak, Phys. Rev. Lett. **74**, 1135 (1995)
229. G. Zwicknagel, D. Deutsch, Phys. Rev. E **56**, 970 (1997)
230. R. Collins, Radiation Effects **37**, 13 (1978)
231. P.S. Ho, Surf. Science **72**, 253 (1978)
232. W. Eckstein, J.P. Biersack, Appl. Phys. **A 37**, 95 (1985)
233. W. Eckstein, J. Roth, E. Gauthier, J. Laszlo, Fusion Technol. **19**, 2076 (1991)
234. W. Williamson, A.J. Antolak, R.J. Meredith, J. Appl. Phys. **61**(9), 4612 (1987)
235. T.E. Everhardt, J. Appl. Phys. **31**(8), 1483 (1960)
236. M. Yasuda, H. Kawata, K. Murata, Phys. Stat. Sol. **(a) 153**, 133 (1996)
237. W. Eckstein, J.P. Biersack, Z. Phys. **A 310**, 1 (1983)
238. R. Wedell, Appl. Phys. **A 35**, 91 (1984)
239. E.S. Parilis, V.K. Verleger, J. Nucl. Mater. **93 & 94**, 512 (1980)
240. A. Weber, H. Mommsen, W. Sarter, A. Weller, Nucl. Instr. and Meth. **198**, 527 (1982)
241. A. Weber, H. Mommsen, Nucl. Instr. and Meth. **204**, 559 (1983)
242. G.D. Archard, J. Appl. Phys. **32**, 1505 (1961)
243. S. Rogaschewski, Phys. Stat. Sol. **(a) 79**, 149 (1983)
244. O.B. Firsov, E.S. Mashkova, V.A. Molchanov, V.A. Snisar, Nucl. Instr. and Meth. **132**, 695 (1976)
245. J.V. Vukanic, R.K. Janev, Nucl. Instr. and Meth. **B16**, 22 (1986)
246. J.V. Vukanic, R.K. Janev, D. Heifetz, Nucl. Instr. and Meth. **B18**, 131 (1987)
247. J.V. Vukanic, R.K. Janev, F.W. Meyer, Phys. Lett. **A 131**(4,5), 294 (1988)
248. S. Tougaard, P. Sigmund, Phys. Rev. **B 25**, 4452 (1982)
249. S. Tougaard, Surf. Science **139**, 208 (1984)
250. S. Tougaard, I. Chorkendorff, Phys. Rev. **B 35**, 6570 (1987)
251. F. Yubero, S. Tougaard, Surface and Interface Analysis **19**, 269 (1992)
252. A.L. Tofterup, Phys. Rev. **B 32**, 2808 (1985)
253. Z. Luo, Nucl. Instr. and Meth. **B33**, 125 (1988)
254. Z. Luo, Nucl. Instr. and Meth. **B33**, 102 (1988)
255. Z. Luo, R. Bai, S. Wang, Nucl. Instr. and Meth. **B48**, 435 (1990)
256. Z. Luo, Nucl. Instr. and Meth. **B48**, 444 (1990)

257. J. Bottiger, J.A. Davies, P. Sigmund, K.B. Winterbon, Radiation Effects **11**, 69 (1971)
258. R. Weissmann, P. Sigmund, Radiation Effects **19**, 7 (1973)
259. R.F. Dashen, Phys. Rev. **134**(4), A1025 (1964)
260. V.P. Afanas'ev, D. Naujoks, Z. Phys. **B 86**, 39 (1992)
261. V.P. Afanas'ev, D. Naujoks, Phys. Stat. Sol. **(b) 164**, 133 (1991)
262. C.A. Ordonez, R.E. Peterkin, J. Nucl. Mater. **228**, 201 (1996)
263. W. Eckstein, J. Nucl. Mater. **248**, 1 (1997)
264. A.V. Nedospasov, M.Z. Tokar, Review of Plasma Physics **18**, 77 (1993)
265. M.I. Baskes, J. Nucl. Mater. **128 & 129**, 676 (1984)
266. J.P. Biersack, L.G. Haggmark, Nucl. Instr. and Meth. **174**, 257 (1980)
267. M.M. Jakas, J.P. Biersack, Z. Phys. **A 315**, 29 (1984)
268. P.L. Grande, M. Behar, J.P. Biersack, F.C. Zawislak, Nucl. Instr. and Meth. **B45**, 689 (1990)
269. W. Eckstein, J.P. Biersack, Z. Phys. **B 63**, 471 (1986)
270. M.T. Robinson, Phys. Rev. **40**(16), 10717 (1989)
271. M.T. Robinson, Nucl. Instr. and Meth. **B48**, 408 (1990)
272. R. Kollath, *Handbuch der Physik.* Review of Plasma Physics, Vol. 21 (Ed. S. Flugge, Springer Verlag, Berlin, 1956)
273. E.W. Thomas, Atomic and Plasma–Material Interaction Data for Fusion: Supplement to the Journal Nuclear Fusion **1**, 79 (1991)
274. I.M. Bronshtein, S.S. Denissov, Sov. Phys. Solid State **9**, 731 (1967)
275. V. Philipps, U. Samm, M.Z. Tokar, B. Unterberg, A. Pospieszczyk, B. Schweer, Nucl. Fusion **33**(6), 953 (1993)
276. L. Verlet, Phys. Rev. **165**, 201 (1968)
277. W.C. Swope, H.C. Andersen, P.H. Berens, K.R. Wilson, J. Chem. Phys. **76**, 637 (1982)
278. W. Takeuchi, Y. Yamamura, Radiation Effects **71**, 53 (1983)
279. D. Naujoks, *Theoretical Study of the Processes of Electron and Ion Interaction with Multicomponent Materials* (PhD Thesis, Moscow State University, Institute of Nuclear Physics, Moscow, 1991)
280. W. Möller, W. Eckstein, Nucl. Instr. and Meth. **B2**, 814 (1984)
281. I. Langmuir, Phys. Rev. **33**, 954 (1929)
282. L. Tonks, I. Langmuir, Phys. Rev. **34**, 876 (1929)
283. M. Laux, H. Grote, K. Günther, A. Herrmann, D. Hildebrandt, P. Pech, H.D. Reiner, H. Wolff, G. Ziegenhagen, J. Nucl. Mater. **162–164**, 200 (1989)
284. M. Weinlich, A. Carlson, Phys. Plasmas **4**, 2151 (1997)
285. L.A. Schwager, C.K. Birdsall, Phys. Fluids **B 2**(5), 1057 (1990)
286. B. Koch, W. Bohmeyer, G. Fussmann, P. Kornejew, H.D. Reiner, J. Nucl. Mater. **290–293**, 653 (2001)
287. K.U. Riemann, J. Phys. D: Appl. Phys. **D 24**, 493 (1991)
288. K.U. Riemann, Phys. Plasmas **1**(3), 552 (1994)
289. K.U. Riemann, Contrib. to Plasma Phys. **36**, 19 (1996)
290. P.C. Stangeby, Phys. Plasmas **2**(3), 702 (1995)
291. P.C. Stangeby, A.V. Chankin, Phys. Plasmas **2**(3), 707 (1995)
292. G.A. Emmert, R.M. Wieland, A.T. Mense, J.N. Davidson, Phys. Fluids **23**(4), 803 (1980)
293. K.U. Riemann, Phys. Plasmas **4**, 4158 (1997)
294. Y. Igitkhanov, D. Naujoks, Contrib. to Plasma Phys. **36**, 67 (1996)

295. G.D. Hobbs, J.A. Wesson, Phys. Plasmas **9**, 85 (1967)
296. J.M. Pedgley, G.M. McCracken, Plasma Phys. Control. Fusion **35**, 357 (1993)
297. K. Asano, N. Ohno, M.Y. Ye, S. Fukuta, S. Takamura, Contrib. to Plasma Phys. **3–4**, 478 (2000)
298. S. Takamura, N. Ohno, M.Y. Ye, T. Kuwabara, Contrib. Plasma Phys. **44**, 126 (2004)
299. D. Naujoks, G. Fussmann, H. Meyer, Contrib. to Plasma Phys. **38**, 127 (1998)
300. R. Chodura, Phys. Fluids **25**(9), 1628 (1982)
301. G.H. Kim, N. Hershkowitz, D.A. Diebold, M.H. Cho, Phys. Plasmas **2**, 3222 (1995)
302. D. Tskhakaya, S. Kuhn, Plasma Phys. Control. Fusion **47**, A327 (2005)
303. B. Koch, W. Bohmeyer, G. Fussmann, J. Nucl. Mater. **313–316**, 1114 (2003)
304. B. Koch, W. Bohmeyer, G. Fussmann, J. Nucl. Mater. **337–339**, 211 (2005)
305. U. Daybelge, B. Bein, Phys. Fluids **24**(6), 1190 (1981)
306. R. Chodura, J. Nucl. Mater. **11 & 112**, 420 (1982)
307. A.B. DeWald, A.W. Bailey, J.N. Brooks, Trajectories of Charged Particles Traversing a Plasma Sheath in Oblique Magnetic Field. Tech. Rep., Georgia Institute of Technology, Atlanta, Georgia 30332, U. S.A. (1984)
308. D. Tskhakaya, S. Kuhn, J. Nucl. Mater. **313–316**, 1119 (2003)
309. C.K. Birdsall, A.B. Langdon, *Plasma Physics via Computer Simulation* (IOP Publishing LTD, London, 1991)
310. R.W. Hockney, J.W. Eastwood, *Computer Simulation using Particles* (IOP Publishing Ltd, Bristol, London, 1988)
311. R. Courant, K.O. Friedrichs, H. Lewy, Math. Ann. **100**, 32 (1928)
312. G.F. Matthews, D.N. Hill, M.A. Mahdavi, Nucl. Fusion **31**(7), 1383 (1991)
313. B. Koch, W. Bohmeyer, G. Fussmann, P. Kornejew, H.D. Reiner, J. Nucl. Mater. **290–293**, 653 (2001)
314. I.V. Tsvetkov, T. Tanabe, J. Nucl. Mater. **266–269**, 714 (1999)
315. S. Takamura, M.Y. Ye, N. Ohno, Y. Uesegi, Recent Res. Devel. Plasmas **1**, 41 (2000)
316. R. Behrisch, G. Venus, J. Nucl. Mater. **202**, 1 (1993)
317. J.B. Withley, W.B. Gauster, R.D. Watson, J.A. Koski, A.J. Russo, Atomic and Plasma–Material Interaction Data for Fusion: Supplement to the Journal Nuclear Fusion **1**, 109 (1991)
318. G. Fussmann, Nucl. Fusion **26**(8), 983 (1986)
319. G.M. McCracken, P.C. Stangeby, Plasma Phys. Control. Fusion **27**(12A), 1411 (1985)
320. C. Ferro, R. Zanino, J. Nucl. Mater. **176 & 177**, 543 (1990)
321. H. Vernickel, J. Bohdansky, Nucl. Fusion **18**, 1467 (1978)
322. D.E. Post, R.V. Jensen, Atomic Data and Nuclear Data **20**, 397 (1977)
323. M.Z. Tokar, Plasma Phys. Control. Fusion **36**, 1819 (1994)
324. J. Kessner, J.P. Freiberg, Nucl. Fusion **35**(2), 115 (1995)
325. B. Lipshultz, J. Nucl. Mater. **145–147**, 15 (1987)
326. M.Z. Tokar, T. Baelmans, V. Philipps, D. Reiter, U. Samm, B. Unterberg, H.A. Claasen, H. Gerhauser, J.D. Hey, A. Huber, P. Mertens, Y.T. Lie, J. Ongena, A. Pospieszczyk, T. Pütz, J. Rapp, D. Rusbüldt, R.P. Schorn, B. Schweer, Plasma Phys. Control. Fusion **37**, A241 (1995)
327. K. Lackner, the ASDEX-Upgrade team, Plasma Phys. Control. Fusion **36**, B79 (1994)

328. V. Abramov, V.I. Pistunovich, Contrib. to Plasma Phys. **30**, 59 (1990)
329. P. Bachmann, D. Sünder, U. Wenzel, Contrib. to Plasma Phys. **36**(4), 519 (1996)
330. D.P. Brennan, R.J. LaHaye, A.D. Turnball, M.S. Chu, T.H. Jensen, L.L. Lao, T.C. Luce, P.A. Politzer, E.J. Strait, S.E. Kruger, D.D. Schnack, Phys. Plasmas **10**, 1643 (2003)
331. A. Sestero, Nucl. Fusion **35**, 919 (1995)
332. H. Bolt, J. Nucl. Mater. **196–198**, 948 (1992)
333. A. Hassanein, I. Konkashbaev, Fusion Engineering and Design **28**, 27 (1995)
334. A. Hassanein, I. Konkashbaev, Atomic and Plasma–Material Interaction Data for Fusion: Supplement to the Journal Nuclear Fusion **5**, 193 (1994)
335. A. Hassanein, I. Konkashbaev, J. Nucl. Mater. **290–293**, 1074 (2001)
336. A. Hassanein, I. Konkashbaev, L. Nikandrov, J. Nucl. Mater. **290–293**, 1079 (2001)
337. A. Hassanein, I. Konkashbaev, J. Nucl. Mater. **313–316**, 664 (2003)
338. H. Würz, S. Pestchanyi, I. Landman, B. Bazylev, V. Tolkach, F. Kappler, Fusion Science and Techn. **40**, 191 (2001)
339. T. Burtseva, A. Hassanein, I. Ovchinnikov, V. Titov, J. Nucl. Mater. **290–293**, 1059 (2001)
340. H. Würz, S. Pestchanyi, B. Bazylev, I. Landman, F. Kappler, J. Nucl. Mater. **290–293**, 1138 (2001)
341. J. Linke, M. Akiba, R. Duwe, A. Lodato, H.J. Penkalla, M. Rödig, K. Schöpflin, J. Nucl. Mater. **290–293**, 1102 (2001)
342. A.W. Leonard, A. Herrmann, K. Itami, J. Lingertat, A. Loarte, T.H. Osborne, W. Suttrop, J. Nucl. Mater. **266–269**, 109 (1999)
343. A.V. Chankin, N. Asakura, T. Fukuda, J. Nucl. Mater. **313–316**, 828 (2003)
344. T. Eich, A. Herrmann, G. Pautasso, P. Andrew, N. Asakura, J.A. Boedo, Y. Corre, M.E. Fenstermacher, J.C. Fuchs, W. Fundamenski, G. Federici, E. Gauthier, B. Goncalves, O. Gruber, A. Kirk, A.W. Leonard, A. Loarte, G.F. Matthews, J. Neuhauser, R.A. Pitts, V. Riccardo, C. Silva, J. Nucl. Mater. **337–339**, 669 (2005)
345. A. Loarte, G. Saibene, R. Sartori, J. Nucl. Mater. **313–316**, 962 (2003)
346. P.T. Lang, J. Neuhauser, L.D. Horton, T. Eich, L. Fattorini, J.C. Fuchs, O. Gehre, A. Herrmann, P. Ignacz, M. Jakobi, S. Kalvin, M. Kaufmann, G. Koscis, B. Kurzan, C. Maggi, M.E. Manso, M. Maraschek, V. Mertens, A.M. anf H. D. Murmann, R. Neu, I. Nunes, D. Reich, M. Reich, S. Saarelma, W. Sandmann, J. Stober, U. Vogl, the ASDEX Upgrade Team, Nucl. Fusion **43**, 1110 (2003)
347. M. Becoulet, G. Huysmans, P. Thomas, E. Joffrin, F. Rimini, P. Monier-Garbet, A. Grosman, P. Ghendrih, V. Parail, P. Lomas, G. Matthews, H. Wilson, M. Gryaznevich, G. Counsell, A. Loarte, G. Saibene, R. Sartori, A. Leonard, P. Snyder, T. Evans, P. Cohil, R. Moyer, Y. Kamada, N. Oyama, T. Hatae, K. Kamiya, A. Degeling, Y. Martin, J. Lister, J. Rapp, C. Perez, P. Lang, A. Chankin, T. Eich, A. Sips, J. Stober, L. Horton, A. Kallenbach, W. Suttrop, S. Saarelma, S. Cowley, J. Lönnroth, M. Shimada, A. Polevoi, G. Federici, J. Nucl. Mater. **337–339**, 677 (2005)
348. H.W. Bartels, T. Kunugi, A.J. Russo, Atomic and Plasma–Material Interaction Data for Fusion: Supplement to the Journal Nuclear Fusion **5**, 225 (1994)
349. G. Maddaluno, G. Maruccia, M. Merola, S. Rollet, J. Nucl. Mater. **313–316**, 651 (2003)

350. R.H. Cohen, Physics of Fluids **19**, 239 (1976)
351. W.B. Gauster, Nucl. Fusion **30**(9), 1897 (1990)
352. D.H. Perkins, *Introduction to High Energy Physics.* 3rd edition (Addison–Wesley Publishing Company Inc., 1987)
353. H. Bolt, A. Miyahara, M. Miyake, J. Nucl. Mater. **155–157**, 256 (1988)
354. H. Bolt, E. Zolti, H. Calen, A. Mörtsell, Proc. 16th Symp. Fusion Techn., London (1990)
355. A.T. Ramsey, C.E. Bush, H.F. Dyla, D.K. Owens, C. Pitcher, M.A. Ulrickson, Nucl. Fusion **31**, 1811 (1991)
356. R. Reichle, Power Handling Capability of C and Be X-Point Target Plates at JET, Termination of High Performance Plasmas in JET. Tech. Rep., JET, Oxford (1992)
357. M.Y. Ye, K. Kudose, T. Kuwabara, N. Ohno, S. Takamura, J. Nucl. Mater. **266–269**, 742 (1999)
358. A.V. Nedospasov, I.V. Bezlyudny, Contrib. to Plasma Phys. **38**(1-2), 337 (1998)
359. K. Reinmüller, Contrib. to Plasma Phys. **38**, 7 (1998)
360. J.N. Brooks, D. Naujoks, Phys. Plasmas **7**, 2565 (2000)
361. D. Naujoks, J.N. Brooks, J. Nucl. Mater. **290–293**, 1123 (2001)
362. P. Chappuis, E. Tsitrone, M. Mayne, X. Armand, H. Linke, H. Bolt, D. Petti, J.P. Sharpe, J. Nucl. Mater. **290–293**, 245 (2001)
363. A.T. Peacock, P. Andrew, P. Cetier, J.P. Coad, G. Federici, F.H. Hurd, M.A. Pick, C.H. Wu, J. Nucl. Mater. **266–269**, 423 (1999)
364. V. Rohde, M. Mayer, the ASDEX-Upgrade team, J. Nucl. Mater. **313–316**, 337 (2003)
365. Y. Watanabe, Plasma Phys. Control. Fusion **39**, A59 (1997)
366. J. Winter, G. Gebauer, J. Nucl. Mater. **266–269**, 228 (1999)
367. J. Winter, V.E. Fortov, A.P. Nefedov, J. Nucl. Mater. **290–293**, 509 (2001)
368. S.I. Krasheninnikov, T.K. Soboleva, Y. Tomita, R.D. Smirnow, R.K. Janev, J. Nucl. Mater. **337–339**, 65 (2005)
369. J.D. Martin, M. Coppins, G.F. Counsell, J. Nucl. Mater. **337–339**, 114 (2005)
370. G.F. Counsell, J.P. Coad, G. Federici, K. Krieger, V. Philipps, C.H. Skinner, D.G. Whyte, J. Nucl. Mater. **290–293**, 255 (2001)
371. M.Z. Tokar, Plasma Phys. Control. Fusion **35**, 1119 (1993)
372. J. Roth, G. Janeschitz, Nucl. Fusion **29**, 915 (1989)
373. A. Kallenbach, R. Neu, W. Poschenrieder, the ASDEX-Upgrade team, Nucl. Fusion **34**(12), 1557 (1994)
374. H. Wolff, Atomic and Plasma–Material Interaction Data for Fusion: Supplement to the Journal Nuclear Fusion **1**, 93 (1991)
375. B. Jüttner, K. Büchl, M. Weinlich, the ASDEX-Upgrade team, Contrib. to Plasma Phys. **34**(2//3), 472 (1994)
376. R. Behrisch, Contrib. to Plasma Phys. **42**, 431 (2002)
377. B. Jüttner, I. Kleberg, J. Phys. D: Appl. Phys. **33**, 2025 (2000)
378. D. Naujoks, R. Behrisch, J.P. Coad, L. de Kock, Nucl. Fusion **33**(4), 581 (1993)
379. D. Naujoks, R. Behrisch, V. Philipps, B. Schweer, Plasma Phys. Control. Fusion **36**, 2021 (1994)
380. Y. Hirooka, Dynamic Materials Compatibility Modeling Between Beryllium, Carbon and Tungsten under Deuterium-Tritium Plasma Bombardment. Tech. Rep. UCSD-ENG-025, University of California, San Diego (1996)

381. Y. Hirooka, Review of Beryllium and Tungsten Erosion Behavior and Universal Modeling of Impurity Effects Observed in Recent PISCES Experiments. Tech. Rep. UCSD-ENG-022, University of California, San Diego (1996)
382. K. Krieger, J. Roth, J. Nucl. Mater. **290–293**, 107 (2001)
383. K. Schmid, J. Roth, J. Nucl. Mater. **313–316**, 302 (2003)
384. P.C. Stangeby, J.D. Elder, Plasma Phys. Control. Fusion **33**(2), 1435 (1991)
385. D. Naujoks, J. Steinbrink, U. Wenzel, the PSI-group, J. Nucl. Mater. **241–243**, 707 (1997)
386. K. Behringer, Description of the Impurity Transport Code STRAHL. Tech. Rep. JET-R(87)08, JET, Culham, U. K. (1987)
387. K.H. Behringer, J. Nucl. Mater. **145–147**, 145 (1987)
388. M.J. Seaton, Plant. Space Sci. **12**, 55 (1964)
389. W. Lotz, Zeitschrift für Physik **216**, 241 (1968)
390. W. Lotz, Electron Impact Ionization Cross–Sections and Ionization Rate Coefficients for Atoms and Ions from Scandium to Zinc. Tech. Rep. IPP 1/76, IPP, Garching (1968)
391. K.L. Bell, H.B. Gilbody, J.G. Hughes, A.E. Kingston, F.J. Smith, Atomic and Molecular Data for Fusion, Part I: Recommended Cross Sections and Rates for Electron Ionisation of Light Atoms and Ions. Tech. Rep. CLM-R216, Culham Laboratory, Abingdon Oxfordshire, U. K. (1982)
392. M.A. Lennon, K.L. Bell, H.B. Gilbody, J.G. Hughes, A.E. Kingston, M.J. Murray, F.J. Smith, Atomic and Molecular Data for Fusion, Part II: Recommended Cross Sections and Rates for Electron Ionisation of Light Atoms and Ions: Fluorine to Nickel. Tech. Rep. CLM-R270, Culham Laboratory, Abingdon Oxfordshire, U. K. (1986)
393. M. Higgins, J.G. Hughes, H.B. Gilbody, F.J. Smith, M.A. Lennon, K.L. Bell, A.E. Kingston, Atomic and Molecular Data for Fusion, Part III: Recommended Cross Sections and Rates for Electron Ionisation of Light Atoms and Ions: Copper to Uranium. Tech. Rep. CLM-R294, Culham Laboratory, Abingdon Oxfordshire, U. K. (1989)
394. T. Kato, K. Masai, M. Arnaund, Comparison of Ionization Rate Coeffients of Ions from Hydrogen through Nickel. Tech. Rep. NIFS-DATA-14, National Institute for Fusion Science, Nagoya, Japan (1991)
395. A. Burgess, Astrophys. J. **139**, 776 (1964)
396. A. Burgess, Astrophys. J. **141**, 1588 (1965)
397. D. Naujoks, K. Asmussen, M. Bessenrodt-Weberpals, S. Deschka, R. Dux, W. Engelhardt, A. Field, G. Fussmann, J.C. Fuchs, C. Garcia-Rosales, S. Hirsch, P.N. Ignacz, G. Lieder, F. Mast, R. Neu, R. Radtke, J. Roth, the ASDEX-Upgrade team, Nucl. Fusion **36**(6), 671 (1996)
398. G. Fussmann, W. Engelhardt, D. Naujoks, K. Asmussen, S. Deschka, A. Field, J.C. Fuchs, C. Garcia-Rosales, S. Hirsch, P. Ignacz, G. Lieder, R. Neu, R. Radtke, J. Roth, U. Wenzel, Plasma Physics and Controlled Nuclear Fusion Research 1995 (Proc. 15th Int. Conf. Seville, IEAE **3**, 143 (1995)
399. J. Roth, D. Naujoks, K. Krieger, A. Field, G. Lieder, S. Hirsch, J. Nucl. Mater. **220–222**, 231 (1995)
400. A.R. Field, G. Fussmann, C. Garcia-Rosales, S. Hirsch, G. Lieder, D. Naujoks, R. Neu, C.S. Pitcher, R. Radtke, U. Wenzel, the ASDEX-Upgrade team, J. Nucl. Mater. **220–222**, 553 (1995)
401. P.C. Stangeby, J.D. Elder, Contrib. to Plasma Phys. **32**(3/4), 353 (1992)

402. D. Hildebrandt, Contrib. to Plasma Phys. **26**(4), 231 (1986)
403. C. Ferro, R. Zanino, J. Nucl. Mater. **196–198**, 321 (1992)
404. R. Bartiromo, I. Condrea, R. de Angelis, M. Leigheb, V. Pericoli-Ridolfnin, F. Romanelli, R. Zagorski, Nucl. Fusion **35**, 1161 (1995)
405. P.C. Stangeby, C. Farrell, L. Wood, Contrib. to Plasma Phys. **28**(4/5), 501 (1988)
406. P.C. Stangeby, C. Farrell, Contrib. to Plasma Phys. **28**(4/5), 495 (1988)
407. P.C. Stangeby, J. Nucl. Mater. **176 & 177**, 51 (1990)
408. P.C. Stangeby, C. Farrell, Plasma Phys. Control. Fusion **32**(9), 677 (1990)
409. G.M. McCracken, U. Samm, P.C. Stangeby, G. Bertschinger, J.A. Boedo, S.J. Davies, D.S. Gray, V. Philipps, R.A. Pitts, D.H.J. Goodall, A. Pospieszczyk, R.P. Schorn, B. Schweer, G. Telesca, B. Unterberg, G. Waidmann, Nucl. Fusion **33**(10), 1409 (1993)
410. J.B. Taylor, The Impurity Problem in Steady–State Toroidal Devices. Tech. Rep., Culham Laboratory, Abingdon (1974)
411. M.Z. Tokar, J. Rapp, G. Bertschinger, L. Könen, H.R. Koslowski, A. Krämer-Flecken, V. Philipps, U. Samm, B. Unterberg, Nucl. Fusion **37**(12), 1691 (1997)
412. F. Wagner, G. Becker, K. Behringer, Phys. Rev. Lett. **49**, 1408 (1982)
413. V. Mertens, the ASDEX team, Plasma Phys. Control. Fusion **32**, 965 (1990)
414. K.H. Burrell, E.J. Doyle, P. Gohil, R.J. Groebner, J. Kim, R.J.L. Haye, Phys. Plasmas **1**(5), 1536 (1994)
415. F.L. Hinton, G.M. Staebler, Phys. Fluids **B5**, 1281 (1993)
416. J.A. Wesson, Plasma Phys. Control. Fusion **37**, A337 (1995)
417. Y.I. Kolesnichenko, V.V. Lutsenko, Y.V. Yakovenko, Nucl. Fusion **34**(12), 1619 (1994)
418. M.F. Nave, J. Rapp, T. Bolonella, R. Dux, M.J. Mantsinen, R. Budny, P. Dumortier, M. von Hellermann, S. Jachmich, H.R. Koslowski, G. Maddison, A. Messiaen, P. Monier-Garbet, K. Ongena, M.E. Puiatti, J. Strachan, G. Telesca, B. Unterberg, M. Valisa, P. de Vries, Nucl. Fusion **43**, 1204 (2003)
419. C. Hopf, T. Schwarz-Selinger, W. Jacob, A. von Keudell, J. Appl. Phys. **87**(6), 2719 (2000)
420. A. von Keudell, Thin Solid Films **402**, 1 (2002)
421. J.N. Brooks, Nuclear Technology/Fusion **4**, 33 (1983)
422. P.C. Stangeby, C. Farrell, S. Hoskins, L. Wood, Nucl. Fusion **28**(11), 1945 (1988)
423. P.C. Stangeby, J.D. Elder, J. Nucl. Mater. **196–198**, 258 (1992)
424. J.N. Brooks, Phys. Fluids **B 2**(8), 1858 (1990)
425. J.N. Brooks, J.P. Allain, D.A. Alman, D.N. Ruzic, Fusion Engineering and Design **72**, 363 (2005)
426. U. Kögler, J. Winter, ERO-TEXTOR 3D–Monte–Carlo Code for Local Impurity Modelling in the Scrape–off–Layer of TEXTOR. Tech. Rep. Jül-3361, KFA Jülich, Jülich (1997)
427. A. Kirschner, A. Huber, V. Philipps, A. Pospieszczyk, P. Wienhold, J. Winter, J. Nucl. Mater. **290–293**, 238 (2001)
428. R. Kawakami, Jpn. J. Appl. Phys. **43**, 785 (2004)
429. R. Kawakami, T. Mitani, J. Nucl. Mater. **337–339**, 45 (2005)
430. R. Schneider, H.S. Bosch, J. Neuhauser, D. Coster, K. Lackner, M. Kaufmann, J. Nucl. Mater. **241–243**, 701 (1997)
431. R.V. Jensen, D.E. Post, D.L. Jassby, Nucl. Sci. and Eng. **65**, 282 (1978)

432. V.M. Prozesky, Plasma Phys. Control. Fusion **35**, 1717 (1993)
433. J. Ehrenberg, J. Nucl. Mater. **162–164**, 63 (1989)
434. J. Ehrenberg, P.J. Harbour, Nucl. Fusion **31**(2), 287 (1991)
435. J. Ehrenberg, P. Andrew, L. Horton, G. Janeschitz, L.D. Kock, V. Philipps, J. Nucl. Mater. **196–198**, 992 (1992)
436. D. Larsson, H. Bergsaker, A. Hedqvist, J. Nucl. Mater. **266–269**, 856 (1999)
437. M. Kikuchi, Plasma Phys. Control. Fusion **35**, B39 (1993)
438. N. Noda, *Calculation of Burning Condition in DT Reactors in Contributions to High-Temperature Plasma Physics* (Ed. K. H. Spatschek, Akademie-Verlag, 1995)
439. M. Mayer, J. Nucl. Mater. **240**, 164 (1997)
440. R. Behrisch, J. Ehrenberg, M. Wielinski, A.P. Martinelli, H. Bergsaker, B. Emmoth, L. de Kock, J.P. Coad, J. Nucl. Mater. **145–147**, 723 (1987)
441. J. Winter, J. Nucl. Mater. **145–147**, 131 (1987)
442. J. Winter, H.G. Esser, L. Könen, V. Philipps, H. Reimer, J. v. Seggern, J. Schlüter, E. Vietzke, F. Waelbroeck, P. Wienhold, T. Banno, D. Ringer, S. Verprek, J. Nucl. Mater. **162–164**, 713 (1989)
443. N. Noda, K. Tsuzuki, A. Sagara, N. Inoue, T. Muroga, J. Nucl. Mater. **266–269**, 234 (1999)
444. J. Rapp, W. Biel, H. Gerhauser, A. Huber, H.R. Koslowski, M. Lehnen, V. Philipps, A. Pospieszczyk, D. Reiser, U. Samm, G. Sergienko, M.Z. Tokar, R. Zagorski, J. Nucl. Mater. **290–293**, 1148 (2001)
445. M. Greenwald, J.L. Terry, S.M. Wolfe, S. Ejima, M.G. Bell, S.M. Kaye, G.H. Neilson, Nucl. Fusion **28**, 2199 (1988)
446. N. Noda, S. Okamura, T. Aoki, H. Yamada, K. Tsuzuki, K. Matsuoka, H. Iguchi, M. Hosokawa, O. Kaneko, S. Kubo, S. Morita, K. Nishimura, A. Sagara, T. Shoji, C. Takahashi, Y. Takeiri, Y. Takita, H. Amemiya, K. Okazaki, Y. Oyama, K. Shimizu, K. Yano, J. Nucl. Mater. **176 & 177**, 640 (1990)
447. P. Grigull, R. Behrisch, R. Brakel, A. Elsner, H. Hacker, H.J. Hartfuss, G. Herre, D. Hildebrandt, R. Jaenicke, J. Kisslinger, H. Maassberg, C. Mahn, H. Niedermeyer, P. Pech, H. Renner, H. Ringler, F. Rau, J. Roth, F. Sardei, U. Schneider, F. Wagner, A. Weller, H. Wobig, H. Wolff, the W7-AS team, the NBI team, J. Nucl. Mater. **196–198**, 101 (1992)
448. S. Kato, M. Watanabe, H. Toyoda, H. Sugai, J. Nucl. Mater. **266–269**, 406 (1999)
449. C.H. Skinner, J. Nucl. Mater. **241–243**, 214 (1997)
450. N.M. Zhang, E.Y. Wang, M.X. Wang, W.Y. Hong, C.H. Cui, Z.W. Wang, D.H. Yan, J. Nucl. Mater. **266–269**, 747 (1999)
451. N.M. Zhang, M.X. Wang, B. Li, J. Nucl. Mater. **313–316**, 255 (2003)
452. J.K. Xie, Y.P. Zhao, J. Li, B.N. Wan, X.Z. Gong, J.S. Hu, X. Gao, X.M. Gu, S.D. Zhang, X.M. Wang, Y.Z. Mao, X.K. Yang, M. Zhen, S.Y. Zhang, the HT-7 team, J. Nucl. Mater. **290–293**, 1155 (2001)
453. R. Brakel, D. Hartmann, P. Grigull, the W7-AS team, J. Nucl. Mater. **290–293**, 1160 (2001)
454. J.P. Gunn, Phys. Plasmas **4**, 4435 (1997)
455. R. Granetz, Phys. Rev. Lett. **49**, 658 (1982)
456. J.L. Terry, S.J. Zweben, K. Hallatschek, B. LaBombard, R.J. Maqueda, B. Bai, C.J. Boswell, M. Greenwald, D. Kopon, W.M. Nevins, C.S. Pitcher, B.N. Rogers, D.P. Stotler, X.Q. Xu, Phys. Plasmas **10**, 1739 (2003)

457. F. Troyan, R. Gruber, H. Saurenmann, Plasma Phys. Control. Fusion **26**, 209 (1984)
458. M. Murakami, J.D. Callen, L.A. Berry, Nucl. Fusion **16**, 347 (1976)
459. J. Hugill, P.J. Lomas, A.J. Wotton, High Density Operation in DITE with Neutral Beam Injection. Tech. Rep. CLM-R239, Culham Laboratory, Abingdon, Oxfordshire (1983)
460. R. Wolf, *Internal Transport Barriers in Tokamak Plasmas* (University de Mons-Hainaut, Belgium, 2002)
461. R. Behrisch, J. Roth, G. Staudenmaier, H. Verbeek, Nucl. Instr. and Meth. **B18**, 629 (1987)
462. W. Engelhardt, W. Feneberg, J. Nucl. Mater. **76 & 77**, 518 (1978)
463. P.C. Stangeby, Contrib. to Plasma Phys. **28**(4/5), 507 (1988)
464. H. Wolff, Contrib. to Plasma Phys. **28**, 131 (1988)
465. G.M. McCracken, Plasma Phys. Control. Fusion **29**(10A), 1273 (1987)
466. R. Behrisch, J.P. Coad, J. Ehrenberg, L. de Kock, J. Roth, M. Wielunski, J.A. Tagle, J. Nucl. Mater. **162–164**, 598 (1989)
467. H. Bergsaker, J.P. Coad, R. Behrisch, S. Clement, F. Lama, A.P. Martinelli, V.M. Prozesky, G. Röver, J.C.B. Simpson, J. Nucl. Mater. **176 & 177**, 941 (1990)
468. J.P. Coad, L. de Kock, A. Koch, F. Weschenfelder, P. Wienhold, R. Wilhelm, Assessment Techniques for Measurement of Target Erosion/Redeposition in Large Tokamaks. Tech. Rep. JET-P(92)61, JET, Abingdon (1992)
469. M. Kaufmann, H.S. Bosch, A. Field, G. Fussmann, O. Gruber, A. Herrmann, W. Junker, A. Kallenbach, W. Köppendörfer, K. Krieger, K. Lackner, M. Laux, V. Mertens, B. Napointek, D. Naujoks, J. Neuhauser, J. Poschenrieder, J. Roth, F. Ryter, H. Zohm, the ASDEX-Upgrade team, Plasma Phys. Control. Fusion **35**(4), B205 (1993)
470. V. Philipps, T. Tanabe, Y. Ueda, A. Pospieszczyk, M.Z. Tokar, B. Unterberg, L. Könen, B. Schweer, U. Samm, P. Wienhold, J. Winter, M. Rubel, B. Emmoth, N.C. Hawkes, the TEXTOR team, Nucl. Fusion **34**(11), 1417 (1994)
471. R.B. et al., J. Nucl. Mater. **220–222**, 310 (1995)
472. V. Philipps, et. al., J. Nucl. Mater. **241–243**, 105 (1997)
473. H.Y. Guo, J.P. Coad, S.J. Davies, J.D. Elder, L.D. Horton, X.L. Li, J. Lingertat, A. Loarte, G.F. Matthews, R.D. Monk, R. Simonini, M.F. Stamp, P.C. Stangeby, A. Tabasso, J. Nucl. Mater. **241–243**, 385 (1997)
474. D.G. Whyte, R. Bastasz, J.N. Brooks, W.R. Wampler, W.P. West, C.P.C. Wong, O.I. Buzhinskij, I. Opimach, J. Nucl. Mater. **266–269**, 67 (1999)
475. K. Krieger, H. Maier, R. Neu, the ASDEX-Upgrade team, J. Nucl. Mater. **266–269**, 207 (1999)
476. J.N. Brooks, D.N. Ruzic, D.B. Hayden, R.B. Turkot, J. Nucl. Mater. **220–222**, 269 (1995)
477. M.F.A. Harrison, E.S. Hotston, J. Nucl. Mater. **176 & 177**, 256 (1990)
478. S.A. Cohen, R.F. Mattas, K.A. Werley, Plasma-Materials Interaction Issues for ITER. Tech. Rep. PPPL-2823, PPPL, Princeton (1992)
479. A. Sagara, N. Noda, R. Akiyama, H. Arimoto, H. Idei, H. Iguchi, O. Kaneko, T. Kohmoto, S. Kubo, N. Matsunami, K. Matsuoka, S. Morita, O. Motojima, K. Nishimura, S. Okamura, J. Rice, T. Shoji, C. Takahasi, Y. Takita, M. Ueda, H. Yamada, I. Yamada, J. Nucl. Mater. **196–198**, 271 (1992)
480. G.M. McCracken, J. Ehrenberg, P.E. Stott, R. Behrisch, L.D. Kock, J. Nucl. Mater. **145–147**, 621 (1987)

481. G.M. McCracken, D.H.J. Goodall, P.C. Stangeby, J.P. Coad, J. Roth, B. Denne, R. Behrisch, J. Nucl. Mater. **162–164**, 356 (1989)
482. J.N. Brooks, Fusion Technol. **18**, 239 (1990)
483. D. Naujoks, Nucl. Fusion **37**(9), 1193 (1997)
484. W.R. Wampler, D.G. Whyte, C.P.C. Wong, W.P. West, J. Nucl. Mater. **290–293**, 346 (2001)
485. J.P. Coad, M. Rubel, C.H. Wu, J. Nucl. Mater. **241–243**, 408 (1997)
486. G.E. Lucas, J. Nucl. Mater. **216**, 322 (1994)
487. V. Barabash, G. Federici, J. Linke, C.H. Wu, J. Nucl. Mater. **313–316**, 42 (2003)
488. C.H. Wu, C. Alessandrini, J.P. Bonal, Fusion Engineering and Design **56–57**, 179 (2001)
489. P. Schiller, J. Nucl. Mater. **206**, 113 (1993)
490. M. Shimada, A.E. Costley, G. Federici, K. Ioki, A.S. Kukushkin, V. Mukhovatov, A. Polevoi, M. Sugihara, J. Nucl. Mater. **337–339**, 808 (2005)
491. A.F. Bardamid, A.I. Belyayeva, V.N. Bondarenko, J. Nucl. Mater. **313–316**, 112 (2003)
492. V.S. Voitsenya, A.F. Bardamid, V.N. Bondarenko, W. Jacob, V. Konovalov, S. Masuzaki, O. Motojima, D.V. Orlinskij, V.L. Poperenko, I.V. Ryzhkov, A. Sagara, A.F. Shtan, S.I. Solodovchenko, M.V. Vinnichenko, J. Nucl. Mater. **290–293**, 336 (2001)
493. M.Z. Tokar, B.J. Peterson, Y. Nakamura, Y. Takeiri, N. Noda, Phys. Plasmas **7**(11), 4357 (2000)
494. B.J. Peterson, S. Masuzaki, R. Sakamoto, J. Nucl. Mater. **290–293**, 930 (2001)
495. P.H. Rebut, K.J. Dietz, P.P. Lallia, J. Nucl. Mater. **162–164**, 172 (1989)
496. M. Keilhacker, the JET team, Phys. Fluids **B 2**(6), 1291 (1990)
497. D.J. Campbell, J. Nucl. Mater. **241–243**, 379 (1997)
498. J. Roth, W. Eckstein, M. Guseva, Fusion Engineering and Design **37**, 465 (1997)
499. J. Roth, W. Eckstein, J. Bohdansky, J. Nucl. Mater. **165**, 199 (1989)
500. T. Tanabe, N. Noda, H. Nakamura, J. Nucl. Mater. **99**, 99 (1992)
501. N. Noda, V. Philipps, R. Neu, J. Nucl. Mater. **241–243**, 227 (1997)
502. W.R. Wampler, B. LaBombard, B. Lipschultz, G.M. McCracken, D.A. Pappas, C.S. Pitcher, J. Nucl. Mater. **266–269**, 217 (1999)
503. T. Burtseva, V. Barabash, I. Mazul, C. Garcia-Rosales, S. Deschka, R. Behrisch, A. Herrmann, J. Nucl. Mater. **241–243**, 716 (1997)
504. J. Li, X. Gong, L. Luo, F.X. Yin, N. Noda, B. Wan, W. Xu, X. Gao, F. Yin, J.G. Jiang, Z. Wu, J.Y. Zhao, M. Wu, S. Liu, Y. Han, J. Nucl. Mater. **241–243**, 878 (1997)
505. G. Maddaluno, F. Pierdominici, M. Vittori, J. Nucl. Mater. **241–243**, 908 (1997)
506. K. Krieger, A. Geier, Y. Gong, J. Nucl. Mater. **313–316**, 327 (2003)
507. R. Neu, R. Dux, A. Geier, J. Nucl. Mater. **313–316**, 116 (2003)
508. H. Bolt, V. Barabash, W. Kraus, J. Linke, R. Neu, S. Suzuki, N. Yoshida, J. Nucl. Mater. **329–333**, 66 (2004)
509. K. Krieger, J. Likonen, M. Mayer, R. Pugno, V. Rohde, E. Vainonen-Ahlgren, the ASEX Upgrade Team, J. Nucl. Mater. **337–339**, 10 (2005)
510. E. Hechtl, W. Eckstein, J. Roth, J. Laszlo, J. Nucl. Mater. **179–181**, 290 (1991)
511. I.S. Landman, H. Wuerz, J. Nucl. Mater. **313–316**, 77 (2003)

512. A. Kurumada, Y. Imamura, Y. Tomota, J. Nucl. Mater. **313–316**, 245 (2003)
513. T. Hirai, V. Philipps, A. Huber, J. Nucl. Mater. **313–316**, 67 (2003)
514. R.A. Neiser, G.R. Smolik, K.J. Hollis, R.D. Watson, J. Thermal Spray Technology **2**(4), 393 (1993)
515. K. Tokunaga, R.P. Doerner, R.S. R., J. Nucl. Mater. **313–316**, 92 (2003)
516. N. Yoshida, J. Nucl. Mater. **266–269**, 197 (1999)
517. R. Behrisch, M. Mayer, C. Garcia-Rosales, J. Nucl. Mater. **233–237**, 673 (1996)
518. R. Andreani, E. Diegele, R. Laesser, B. van der Schaaf, J. Nucl. Mater. **329–333**, 20 (2004)
519. L.K. Mansur, A.F. Rowcliffe, R.K. Nanstad, S.J. Zinkle, W.R. Corwin, R.E. Stoller, J. Nucl. Mater. **329–333**, 166 (2004)
520. R.F. Mattas, D.L. Smith, C.H. Wu, T. Kuroda, G. Shatalov, J. Nucl. Mater. **191–194**, 139 (1992)
521. C.H. Wu, U. Mszanowski, J. Nucl. Mater. **99**(99), 99 (1994)
522. E.E. Bloom, S.J. Zinkle, F.W. Wiffen, J. Nucl. Mater. **329–333**, 12 (2004)
523. A. Kimura, T.T. Sawai, K. Shiba, A. Hishinuma, S. Jitsukawa, S. Ukai, A. Kohyama, Nucl. Fusion **43**, 1246 (2003)
524. E.B. Bloom, Nucl. Fusion **30**(9), 1879 (1990)
525. A. Brendel, C. Popescu, C. Leyens, J. Woltersdorf, E. Pippel, H. Bolt, J. Nucl. Mater. **329–333**, 804 (2004)
526. B.Q. Deng, J.H. Huang, J.C. Yan, J. Nucl. Mater. **313–316**, 630 (2003)
527. A. Grossman, R.P. Doerner, S. Luckhardt, J. Nucl. Mater. **290–293**, 80 (2001)
528. R.P. Doerner, M.J. Baldwin, S.I. Krasheninnikov, D.G. Whyte, J. Nucl. Mater. **313–316**, 383 (2003)
529. J.P. Allain, J.N. Brooks, D.A. Alman, L.E. Gonzalez, J. Nucl. Mater. **337–339**, 94 (2005)
530. R.P. Schorn, E. Hintz, B. Baretzky, J. Bohdansky, W. Eckstein, J. Roth, E. Taglauer, J. Nucl. Mater. **162–164**, 924 (1989)
531. B. Baretzky, W. Eckstein, R.P. Schorn, J. Nucl. Mater. **224**, 50 (1995)
532. N.V. Antonov, V.G. Belan, V.A. Evtihin, L.G. Golubchikov, V.I. Khripunov, V.M. Korjavin, I.E. Lyublinki, J. Nucl. Mater. **241–243**, 1190 (1997)
533. A.V. Zhmendak, A. Huber, V.A. Kvitcinsky, E.V. Mudretskaya, A.V. Nedospasov, V.V. Panechkina, S.N. Pavlov, A. Pospieszczyk, G.V. Sergienko, V.F. Virko, J. Nucl. Mater. **290–293**, 220 (2001)
534. B.I. Khripunov, V.B. Petrov, V.V. Shapkin, J. Nucl. Mater. **313–316**, 619 (2003)
535. O.I. Buzhinskij, V.G. Otroshchenko, D.G. Whyte, J. Nucl. Mater. **313–316**, 214 (2003)
536. Y. Hirooka, A Proposal for Experimental Demonstration of Steady–State Impurity Control by Moving–Belt Plasma–Facing Components in the PISCES Plasma Device. Tech. Rep. UCSD-ENG-047, Fusion Energy Research Program, University of California, San Diego (1997)
537. G. Federici, J. Nucl. Mater. **313–316**, 11 (2003)
538. J.B. Roberto, R. Behrisch, J. Nucl. Mater. **128 & 129**, 764 (1984)
539. D.A. Alman, D.N. Ruzic, J. Nucl. Mater. **313–316**, 182 (2003)
540. A. Kirschner, V. Philipps, D.P. Coster, S.K. Erents, H.G. Esser, G. Federici, A.S. Kukushkin, A. Loarte, G.F. Matthews, J. Roth, U. Samm, J. Nucl. Mater. **337–339**, 17 (2005)

541. D. Naujoks, W. Bohmeyer, A. Markin, I. Arkhipov, P. Carl, B. Koch, H.D. Reiner, D. Schröder, Physica Scripta **111**, 80 (2004)
542. D.G. Whyte, J.W. Davis, J. Nucl. Mater. **337–339**, 560 (2005)
543. J.P. Sharpe, V. Rohde, the ASDEX-Upgrade team, J. Nucl. Mater. **313–316**, 455 (2003)
544. R.A. Jameson, R. Ferdinand, H. Klein, J. Rathke, J. Sredniawski, M. Sugimoto, J. Nucl. Mater. **329–333**, 193 (2004)
545. G.S. Bauer, H. Ullmaier, J. Nucl. Mater. **318**, 26 (2003)
546. P. Vladimirov, A. Möslang, J. Nucl. Mater. **329–333**, 233 (2004)
547. D. Potter, *Computational Physics* (John Wiley & Sons, London, 1972)
548. D. Naujoks, R. Behrisch, J. Nucl. Mater. **220–222**, 227 (1995)
549. W. Eckstein, J.P. Biersack, Nucl. Instr. and Meth. **B2**, 550 (1984)
550. P.J. Mohr, B.N. Taylor, J. of Physical and Chemical Reference **28**(6), 1 (1999)
551. P.J. Mohr, B.N. Taylor, Reviews of Modern Physics **72**(2), 1 (2000)
552. E.R. Cohen, K.M. Crowe, J.W.M. DuMond, *Fundamental Constants of Physics* (Interscience, New York, 1957)
553. K.D. Froome, L. Essen, *Velocity of Light and Radio Waves* (Academic Press, London, 1969)
554. B.W. Petley, *Fundamental Constants and the Frontier of Measurement* (Adam Hilger, Bristol, 1988)
555. J. Bordtfeldt, B. Kramer (eds.), *Units and Fundamental Constants in Physics and Chemistry* (Springer Verlag, Berlin, 1992)

Index

Springer Series on

ATOMIC, OPTICAL, AND PLASMA PHYSICS

Springer Series on

ATOMIC, OPTICAL, AND PLASMA PHYSICS

1 **Polarized Electrons**
2nd Edition
By J. Kessler

2 **Multiphoton Processes**
Editors: P. Lambropoulos and S.J. Smith

3 **Atomic Many-Body Theory**
2nd Edition
By I. Lindgren and J. Morrison

4 **Elementary Processes in Hydrogen-Helium Plasmas**
Cross Sections and Reaction Rate Coefficients
By R.K. Janev, W.D. Langer, K. Evans Jr., and D.E. Post Jr.

5 **Pulsed Electrical Discharge in Vacuum**
By G.A. Mesyats and D.I. Proskurovsky

6 **Atomic and Molecular Spectroscopy**
Basic Aspects and Practical Applications
3rd Edition
By S. Svanberg

7 **Interference of Atomic States**
By E.B. Alexandrov, M.P. Chaika, and G.I. Khvostenko

8 **Plasma Physics**
Basic Theory with Fusion Applications
3rd Edition
By K. Nishikawa and M. Wakatani

9 **Plasma Spectroscopy**
The Influence of Microwave and Laser Fields
By E. Oks

10 **Film Deposition by Plasma Techniques**
By M. Konuma

11 **Resonance Phenomena in Electron–Atom Collisions**
By V.I. Lengyel, V.T. Navrotsky, and E.P. Sabad

12 **Atomic Spectra and Radiative Transitions**
2nd Edition
By I.I. Sobel'man

13 **Multiphoton Processes in Atoms**
2nd Edition
By N.B. Delone and V.P. Krainov

14 **Atoms in Plasmas**
By V.S. Lisitsa

15 **Excitation of Atoms and Broadening of Spectral Lines**
2nd Edition, By I.I. Sobel'man, L. Vainshtein, and E. Yukov

16 **Reference Data on Multicharged Ions**
By V.G. Pal'chikov and V.P. Shevelko

17 **Lectures on Non-linear Plasma Kinetics**
By V.N. Tsytovich

18 **Atoms and Their Spectroscopic Properties**
By V.P. Shevelko

19 **X-Ray Radiation of Highly Charged Ions**
By H.F. Beyer, H.-J. Kluge, and V.P. Shevelko

20 **Electron Emission in Heavy Ion–Atom Collision**
By N. Stolterfoht, R.D. DuBois, and R.D. Rivarola

21 **Molecules and Their Spectroscopic Properties**
By S.V. Khristenko, A.I. Maslov, and V.P. Shevelko

22 **Physics of Highly Excited Atoms and Ions**
By V.S. Lebedev and I.L. Beigman

23 **Atomic Multielectron Processes**
By V.P. Shevelko and H. Tawara

24 **Guided-Wave-Produced Plasmas**
By Yu.M. Aliev, H. Schlüter, and A. Shivarova

25 **Quantum Statistics of Nonideal Plasmas**
By D. Kremp, M. Schlanges, and W.-D. Kraeft

26 **Atomic Physics with Heavy Ions**
By H.F. Beyer and V.P. Shevelko

MIX
Papier aus verantwortungsvollen Quellen
Paper from responsible sources
FSC® C105338

If you have any concerns about our products,
you can contact us on
ProductSafety@springernature.com

In case Publisher is established outside the EU,
the EU authorized representative is:
Springer Nature Customer Service Center GmbH
Europaplatz 3, 69115 Heidelberg, Germany

Printed by Libri Plureos GmbH
in Hamburg, Germany